Progress in Optical Science and Photonics

Volume 2

Series editor

Javid Atai

The purpose of the series Progress in Optical Science and Photonics is to provide a forum to disseminate the latest research findings in various areas of Optics and its applications. The intended audience is physicists, electrical and electronic engineers, applied mathematicians, and advanced graduate students.

More information about this series at http://www.springer.com/series/10091

Prabhakar Misra
Editor

Applied Spectroscopy and the Science of Nanomaterials

Springer

Editor
Prabhakar Misra
Department of Physics and Astronomy
Howard University
Washington, DC
USA

ISSN 2363-5096 ISSN 2363-510X (electronic)
ISBN 978-981-10-1189-4 ISBN 978-981-287-242-5 (eBook)
DOI 10.1007/978-981-287-242-5

Springer Singapore Heidelberg New York Dordrecht London

Printed on acid-free paper

Springer is part of Springer Science+Business Media (www.springer.com)

Preface

The present monograph *Applied Spectroscopy and the Science of Nanomaterials* is a synergistic compendium of 11 informative and cutting-edge chapters written by leading researchers in the fields of spectroscopy and condensed matter physics as applied to a variety of materials at the nanoscale.

"Raman Spectroscopy, Modeling and Simulation Studies of Carbon Nanotubes" authored by Daniel Casimir, Raul Garcia-Sanchez and Prabhakar Misra focuses on carbon nanotubes (CNTs) that are allotropes of carbon and consist of a single layer of sp^2-hybridized carbon atoms. CNTs have a hollow cylindrical shape and are of two types: single-walled carbon nanotubes (SWCNTs) and multi-walled carbon nanotubes (MWCNTs). The majority of SWCNT have diameters of the order of ~ 1 nm and lengths of the order of microns to centimeters. MWCNTs are composed of concentric layers of SWCNTs nested inside one another, giving it a layered cylindrical shape. In this chapter, the authors provide a historical overview of CNTs and examine specifically their thermal properties and describe in-depth the use of Raman spectroscopy and Molecular Dynamics (MD) simulations to characterize and investigate the thermal characteristics of SWCNTs. Such investigations lead to a clearer understanding of the chirality of nanotubes through the visualization of rolled-up graphene sheets, and in turn provide insight into the versatility and myriad thermomechanical and electrical properties that make CNTs excellent candidates for a wide variety of applications in nanoelectronics and the semiconductor industry.

In "Laser Optogalvanic Spectroscopy and Collisional State Dynamics Associated with Hollow Cathode Discharge Plasmas" Michael Blosser, Xianming Han, Raul Garcia-Sanchez, and Prabhakar Misra discuss the laser optogalvanic effect in a discharge plasma environment, specifically associated with an iron–neon (Fe–Ne) hollow cathode lamp. The theoretical model behind the optogalvanic effect provides insight into the importance of laser optogalvanic spectroscopy as a tool for spectral characterization of plasma processes and enhanced understanding of the collisional state dynamics associated with the discharge species in hollow cathode lamps. The chapter focuses on transition states of neon in the Fe–Ne hollow cathode lamp and for illustrative purposes analyzes in detail the waveforms associated with the optogalvanic transitions of neon: $1s_4$–$2p_3$ (607.4 nm), $1s_5$–$2p_7$ (621.7 nm), $1s_3$–$2p_5$ (626.6 nm), $1s_5$–$2p_8$ (633.4 nm) and $1s_5$–$2p_9$ (640.2 nm). A comparison between the experimentally recorded waveforms and the Monte Carlo fitting routine, along with a discussion related to the variation of the fitting coefficients as a function of the discharge current, illustrates the success of the theoretical model in elucidating collisional state dynamics associated with discharge species in a plasma environment. The authors also describe the potential applications of the optogalvanic effect at the nanoscale in fields such as graphene-based nanoelectronics and nanoplasmonics.

"Applications of Fluorescence Anisotropy in Understanding Protein Conformational Disorder and Aggregation" written by Neha Jain and Samrat Mukhopadhyay focuses on the applications of fluorescence anisotropy in protein conformational disorder and aggregation. Fluorescence spectroscopy is an ultra-sensitive multiparametric technique that provides key insights into protein conformational dynamics and size changes simultaneously. Fluorescence polarization (anisotropy) is one of the parameters related to the rotational dynamics of a fluorophore either intrinsic to the molecule or attached to a biomolecule. The anisotropy measurements can be utilized to unravel the structural and dynamical properties of biomolecules. The advantage of fluorescence anisotropy measurements is that it is a concentration-independent parameter; it can be measured either in the steady-state or in the time-resolved format. Steady-state fluorescence anisotropy provides important information about the overall size/dynamics of biomolecules, whereas the time-resolved fluorescence anisotropy can distinguish between the local and the global dynamics of a fluorophore. Therefore, the time-resolved anisotropy measurements allow one to determine the conformational flexibility of biomolecules. In recent years, it has been demonstrated that fluorescence anisotropy can be effectively utilized to obtain structural and dynamical information about protein-based assemblies such as aggregates, protein-lipid complexes, etc. This chapter provides an overview of the applications of fluorescence anisotropy to study protein conformational disorder, misfolding, and aggregation, leading to the formation of nanoscopic amyloid fibrils that are implicated in a range of human diseases.

"Nuclear Magnetic Resonance Spectroscopy in Nanomedicine" authored by Ping-Chang Lin talks about the essentials of noninvasive measurements and deep penetration through living bodies using nuclear magnetic resonance (NMR) spectroscopy as a useful adjunct tool that complements *in vitro* and *in vivo* centered investigations in nanomedicine. This chapter introduces basic NMR principles and important NMR techniques that are especially relevant to the field of nanomedicine, followed by exploration of physicochemical characterization and metabolic profiles germane to the treatment and understanding of serious diseases. Conventional medicine generally relies on certain expression of disease symptoms (eg., metabolic responses, infections, and cancers) detected or diagnosed through the use of sophisticated and usually expensive medical instruments, devices, or implants, while appropriate therapy is subsequently developed based on clinical trials. On the contrary, nanomedicine, which focuses more attention on nanoscale interactions within biomolecules, cell organelles, and cells, is aimed at achieving early diagnosis and to ensure accurate dosage for treatment at the molecular or cellular level, prior to the appearance of traditional symptoms. Two major challenges remain at the forefront of nanomedicine, namely: (1) how to reliably regulate the assembly of more than one active component in a nanoscale vector and (2) how well to control the forms and dosages of the active agents that are spatially and temporally released.

"An Efficient Coupled Dipole Method for the Accurate Calculation of van der Waals Interactions at the Nanoscale" is written by Hye-Young Kim and details the powerful and efficient coupled dipole technique for understanding van der Waals (VDW) interactions at the nanoscale. The VDW force arises from purely quantum mechanical charge fluctuations and is variously called a dispersion or London or Casimir force. This often considered as weak, yet ubiquitous, attractive interaction is important in many nanoscale systems. This chapter provides an overview of the Coupled Dipole Method (CDM), an atomistic and accurate computational method widely adopted to predict the VDW forces between dielectric nanomaterials. There is a concern about the burden of memory and computing time needed to solve eigenvalue problems by either diagonalization or iteration, which have hindered the implementation of CDM for large systems. Here, an efficient way, named trace-CDM (TCDM), is presented. TCDM uses the simple fact that the trace of a square matrix is equal to the sum of its eigenvalues and thus calculates the accurate VDW energies without solving for the eigenvalues. Four examples are presented to demonstrate the advantages of the method as a time and effort saver computational technique for complex eigenvalue problems in nanoscience.

"Adsorption of Gases in Nanomaterials: Theory and Simulations" by Mamadou T. Mbaye, Sidi M. Maiga and Silvina M. Gatica describes the theory and simulations associated with the adsorption of gases in nanomaterials. "Physical adsorption" (physisorption) is a term applied to atoms or molecules that are weakly bound to surfaces and has been extensively explored for more than half of a century because of interest in both potential applications and basic science. These applications include the separation of cryogenic gases, their storage, and their use as surface characterization tools, such as the measurement of surface area of porous media by nitrogen adsorption. The science of physisorption encompasses a wide variety of fundamental questions, including many related to phase transitions. The remarkable capability of physical adsorption to explore and analyze these diverse phenomena *quantitatively* derives from the weak binding energy <0.3 eV, which has several significant consequences. One is that the adsorbed film represents just a small perturbation of the underlying surface, simplifying theoretical analysis enormously. Often the substrate is assumed to be perfectly rigid, unaffected by the film. Another consequence is that the adsorbed gases are not significantly altered from their 3D gas phase electronic state, meaning that their mutual interactions are relatively well known. A particularly valuable consequence of the small binding energy is that a vapor coexists with the film (except at very low T), so that one can experimentally determine the chemical potential of both phases. This chapter focuses on adsorption in nanomaterials. In particular, the authors describe equilibrium properties of gases adsorbed in CNTs, graphene and Metal Organic Frameworks (MOFs). The behavior of adsorbed matter at the nanoscale is dramatically different from the corresponding bulk material, due in part to the reduced dimensionality. The chapter describes the main results of simulations of adsorbed gases in nanomaterials and predicts novel phases that have not been observed yet, presenting challenges and opportunities to make significant breakthroughs in understanding intriguing phenomena linked with the behavior of adsorbed molecules at the nanoscale.

"Atom-Precise Metal Nanoclusters" penned by Anu George and Sukhendu Mandal looks at the world of atom-precise metallic nanoclusters. Nanoscale metals are generally classified based on three length scales into nanoparticles, nanocrystals, and nanoclusters. Nanoparticles are broadly defined as nanoscale particles in the range 1–100 nm, while the term "nanocrystal" is defined as a nanoscale crystallite of size >2 nm and nanoclusters (NCs) are those noncrystalline nanoparticles that are typically very small and composed of a specific number of metal atoms in the metal core, which are protected by shell of ligands. NCs are noncrystallographic in nature due to lack of translational symmetry. Optical properties of large metal nanoparticles to external electromagnetic fields depend on their sizes, free-electron density, and therefore their nearly bulk-like dielectric function relative to that of the surrounding medium. When particle size approaches the subnanometer size regime, the optical, electronic, and chemical properties of metal NCs differ dramatically from the other two size regimes. The ultra-small size of these NCs induces distinctive quantum confinement effects, which result in discrete electronic structure and molecular-like properties, such as HOMO–LUMO electronic transition, enhanced photoluminescence, intrinsic magnetism, to name a few. These NCs bridge small organometallic complexes and larger crystalline metal nanoparticles. Understanding the evolution as a function of cluster size is one of the grand challenges in chemistry and physics at the present time. In this chapter, the authors discuss the bridging roles of metal NCs between molecular chemistry and nanochemistry. Fundamental understanding of the evolution of the structure, electronic, and optical properties, as the materials evolve from the atomic state to NCs to fcc-structured nanocrystals, is of paramount importance and constitutes a major task in nano and materials science research. The evolution across length scales holds promise to yield fundamental insights into correlation between structure and the important quantum mechanical properties of NCs.

"Plasmonic Properties of Metallic Nanostructures, Two Dimensional Materials, and Their Composites" is authored by Lauren Rast and describes the intense and highly tunable optical field enhancement provided by nanomaterials supporting plasmon resonances, which has diverse applications in the areas of biophotonics, terahertz spectroscopy, and subwavelength microscopy. This chapter compares plasmon resonance behavior and tunability in noble metal nanostructures with that of two-dimensional and quasi-two dimensional materials including graphene, silicene, germanene, and transition metal dichalcogenides. Plasmonic optical behavior and related advancements in two-dimensional materials functionalized by metallic nanostructures are discussed, as well as possibilities for new directions for work on similar composite plasmonic systems are outlined.

"Application of Graphene Within Optoelectronic Devices and Transistors" written by F.V. Kusmartsev, W.M. Wu, M.P. Pierpoint, and K.C. Yung talks about how scientists are always yearning for new and exciting ways to unlock graphene's true potential. There are reports that suggest this two-dimensional material may possess some unique properties that make it a viable candidate for use in optoelectronic and semiconducting devices. Although graphene is highly transparent due to its atomic thickness, the material does exhibit strong interaction with photons, which provides clear advantages over existing materials when used in photonic devices. Moreover, the material can be used to *trap* light and alter the incident wavelength, forming the basis for a plasmonic device. The authors highlight graphene's nonlinear optical response to an applied electric field and the phenomenon of saturable absorption. Within the context of logical devices, graphene has no discernible band-gap. Therefore, generating one will be of utmost importance. Among many others, some existing methods to open this band-gap include chemical doping, deformation of the honeycomb structure, or the use of CNTs. The authors also discuss various designs of transistors, including those which incorporate CNTs, and others which exploit the idea of quantum tunneling. A key advantage of the CNT transistor is that ballistic transport occurs throughout the CNT channel, with short channel effects being minimized. The authors go on to discuss recent developments associated with the graphene tunneling transistor, with emphasis being placed upon its operational mechanism. Kusmartsev and colleagues also discuss the incorporation of graphene within high frequency devices that do not require a predefined band-gap.

"The Versatile Roles of Graphene in Organic Photovoltaic Device Technology" authored by S. Jayalekshmi and Sreekanth J. Varma builds on the wonder material graphene's potential applications in the realization of efficient and stable organic optoelectronic devices, especially flexible solar cells, and assesses the prospect of realizing all carbon photovoltaic devices. The combination of uniqueness and fine tuning of electrical and optical properties of graphene, makes it a highly sought after candidate for various technologically important applications in optoelectronics. Graphene has been identified as a suitable replacement for the highly expensive, brittle and less abundant indium tin oxide, as the transparent electrode material for optoelectronic device applications. The absence of energy band-gap in graphene had originally limited its applications in optoelectronic devices. This problem has since been solved with the advent of functionalized graphene and graphene nanoribbons (GNRs). Functionalized graphene and GNRs have extended its use as hole and electron transport layers in organic/polymer light emitting diodes and organic solar cells by appropriate tuning of the band-gap energy. Blending dispersions of functionalized graphene with the active layers in photovoltaic devices has been found to enhance light absorption and to enable carrier transport efficiently. Graphene layers with absorption in the entire visible region can be fine-tuned to be incorporated into the active layers of organic solar cells. Realization of optimal conditions for the synthesis of GNRs and functionalized graphene hold promise for achieving the requisite structural, optical and electrical properties for the development of all carbon-based cost-effective organic solar cells with improved efficiency in the not-too-distant future.

"Nanomaterials in Nanomedicine" written by Francis Mensah, Hailemichael Seyoum, and Prabhakar Misra focuses on the applications of nanomaterials in nanomedicine. Nanotechnology has led to the construction of new devices for diagnostics and therapeutics, which permits early detection and treatment of malignant tumors and other serious diseases. It has also led to the improvement and efficiency of drug, gene, and protein delivery. Nanoparticles are being used to target the central nervous system (CNS), which can be challenging because of the necessity to cross the blood–brain barrier (BBB). They are also used in the treatment of Alzheimer's Disease (AD), which is a neurodegenerative disorder prevalent in the aging population. It is characterized by severe neuronal loss and proliferation of plaques composed of β-amyloid peptide (A_β) and τ-protein deposits. An imbalance between production and clearance leads to the aggregation of A_β peptides, especially in neurotoxic forms, and may be the initiating factor in AD. Nanobiotechnology forms the basis of many new devices being developed for medicine and surgery, such as nanorobots. It has applications in most branches of medicine, such as: nanooncology—concerning cancer research; nanoneurology—the study of neurological disorders; nanocardiology, which studies cardiovascular disorders; nanoorthopedics, which studies diseases of bones and joints; nanoophthalmology, which studies eye's diseases; and also for other infectious diseases. This chapter illustrates the use of nanoparticles as effective drug carriers for localized treatment of tumors via the use of liposomes, which are viscoelastic and deformable materials, akin to red cells. In this chapter, the authors also present an overview of the viscoelastic properties of liposomes in connection with the use of red blood cells evolution equation in mathematical biology. Recent advances have shown that liposomes can be used to incorporate red blood cells for delivery in the human body, and understanding rheological properties of liposomes is vital to their effective therapeutic utilization.

I would like to thank all the chapter authors for their excellent contributions and for their patience as the book project evolved from the planning to the implementation stage over a period of several months and a number of iteration cycles. Also, I would like to thank Mr. Loyola D'Silva of Springer for making the initial contact with me at OPAP 2013, the Annual International Conference in Optoelectronics, Photonics and Applied Physics, held in Singapore in February 2013, when this monograph took initial shape, and for his persistent, guidance and support in making this book project a reality.

In closing, I would like to acknowledge my wife, Ms. Suneeta Misra, and my children, Dr. Isha Misra and Mr. Uday Misra, without whose love and support this book may never have been completed in a timely manner.

I would like to dedicate this book in fond memory of younger brother, Dr. Sudhakar Misra, who passed away prematurely on December 15, 2013.

Washington, DC, August 2014 Prabhakar Misra

Contents

Raman Spectroscopy, Modeling and Simulation Studies of Carbon Nanotubes 1
Daniel Casimir, Raul Garcia-Sanchez and Prabhakar Misra

Laser Optogalvanic Spectroscopy and Collisional State Dynamics Associated with Hollow Cathode Discharge Plasmas 21
Michael Blosser, Xianming L. Han, Raul F. Garcia-Sanchez and Prabhakar Misra

Applications of Fluorescence Anisotropy in Understanding Protein Conformational Disorder and Aggregation 41
Neha Jain and Samrat Mukhopadhyay

Nuclear Magnetic Resonance Spectroscopy in Nanomedicine 59
Ping-Chang Lin

An Efficient Coupled Dipole Method for the Accurate Calculation of van der Waals Interactions at the Nanoscale 85
Hye-Young Kim

Adsorption of Gases in Nanomaterials: Theory and Simulations 121
M.T. Mbaye, S.M. Maiga and S.M. Gatica

Atom-Precise Metal Nanoclusters 141
Anu George and Sukhendu Mandal

Plasmonic Properties of Metallic Nanostructures, Two Dimensional Materials, and Their Composites 165
Lauren Rast

Application of Graphene Within Optoelectronic Devices and Transistors 191
F.V. Kusmartsev, W.M. Wu, M.P. Pierpoint and K.C. Yung

The Versatile Roles of Graphene in Organic Photovoltaic Device Technology 223
Jayalekshmi Sankaran and Sreekanth J. Varma

Nanomaterials in Nanomedicine 253
Francis Mensah, Hailemichael Seyoum and Prabhakar Misra

Raman Spectroscopy, Modeling and Simulation Studies of Carbon Nanotubes

Daniel Casimir, Raul Garcia-Sanchez and Prabhakar Misra

Abstract This chapter focuses on two types of carbon nanotubes (CNTs): single-walled carbon nanotubes (SWCNTs) and multi-walled carbon nanotubes (MWCNTs). CNTs are cylindrically-shaped carbon allotropes. They consist of a single layer of sp^2-hybridized carbon atoms, giving it a hollow cylindrical shape. The majority of SWCNT samples have diameters on the order of ~ 1 nm and lengths on the order of microns to centimeters. MWCNTs are composed of concentric layers of SWCNTs nested inside one another, giving it a layered cylindrical shape. In the present chapter, we will provide a historical overview of CNTs and examine specifically their thermal properties as it relates to their applications to the semiconductor industry and nanoelectronics. The understanding of CNT chirality through the visualization of rolled-up graphene sheets will provide insight into the versatility and myriad thermo-mechanical and electrical properties of CNTs. We will focus on the use of Raman spectroscopy and Molecular Dynamics (MD) simulations to characterize and investigate the thermal characteristics of SWCNTs.

Keywords Carbon nanotubes · Single-walled · Multi-walled · Chirality · Thermal properties · Thermo-mechanical properties · Electrical properties · Graphene · Raman spectroscopy · Modeling · Molecular Dynamics (MD) simulation

1 Introduction

In order to provide an overview of carbon nanotubes, we will take the historical perspective, although the reported timeline has been mired in controversy. In 1952, Soviet scientists, Radushkevich and Lukyanovich reported the first Transmission Electron Microscope (TEM) images of MWCNTs [1]. There was a period of lull

D. Casimir · R. Garcia-Sanchez · P. Misra (✉)
Department of Physics and Astronomy, Howard University, Washington, DC 20059, USA
e-mail: pmisra@howard.edu

P. Misra (ed.), *Applied Spectroscopy and the Science of Nanomaterials*,
Progress in Optical Science and Photonics 2, DOI 10.1007/978-981-287-242-5_1

regarding landmark discoveries relating to CNTs until 1976, when Oberlin et al. generated single- and multi-walled CNTs while trying to grow carbon fibers by pyrolyzing a mixture of benzene and hydrogen around 1,100 °C [2]. Such filamentous growth of carbon was followed by the observation of so-called carbon fibers in an arc discharge by Abrahamson, who reported this at the biennial carbon conference held at The Pennsylvania State University in 1979 and later reported their characterization in 1999 [3]. Circular shaped (armchair configuration) and helical shaped (chiral form) CNTs were suggested by Soviet scientists in 1982 following Transmission Electron Microscopy (TEM) imagery and X-ray Diffraction (XRD) pattern recording of carbon nanoparticles [4]. In 1987, Tennent received a U.S. patent for producing and characterizing the so-called carbon fibrils [5]. This was followed by Iijima's discovery of MWNTs in 1991, while attempting to understand the growth mechanisms associated with another carbon allotrope, namely buckminsterfullerene (or bucky-ball) [6] and Mintmire et al.'s publication in 1992 [7] regarding the promise of excellent conducting properties of fullerene tubules and their zero band gap at room temperature. Both Iijima and Mintmire et al.'s publications in the early 1990s helped bring significant visibility to CNTs and energized research into their characterization and properties at an astonishing pace all around the globe. Such investigations were spurred greatly by subsequent discoveries related to SWCNTs by Iijima and Iohihashi [8] and Bethune et al. [9].

Owing to their high aspect ratio, studies of CNTs can serve in testing theories of 1-dimensional systems [10]. Since being brought into the scientific spotlight over 20 years ago by Iijima [6], there have been rapid developments regarding the unique physical properties of CNTs and both SWCNTs and MWCNTs have been extensively studied. Investigations into their thermal properties serve as a good illustration of the continued interest in understanding both their behavior and the potential for a number of applications in materials science and engineering that arise from the wide-ranging configurations this material can take. Knowledge regarding the thermo-mechanical properties of CNTs, although not at presently limited, has often been controversial. One such topic lies in the axial and radial thermal expansions of SWCNTs. Kwon et al. [11] reported a negative thermal expansion of SWCNTs in a temperature interval of 0–800 K, while Li [12] reported positive values for the same temperature range. Optical techniques, such as fluorescence and Raman spectroscopy, allow rapid and reliable characterization of CNTs, including their structural imperfections and chirality.

The concept of chirality, or the ability to superpose the mirror image of an object on top of the original, is important for understanding the lattice and structural complexity of SWCNTs. SWCNTs can be grouped via two major classifications into chiral and achiral. By utilizing a folding construction on a flat graphene lattice, the primitive lattice of any SWCNT can be constructed based on the chiral and achiral classifications [13]. The axis of symmetry for SWCNTs is along the longitudinal direction, which in turn determines the arbitrary orientation and angle of the carbon hexagons along the cylindrical wall of the resulting nanotube. A surprisingly simple folding operation applied to the planar graphene lattice can account for the existence of only two achiral varieties (*armchair* and *zigzag*), and also for the helicity of the remaining non-symmetric or chiral tube types. Also significant in this discussion is that the electronic properties of SWCNTs depend exclusively on their geometry [10]. However, it should be emphasized that the description regarding the nanotube structure is merely a mathematical construction, and that actual CNTs grow naturally in the form of cylinders according to complex and not fully understood growth mechanisms [14].

2 Electronic Structure of CNTs

Large-scale Atomic/Molecular Massively Parallel Simulator (LAMMPS) is an open source code that performs Molecular Dynamics (MD) simulations. By creating an input text file that specifies lattice spacing, atom type, atom interactions and other parameters of the nanoscale system we are trying to model, LAMMPS outputs a 'dump' file. This 'dump' file can be run by an MD visualization software, such as The Open Visualization Tool (Ovito) or Visual Molecular Dynamics (VMD) package. The visualization in Ovito shown in Fig. 1 shows a single sheet of graphene and the structure of the carbon atoms. Each atom is bonded to three other atoms forming the hexagonal shaped structure. With LAMMPS, the trajectories of atoms can be calculated and more complicated interactions can be simulated. Interestingly, for example, this structure could be extended to analyze water on graphene sheets.

LAMMPS utilizes four main components that need to be specified in each input text file, which are: *initialization*, *atom definition, settings* and *running simulations*. The 'initialization' dictates units, dimensions, atom styles, etc. For 2D graphene, the units are 'metal', and atom style is simply 'atomic'. The next step, atom definition, dictates the lattice spacing for graphene. A custom lattice is needed to

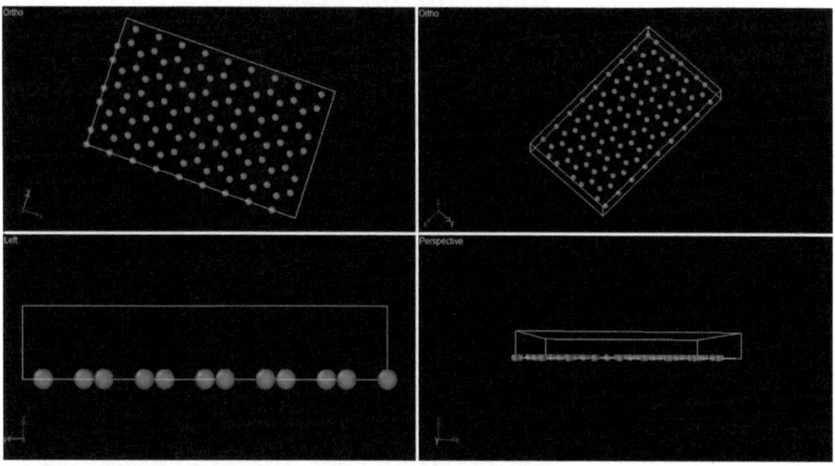

Fig. 1 3-D Visualization of graphene structure using the open visualization tool (*Ovito*) software

replicate the sheet of graphene (shown in Fig. 1). The distance between each atom was defined in angstroms based on literature values of lattice spacing for graphene [15]. The third section, settings, includes properties, such as pair and bond coefficients. This particular simulation required pair interactions, which can be defined as 'pair_style' under the atom definition section, and 'pair_coefficients' under settings. As a first approximation, one can use the Born potential for visualizing graphene; however, one needs to use the refined potential termed the Adaptive Intermolecular Reactive Empirical Bond Order (AIREBO) potential for a more accurate modeling and representation of the structure and its dynamics. Finally, a simulation can be set up for the molecules produced. Figure 1 shows the 3-D visualization generated for the structure of graphene employing a combination of LAMMPS and Ovito.

Let a_1 and a_2 be two graphene basis vectors, as illustrated in Fig. 2, and C_h the so-called chiral vector obtained as a linear combination of a_1 and a_2 in terms of a pair of integers n and m:

$$\boldsymbol{C_h} = n\boldsymbol{a}_1 + m\boldsymbol{a}_2 \tag{1}$$

with magnitude

$$|\boldsymbol{C_h}| = a\sqrt{n^2 + m^2 + nm} \tag{2}$$

where $|\boldsymbol{C_h}|$ can be considered as a measure of the circumference of the nanotube.

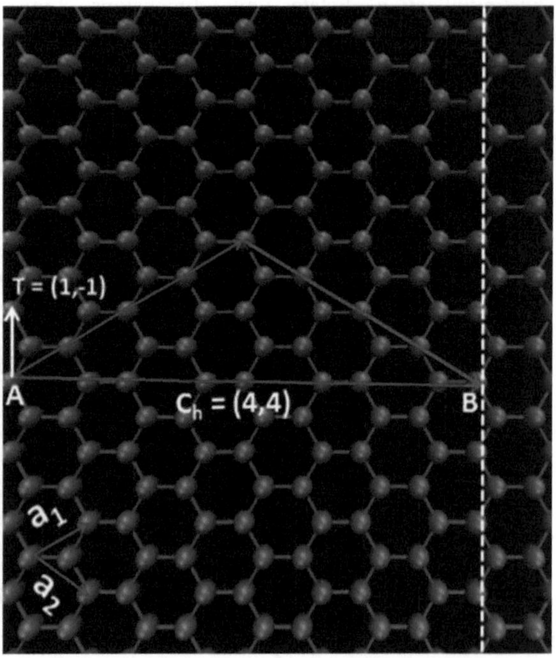

Fig. 2 Figure demonstrating the geometrical construction of a (4, 4) armchair single-walled carbon nanotube from a graphene sheet. The sheet is rolled along the chiral vector C_h until the Carbon atom at point A coincides with the Carbon atom at point B. The *rectangle* formed by the Translation vector T and the chiral vector form the unit cell of the resulting nanotube

The values of n and m are arbitrary with the constraint $0 \leq |m| \leq n$. The Cartesian components of the lattice vectors a_1 and a_2 are $(a\sqrt{3}/2,\ a/2)$ and $(a\sqrt{3}/2,\ -a/2)$, respectively, where $a = a_{C-C}\sqrt{3} = 2.46$ Å.

Figure 2 shows a diagram of a planar hexagonal lattice and the relevant geometrical parameters in the description of the conceptual rolling/folding operation in the construction of a single walled carbon nanotube from this sheet. The two vectors a_1 and a_2 are the basis lattice vectors of the planar sheet, and the vector is rolled along the chiral vector, C_h until point A coincides with point B. The first and second integers in the designation of the nanotube chirality n and m come from the amount of a_1 and a_2 vectors used to construct the two vectors that have the chiral vector as their resultant. The translation vector T, is the shortest possible vector that is perpendicular to the chiral vector that can be constructed from the basis vectors a_1, and a_2.

3 Thermo-Mechanical Properties of SWCNTs

It was anticipated in the 1990s when they were discovered that SWCNTs would have unique remarkable mechanical properties, and that has have proven to be the case. A common denominator among many of these extraordinary mechanical properties is the nature of the intra-planar Carbon-Carbon bond in the graphene lattice, the conceptual parent material of all other carbon allotropes including nanotubes. Overall, the mechanical properties of carbon nanotubes are reasonably well understood. However, research on these thermo-elastic properties is still not at a uniform level of development [11]. This is due to the fact that although continuum/elasticity theory has proven to be applicable in studying the mechanical properties of SWCNTs, its use at the nano-scale requires careful use. A major portion of the research of SWCNT mechanical properties concerns the measurement and calculation of various elastic constants, such as Poisson's ratio, Young's modulus, and the thermal expansion coefficients of bundled and individual SWCNTs.

Figure 3 shows an SEM image of a sample of the purified SWCNT reported on in this chapter and illustrates one of the exceptional mechanical properties of these objects hinted at above. SWCNTs were predicted to have rather "soft" surfaces for externally applied loads perpendicular to their axis, with the ability in some cases to flexibly bend around sharp corners or in circles and not break [1]. The relatively large free standing SWCNT imaged in the center of Fig. 3 shows just this behavior. In sp^2 graphitic materials the intra-planar covalent σ bonds from the base structure resulting in elastic thin film behavior along the axial direction for nanotubes. However, because SWCNTs have hollow cores with no interlayer potential interactions, perpendicular deformations, including unstable ones result in bending of the nanotube without any fracture of the base σ skeleton.

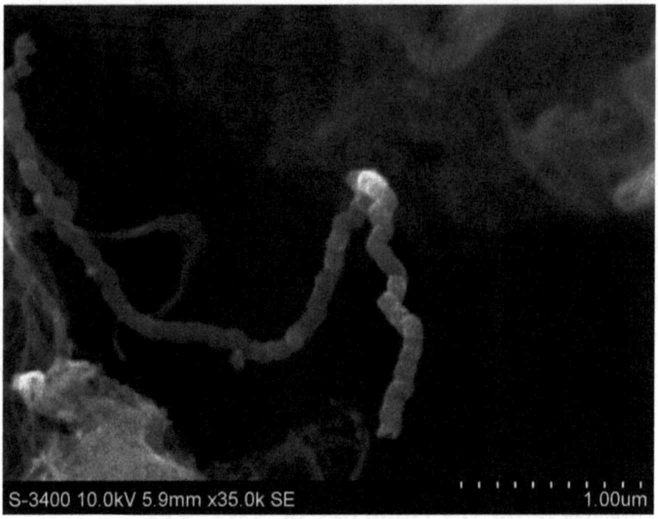

Fig. 3 SEM Image of purified SWCNT sample produced by the HiPco technique

4 Raman Spectroscopy of SWCNTs

We have used Raman spectroscopy for our experimental investigations into the thermal effects associated with SWCNTs. Resonance Raman spectroscopy serves as a sensitive tool for probing the mechanical properties of SWCNTs. Intensities of Raman transitions are typically quite weak, especially from the laser excitation of nano-sized entities, however the unique electronic properties of these quasi one dimensional objects cause a many fold increase in the Raman intensity, known as resonance Raman scattering, allowing in some cases for the collection of Raman spectra of individual SWCNTs.

A schematic of the instrument (Thermo Scientific DXR SmartRaman Spectrometer), employed to carry out the Raman spectroscopy measurements of the CNT samples, is shown in Fig. 4. A 780 nm laser source (frequency stabilized single mode diode with high brightness; frequency stability <1 cm^{-1} over a one-hour period and beam quality with a power setting of 6 mW was used. A full range grating was employed in the triplet spectrograph used (with an average spectral dispersion of 2 cm^{-1} per CCD pixel). The laser beam is focused on the sample using an objective lens and the scattered radiation from the sample is directed into the Imaging Spectrometer through a collection lens and a Raman signature (Stokes transitions only) is recorded using the appropriate Rayleigh filters and automated aperture (20–50 μm) selections as read by the CCD detector. Specifically, for instance, a 25 μm slit provided a resolution ~ 1.93 cm^{-1}. A Ventacon Heated Cell served as a temperature-controlled environmental chamber (in the temperature range room temperature up to 200 °C) used and housed the sample either on its surface (in case of solid samples) or within (as in the case of nanopowders and

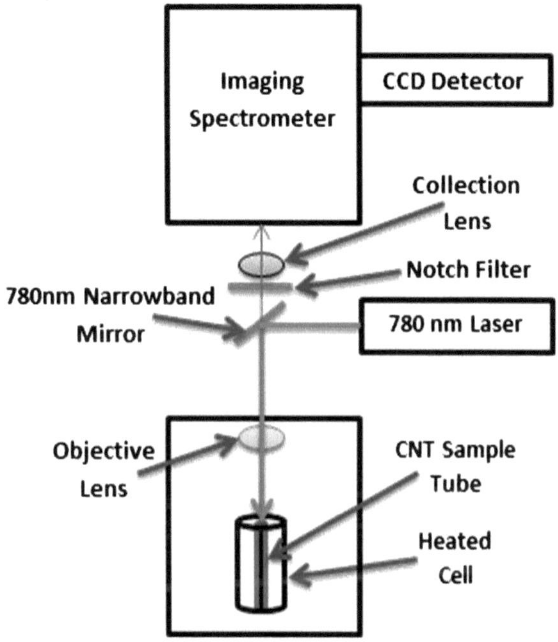

Fig. 4 A schematic of the experimental arrangement used for recording the Raman spectra of the CNT samples

carbon nanotubes). All temperature-dependent data were specifically collected for the purified batch of SWCNTs.

Significant Raman bands recorded for SWCNTs and their relationship to important mechanical properties are discussed below [16].

Figure 5 is an example Raman spectrum collected in the backscattering configuration with 780 nm laser excitation for a sample of SWCNTs produced through the high-pressure carbon monoxide (HiPco) process—a gas-phase method for large-scale production of CNTs—and subsequently purified to 15 % residual Fe catalyst by weight. The pair of peaks (labeled G^+ and G^-) are customarily present in sp^2-hybridized graphitic materials as a single peak characteristically identified as the G (Graphite)-band around $\sim$1,580 cm^{-1} and is related to the *in plane* C–C bond stretching mode for graphitic nanomaterials. Such (G^+ and G^-) peaks occur because of strain induced in rolled-up CNTS that cause the transverse and optical phonon modes to mix and become Raman active and subsequently split into two separate phonon frequency peaks—with peak separation serving as a direct measure of the amount of strain present in the graphitic nanomaterial being analyzed. (Both elasticity theory and ab initio techniques have been used in understanding how the above spectral features can also be linked to the diameters of the nanotubes.)

In Fig. 5, the splitting observed shows two peaks at 1559.88 and 1585.00 cm^{-1}, which correspond to G^- and G^+, respectively. These experimental results agree with the ab initio derived expression for G^- in Ref. [9].

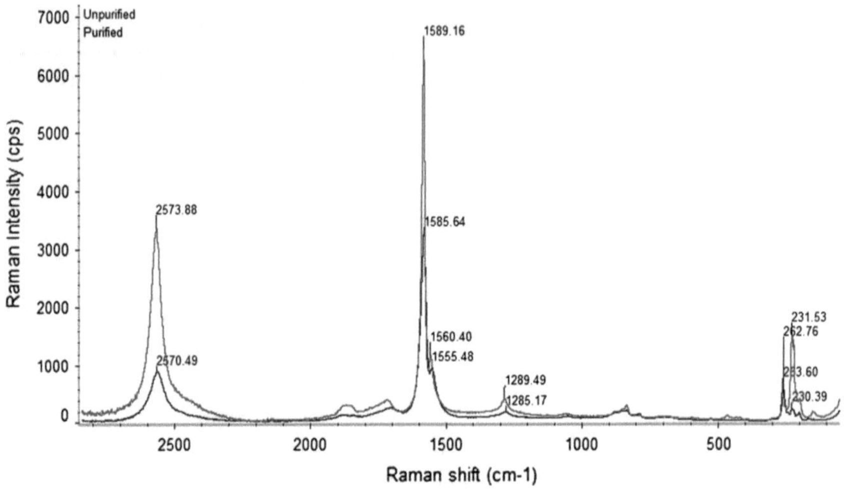

Fig. 5 Comparison of the Raman spectrum of both unpurified and purified SWCNT sample obtained with 780 nm excitation

$$\omega_G = 1592 - \frac{C}{d_t^\beta} \tag{3}$$

The experimental parameters obtained for this expression in Eq. (3) were as follows: $\beta = 1.4$, $C_{A1} = 41.4\ cm^{-1}$ nm, $C_{E1} = 32.6\ cm^{-1}$ nm, and $C_{E2} = 64.6\ cm^{-1}$ nm [9]. Small strain-induced changes in carbon nanotubes are easily measurable by Raman spectroscopy, due to the high frequency of the graphitic G band. For example, a mere 1 % strain-induced change in this Raman feature of sp^2-hybridized CNTs resulted in a measurable peak splitting $\sim 16\ cm^{-1}$, which is larger than the 10 cm^{-1} width of the natural G band. The more than 25 cm^{-1} difference between the two G band features for the semiconducting tubes in resonance, without any external loading, also exhibited this very useful feature of SWCNTs. The next group of bands in the low frequency regime, specifically the 264 and 228 cm^{-1} features, dubbed the radial breathing modes (RBM), are another set of Raman features of SWCNTs that are intimately tied to their mechanical properties (like the previously discussed G band). However, where the G-band is present in all sp^2 graphitic materials, the RBM features are unique to nanotubes. As its name indicates, the RBM refers to the collective radial vibration of the SWCNT, as if it were breathing. This Raman feature is the one usually relied on for the estimation of the average diameter of CNTs, and not the diameter dependence of the G-band feature in Eq. (3). Both elasticity theory and ab initio techniques, in addition to experiment, predict an inverse diameter relationship for this Raman feature, with small deviations from the simple relationship given in Eq. (4) arising from effects due to inter-tube interactions in bundled nanotube samples and the Kohn anomaly.

$$\omega_{RBM} = \frac{A}{d_t} + B \tag{4}$$

In the present case, the pair of fitted parameters (A, B) used in the diameter estimation have the following values: A = 234 cm^{-1} nm and B = 10 cm^{-1}, yielding an average SWCNT diameter of ~1.09 nm for the 228.81 cm^{-1} Raman peak in Fig. 5. Based on the geometrical structure of SWCNTs described earlier, and Eq. (2), this diameter value can be shown to correlate to the specific chiralities (14, 0) and (10, 6). The RBM seen at 264.01 cm^{-1} is also of special interest. It has been shown in Ref. [10], based on an RBM intensity map with 785 nm excitation, that this peak is useful in assessing the extent of aggregation of the different tubes present in a bundled sample, which may prove to be another helpful technique in the ongoing search for simpler robust techniques in determining the diameter distribution of samples compared to scanning probe methods. This "roping" peak, present in the spectra in Fig. 5, corresponded to the chirality (10, 2)—being in resonance for bundled samples, while for individualized nanotubes a chirality of (12, 1) was the dominant one [10].

The two other labeled features in the Raman spectrum shown in Fig. 5 are the D (Defect) band at 1290.22 cm^{-1} and the 2D band at 2578.61 cm^{-1}, respectively. Although the intensity ratio of the D "Defect" band to the G band is used in assessing the quality of SWCNT samples, these results must be used cautiously as it is still uncertain if the D band can be used for defect characterization in all sp^2 hybridized nanocarbon materials. An important wrinkle is that the D band Raman feature of individual SWCNTs is narrow (with line-widths in the range of 7–40 cm^{-1}), and the Raman spectra of bundled samples are made up of broad peaks with superimposed features from separate nanotubes in a given graphitic nano-sample. Another issue is the lower frequency of the D band in SWCNTs, compared to other bulk sp^2 nanocarbons, and in addition the downshift of the D band in nanotubes is also diameter-dependent. In addition, the 2D band results of second-order scattering near the K-point in the planar graphene Brillouin zone occurs via a double resonance process and such a highly dispersive second order feature also proves useful in the characterization of SWCNTs.

Although the focus of the present report is single walled carbon nanotubes (SWCNTs), a sample Raman spectrum of multi walled carbon nanotubes (MWCNTs) is shown in Fig. 6 for comparison, and to illustrate the effectiveness in the use of Raman spectroscopy in the characterization of Carbon allotropes. The major differences between the two sets of spectra of MWCNT and SWCNT, respectively, are the reduction in intensity and the number of RBMs present in the multi-walled sample [17]. Also noticeable is the more prominent defect peak in the multi-walled carbon nanotube sample. The first effect is due to the restriction of the radial vibration by the outer walls of the nested coaxial nanotubes, and the latter effect has been attributed to the stacked multi-layer of MWCNTs leading to more opportunities for disorder [18].

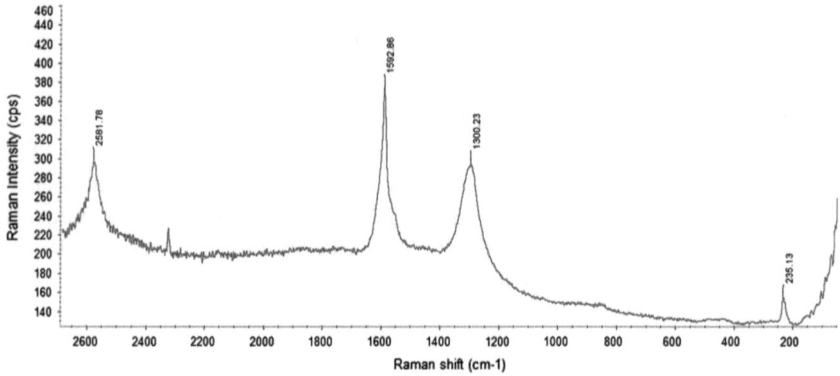

Fig. 6 Raman spectrum of multi walled carbon nanotubes with 780 nm laser excitation

4.1 Raman Spectra of SWCNTs as a Function of Temperature

When heated to higher temperatures the location of the Raman peaks of SWCNT and other carbon allotropes, shift to lower frequencies due to the softening of the inter-atomic bonds [18]. This red-shifting of the Raman frequencies of carbon nanotubes demonstrates the suitability of Raman spectroscopy in the characterization of carbon allotropes and carbon nanotubes in particular. This phenomenon can be decomposed into two effects, one purely thermal and the other a volume/strain effect caused by the thermal expansion of the crystal lattice. The first term in Eq. (5) represents the thermal effect, and the second term the volume effect [19].

$$\Delta\omega = \left(\frac{\partial\omega}{\partial T}\right)_V \Delta T + \left(\frac{\partial\omega}{\partial V}\right)_T \left(\frac{\partial V}{\partial T}\right)_P \Delta T \quad (5)$$

Many studies have shown however that the temperature effect is the dominant one in this phenomenon. For this reason both our results and those from the literature on this effect usually display a linear trend, since the second order term in Eq. (6) associated with the volume effect is usually very small [20].

$$\omega = \omega_0 - a_1 T - a_2 T^2 \quad (6)$$

Our results in reproducing the temperature induced linear red-shift of major Raman peaks in a sample of purified SWCNT ropes, specifically the G^+, G−, D, G′ and RBM peaks are shown in Figs. 7 and 8. Each data point on the plots is the average of two Raman spectra collected at the same sample point at each respective temperature. The linearity of the temperature variation of the G′ peak was not clearly evident in our data, with a significant amount of scattering in contrast to the other Raman peaks, as shown by its low coefficient of determination. The cause of

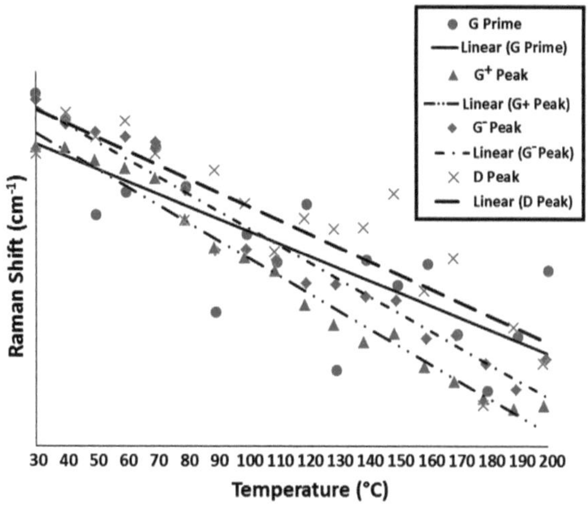

Fig. 7 The Raman Shift for the major SWCNT peaks as a function of temperature

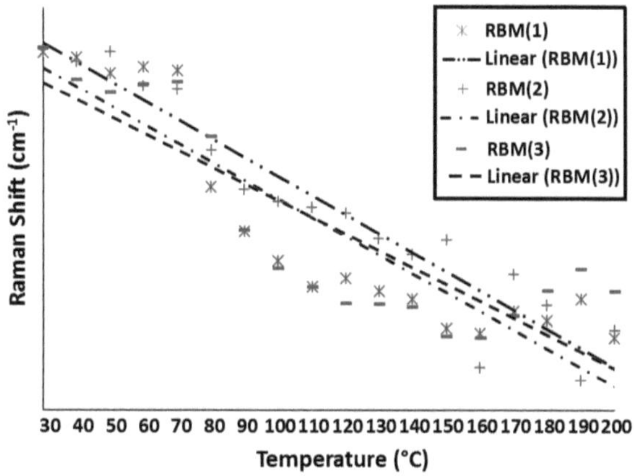

Fig. 8 The Raman Shift for the SWCNT RBM modes as a function of temperature

the departure from linearity in the redshift trend for this second order Raman band in our results is presently unknown. This peak however is very dispersive and sensitive to electronic and other perturbations [21]. This extreme sensitivity to thermal, mechanical and other perturbations was also the reason Deng et al., whose study is mentioned later also chose this Raman band in their thermal expansion study [22]. Raman spectral data were collected on the CNT samples under increasing temperature from room temperature up to 200 °C in 10 °C increments. The HiPco single walled nanotube samples were obtained from Unidym Carbon Nanotubes, and according to the manufacturer had only 15 % by weight of Fe catalyst present. The diameter range of the single walled carbon nanotubes was also stated as 0.8–1.2 nm [16].

Table 1 Values for the Raman shift (cm^{-1}) of the major SWCNT peaks with increasing temperature

Temp (°C)	G′	G^+	G	D	RBM (1)	RBM (2)	RBM (3)
30	2579.53	1594.2	1565.665	1294.405	267.58	233.995	206.425
40	2578.98	1594.19	1565.37	1294.96	267.555	233.935	206.28
50	2576.94	1594.055	1565.275	1294.625	267.475	233.985	206.225
60	2577.425	1593.97	1565.215	1294.845	267.51	233.815	206.2625
70	2578.375	1593.86	1565.155	1294.405	267.49	233.795	206.2725
80	2577.54	1593.42	1564.58	1293.52	266.91	233.495	206.0275
90	2574.86	1593.12	1563.86	1294.18	266.69	233.3	205.6075
100	2576.525	1593.015	1563.865	1293.735	266.54	233.24	205.4375
110	2575.925	1592.87	1563.74	1293.1	266.41	233.21	205.3525
120	2577.16	1592.51	1563.455	1293.545	266.455	233.18	205.28
130	2573.625	1592.3	1563.445	1293.4	266.39	233.055	205.2775
140	2575.975	1592.115	1563.3	1293.42	266.35	232.975	205.2625
150	2575.425	1592.205	1563.25	1293.875	266.205	233.045	205.1325
160	2575.88	1591.845	1562.795	1292.57	266.18	232.415	205.1275
170	2574.385	1591.69	1562.835	1293.005	266.29	232.875	205.225
180	2573.175	1591.51	1562.495	1291.04	266.245	232.725	205.3375
190	2574.335	1591.39	1562.18	1292.08	266.35	232.35	205.4325
200	2575.74	1591.43	1562.545	1291.6	266.16	232.6	205.335

Our tabulated and plotted results are shown in Figs. 7, 8, and Table 1, respectively, and the temperature slopes $\frac{\Delta\omega}{\Delta T}$ presented in Table 2 agree well with Gregan [20] and Raravikar [23].

Raravikar [23] reported temperature slope $\left(\frac{\Delta\omega}{\Delta T}\right)$ values of −0.0189 and −0.0238 cm^{-1}/K for the G^+ and G– peaks, respectively, using 514.4 nm excitation, while Gregan [20] reported similar values of −0.016 cm^{-1}/K for both G^+ and G^- peaks again using 514.5 nm excitation. In addition, in Table 2 we provide the temperature slopes for the G′, D and Radial Breathing Mode (RBM) features as well (with 780 nm excitation).

The temperature-shifted Raman peaks are useful in determining the thermal expansion coefficients associated with SWCNTs. It has been noted in the literature

Table 2 Temperature slopes of the various Raman peaks of bundled SWCNTs illustrated in Figs. 7 and 8

Raman band feature	Temperature slope $\frac{\Delta\omega}{\Delta T}$ (cm^{-1}/°C) using 780 nm excitation wavelength
G^+ Peak	−0.0187
G^- Peak	−0.0204
G′ Peak	−0.0264
"Defect" D Peak	−0.0182
RBM ~ 267 cm^{-1}	−0.0093
RBM ~ 234.31 cm^{-1}	−0.0095
RBM ~ 206.49 cm^{-1}	−0.0075

however, that the determination of the axial thermal expansion coefficient, and many other thermo-mechanical properties of individual carbon nanotubes is extremely difficult owing to their nanoscopic dimensions and also due to their appearance as one of many nanotubes entangled in complex nanotube bundles upon production (as illustrated in the SEM image shown in Fig. 3). Therefore, the majority of the values for the Coefficient of Thermal Expansion (CTE) coefficients are derived from experimental studies done on bundles of carbon nanotubes.

The latter is also true for the novel and practical recent work done by Deng et al. [22]. In Deng et al., the temperature shift in the Raman G′ band of single and double walled carbon nanotubes both in and outside of an epoxy mixture were measured. They chose to use the G′ band for the study because it, "was the most sensitive to external perturbations" [22]. The shift in the G′ band with applied strain with the laser polarization in the same direction as the applied strain was also measured, and finally using the temperature slope of the G′ band under the two previously mentioned conditions in addition to an average over all angular orientations of the strain shifted G` frequencies, Deng was able to calculate the CTE of the carbon nanotubes denoted as α_F in Eq. (7). α_E is the CTE of the epoxy matrix that the carbon nanotubes are embedded in Deng et al. [22].

$$\alpha_F = \alpha_E - {(\chi_C - \chi_F)}/{S_0} \tag{7}$$

Finally in this section, it is also shown that the thermal conductivity of SWCNTs can also be obtained through resonant Raman spectroscopy. Terekhov et al. [24] discusses a robust practical technique to estimate the thermal conductivity of a carbon nanotube sample. Initially while developing a Raman based calibration method of determining the amount of SWCNT content in powdered single walled carbon nanotube samples, they also noted the correlation between the temperature slope of the G^+ band of heated SWCNTs and the amount of SWCNT content present in the sample; using the above relationship, and the fact that the physical interpretation of the slope of the shift in the G^+ peak versus laser power density is the inverse thermal conductivity allowed them to estimate the thermal conductivity in samples with different amounts of SWCNT content or purity levels. Our reproduction of this technique, based on our data is shown in Fig. 9.

5 Molecular Dynamics Simulation

Computational studies regarding the mechanical properties of SWCNTs have often reported contradictory results for elastic moduli and constants. A major factor contributing to the above problem lies in the arbitrariness of the thickness of a nanotube shell, which is usually prescribed the value of the interlayer spacing in graphite ~0.35 nm. The early computational studies of the elastic properties of SWCNTs usually ascribed to Robertson et al. [25] employed the Tersoff and

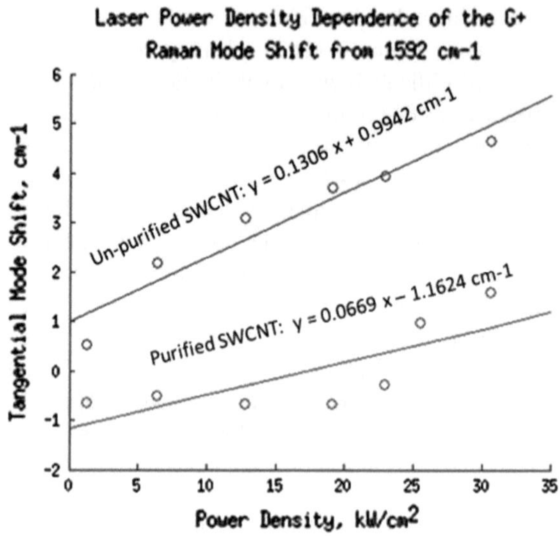

Fig. 9 The shift of the G^+ Raman feature away from 1,592 cm^{-1} upon heating caused by increased laser power density

Brenner empirical potentials, one of the first many-body empirical potentials used in such investigations. This particular study confirmed the elasticity theory-based prediction of the inverse squared diameter dependence of the strain energy, in addition to other interesting predictions, such as a decrease in the Young's modulus for extremely narrow SWCNTs. We have carried out Molecular Dynamics (MD) simulations based on the Adaptive Intermolecular Reactive Empirical Bond Order (AIREBO) potential [26, 27]. The latest iteration of the Tersoff-Brenner empirical potential, specifically developed for hydrocarbons, commonly in use now is the AIREBO potential:

$$E = \frac{1}{2}\sum_{i}\sum_{j\neq i}\left[E_{ij}^{REBO} + E_{ij}^{LJ} + \sum_{k\neq i,j}\sum_{l\neq i,j,k} E_{kijl}^{TORSION}\right] \quad (8)$$

The first term contains the repulsive and attractive interactions, including a special "bond order" term to modify bonded energies, the second term is the Lennard-Jones potential, and the last term calculates angular and torsional interactions. This potential was used to produce the illustrations in Figs. 10 and 11, a snapshot of one unit cell of a (12, 1) SWCNT representative of one of the chiralities present in the actual sample. Both figures are from equilibration runs, of MD simulations with 0.0005 ps time-steps, and a total of 80,000 time-steps. The bottom portion of Fig. 10 shows the distortion of a cross-sectional portion of the SWCNT during the first 3,000 time-steps in 1,000-step intervals. The choice of this empirical potential was also influenced by the extensive table of single-walled carbon nanotube mechanical properties produced by Bialoskorski and Rybicki using the above potential given in Eq. (8) [13].

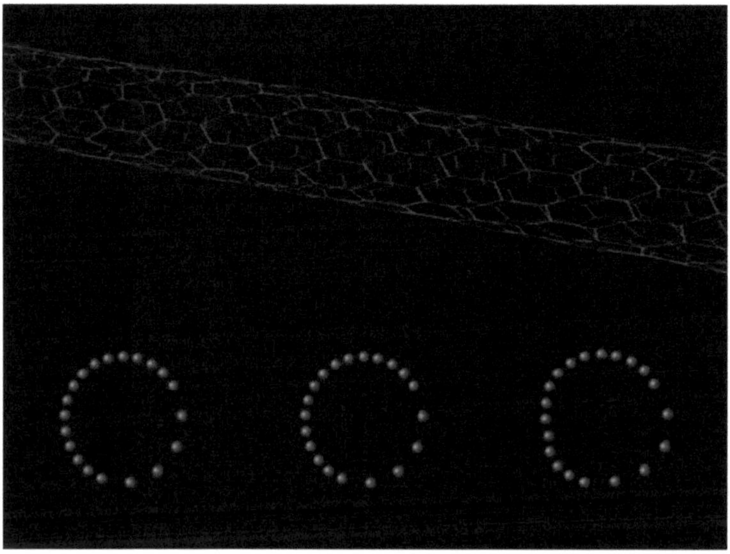

Fig. 10 Illustration of a (12, 1) chirality single-wall nanotube via molecular dynamics simulation using LAMMPS

Fig. 11 The axial sampling geometry of a SWCNT model based on Raman and Scanning Electron Microscope (SEM) analysis

Molecular Dynamics (MD) simulations examining the axial dimension denoted by the white arrow in Fig. 11 were also performed. This "apparent length" dimension, denoted by S, of the model carbon nanotube was constructed by separating the SWCNT into cylindrical sections 2 nm in height, and summing the center of mass distance between consecutive disks along the axial direction of the nanotube. Similar to the radial distortions, Verlet time integration with a 0.5 fs time step was used, and fixed boundary conditions employed in studying the distortion of this particular SWCNT dimension. MD simulations were performed at various temperatures in the range 100–800 K (in increments of 100 K). At each temperature, the apparent length behavior can be used to obtain the length of the SWCNT.

The apparent length variation of CNTs with temperature has been discussed by Cao et al. [28]. Our results also exhibit such variation. Figure 12, for instance, illustrates such variation of the axial dimension over the course of 100 ps during Nose-Hoover thermostated Molecular Dynamics (MD) simulation runs at 400 and 800 K.

Each sample point of the (apparent length) quantity S is the average of 10 samples selected every 100 time-steps, and averaged on every thousandth time-step.

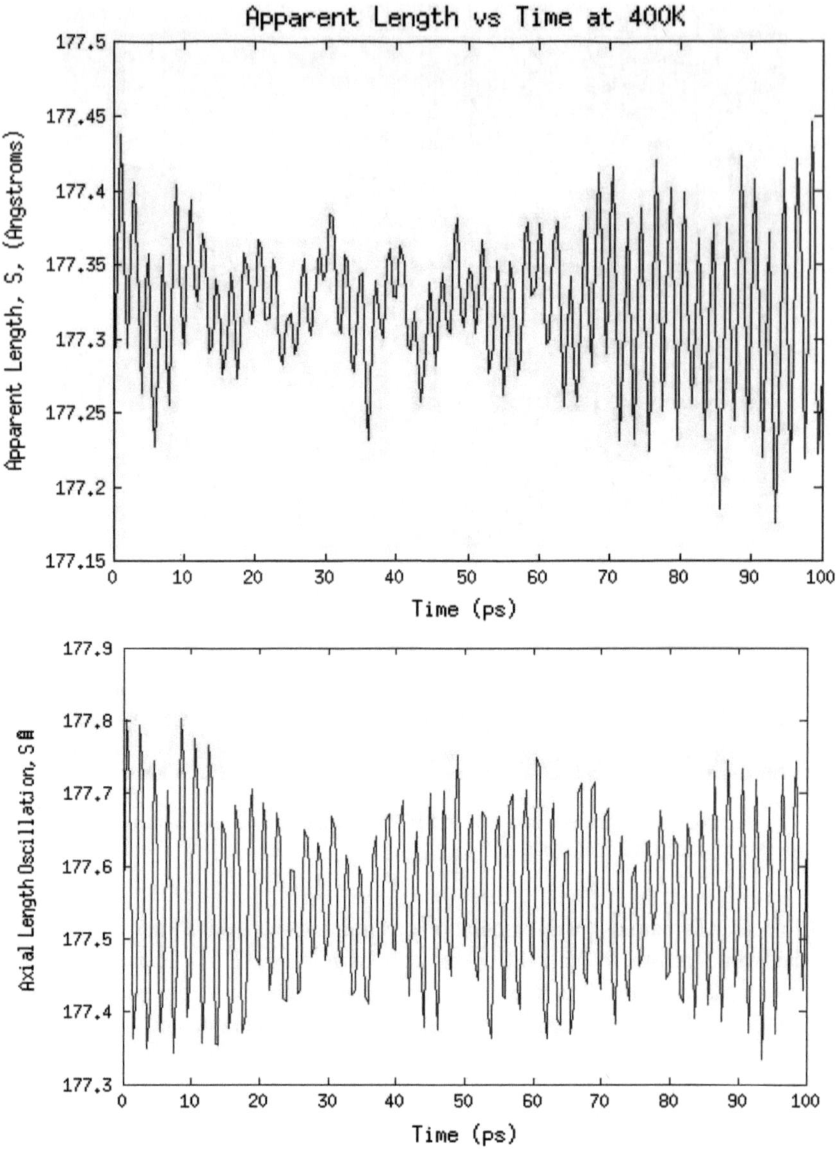

Fig. 12 Apparent length of SWCNTs versus time variation on the ps scale is shown in the plots for temperature runs of 400 and 800 K, respectively

Also, it needs to be noted that, before any sampling of the axial length S had begun, data from the first 100,000 time-steps of this quantity were discarded in order to allow time for thermal equilibrium. The oscillation of the apparent length S (shown in Fig. 12), appears to be a combination of low and high frequency vibration modes (akin to that reported by Cao et al. [28]).

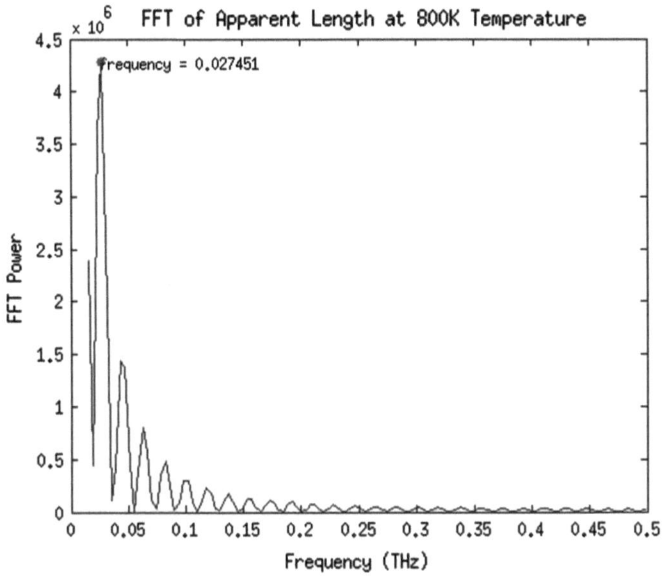

Fig. 13 Fast Fourier Transform (FFT) of the apparent length of SWCNTs at 800 K is illustrated as a FFT Power versus Frequency (in THz) plot

Figure 13 shows a discrete Fourier Transform of the axial length oscillation at a temperature of 800 K; it includes vibrations in both the longitudinal axial and transverse bending directions. Note that in this figure, only the dominant frequency at 0.027451 THz is labeled. As discussed in Cao et al. [28], this Fourier decomposition of the trajectory data obtained from Molecular Dynamics simulations is beneficial to the structural analysis of SWCNTs since it complements the predictions of continuum mechanics, wherein the nanotube is modeled as a smooth cylindrical shell made of an elastic homogeneous material.

6 Concluding Remarks

In this chapter we have discussed the history of carbon nanotube research and the remarkable properties they possess as a result of their varied electronic structure, which allows nanomaterials to have versatile electric and thermal properties [29, 30]. At the center of this versatility is the concept of chirality, which we can visualize by treating carbon nanotubes as if they were rolled-up sheets of graphene; rolling the graphene sheet around a chiral vector (whose direction and size are determined by the n and m chiral coefficients) results in a wide variety of carbon nanotube (n, m) configurations with metallic or semiconductor electrical properties and achiral or chiral structures.

The present chapter also discusses how Raman spectroscopy, a powerful optical technique, has been of increasing use in the study of the properties of nanomaterials.

With Raman spectroscopy, we can determine quantitatively the temperature shifts in frequency of the Raman peaks associated with carbon nanotubes. In addition, several researchers as cited above have shown significant amount of structural characterization based on thermal shifts of Raman spectral characteristics. Specifically, the use of the strain and temperature-induced frequency shift of the 2nd order G′ peak to obtain the axial thermal expansion coefficient was noted and discussed. We have also discussed the method of classical Molecular Dynamics simulation, and its application to the structural characterization of single-walled carbon nanotubes, with pertinent results and lucid demonstrations that compliment continuum based studies of the mechanical behavior of SWCNTs with data obtained from MD simulations. The useful ability to obtain an estimate of the thermal conductivity of SWCNT bundles via the temperature shift of the Raman G^+ mode under laser heating has also been demonstrated. Thanks in part to the electronic density of states of CNTs, Raman intensities are also enhanced several-fold, allowing for the identification of the Raman signal from de-bundled SWCNT samples. An important concept associated with the Raman spectroscopic studies of nanomaterials is the identification of the Raman peaks related to carbon allotropes, resulting in the important G, D and RBM Raman modes, which are used to determine a variety of significant mechanical and electrical properties of CNTs, including important differences between different types of CNTs, specifically, SWCNTs (purified and unpurified) and MWCNTs.

As shown in the reviews and the reported literature results summarized in the current chapter, much still remains to be done in the area of carbon nanotube characterization in general, and also specifically in our on-going investigations relating to thermal effects associated with CNTs. Regarding the latter, our Molecular Dynamics simulation results were primarily focused on the axial dimension of SWCNTs, and not on the radial dimension. Results on the latter based on our ongoing work are anticipated to yield an interesting dimension to the thermal characterization of SWCNTs. Also, the unique specific heat capacity of SWCNTs, the ability to determine the radial CTE of SWCNTs, and be able to probe their 1-dimensional characteristics, as discussed in [31], would add to greater understanding of the thermal behavior of CNTs and in their widespread applications.

Acknowledgments The authors would like to acknowledge the assistance of Ms. Larkin Sayre, a rising sophomore at the Massachusetts Institute of Technology (MIT) and a 2014 Summer Research Experiences for Undergraduates (REU) in Physics participant at Howard University (NSF Grant PHY-1358727), for generating Fig. 1.

References

1. Radushkevich LV, Lukyanovich VM (1952) Structure of the carbon produced in the thermal decomposition of carbon monoxide on an iron catalyst. Russ J Phys Chem 26:88–95
2. Oberlin A, Endo M, Koyama T (1976) Filamentous growth of carbon through benzene decomposition. J Cryst Growth 32:335–349
3. Abrahamson J, Wiles PG, Rhoades BL (1999) Structure of carbon fibers found on carbon arc anodes. Carbon 37(11): 1873

4. Nesterenko AM, Kolesnik NF, Akhmatov YuS, Suhomlin VI, Prilutskii OV (1982) Characteristics of the phase composition and structure of products of the interaction of nickel(II) and iron(III) oxide with carbon monoxide. Izevestiya Akadem ii Nauk, SSSR, Metals 3:12–17
5. Tennent HG (1987) Carbon fibrils, method for producing the same and compositions containing same. US Patent No. 4663230, 1987-05-05
6. Iijima S (1991) Helical microtubules of graphitic carbon. Nature 354:56–58
7. Mintmire JW, Dunlap BI, White CT (1992) Are fullerene tubules metallic? Phys Rev Lett 68:631
8. Iijima S, Iohihashi T (1993) Single-shell carbon nanotubes of 1-nm diameter. Nature 363 (6430):603–605
9. Bethune DS, Kiang CH, De Vries MS, Gorman G, Savoy R, Vazquez J, Beyers R (1993) Cobalt-catalyzed growth of carbon nanotubes with single-atomic—layer walls. Nature 363 (6430):605–607
10. Saito R, Dresselhauss G, Dresselhauss MS (1998) Physical properties of carbon nanotubes. Imperial College Press, London
11. Kwon YK, Berber S, Tomanek D (2004) Thermal contraction of carbon fullerenes and nanotubes. Phys Rev Lett 92:015901
12. Li C, Chan TW (2005) Axial and radial thermal expansions of single-walled carbon nanotubes. Phys Rev B 71:235414
13. Wong P, Akinwande D (2011) Carbon nanotube and graphene device physics. Cambridge University Press, Cambridge
14. Thess et al (1996) Crystalline ropes of metallic carbon Nanotubes. Science 273(5274): 483–487
15. Gray D, McCaughan A, MookerjCrystal B (2009) Structure of graphite, graphene and silicon. http://community.wvu.edu/ ~ mbh039/graphene.pdf. Accessed 14 Jun 2014
16. Misra P, Casimir D, Garcia-Sanchez R (2013) Thermal expansion properties of single-walled carbon nanotubes by raman spectroscopy at 780 nm wavelength. OPAP 2013 proceedings
17. O'Connell MJ, Sivaram S, Doorn SK (2004) Near-infrared resonance Raman excitation profile studies of single-walled carbon nanotube intertube interactions: a direct comparison of bundled and individually dispersed HiPco nanotubes. Phys Rev B 69:235415
18. Thermo scientific characterizing carbon with Raman. http://fscimage.thermoscientific.com/images/D19504 ~ .pdf. Accessed 14 Jun 2014
19. Dresselhaus MS, Eklund PC (2000) Phonons in carbon nanotubes. Adv Phys 49(6):705–814
20. Gregan E (2009) The use of raman spectroscopy in the characterization of single walled carbon nanotubes. Doctoral Dissertation, Dublin Institute of Technology, School of Physics
21. Jorio A, Saito R, Dresselhauss G, Dresselhauss MS (2011) Raman spectroscopy in graphene related systems. Wiley-VCH Verlag GmbH and Co, KGaA, Weinheim, Germany
22. Deng L, Young RJ, Kinloch IA, Sun R, Zhang G, Noe L, Monthioux M (2014) Coefficient of thermal expansion of carbon nanotubes measured by raman spectroscopy. Appl Phys Lett 104:051907
23. Raravikar et al (2002) Temperature dependence of radial breathing mode Raman frequency of single-walled carbon nanotubes. Phys Rev B 66:235424-1–235424-9
24. Terekhov SV, Obraztsova ED, Detltlaff-Weglikowska U, Roth S, Calibration of raman-based method for estimation of carbon nanotube purity. AIP Conf Proc 685:116
25. Robertson DH, Brenner DW, Mintmire JW (1992) Progress on mechanics of carbon nanotubes and derived materials. Phys Rev B 45:12592
26. Stuart SJ, Tutein AB, Harrison JA (2000) A reactive potential for hydrocarbons with intermolecular interactions. J Chem Phys 112:6472
27. Bialoskorski M, Rybicki J (2012) Mechanical properties of single-walled carbon nanotubes simulated with airebo force-field. Comp Meth Sci Tech 18(2):67–77
28. Cao G, Chen X, Kysar JW (2005) Apparent thermal contraction of single-walled carbon nanotubes. Phys Rev B 72:235404

29. Jorio A, Dresselhauss G, Dresselhauss MS (eds) (2008) Carbon nanotubes: advanced topics in the synthesis, structure, properties and applications. Springer, Berlin
30. Rao CNR, Subrahmanyan KS, Ramakrishna Matte HSS, Govindraj A (2011) Graphene: synthesis, functionalization and properties. Mod Phy Lett 25(7):427–451
31. Bagatskii MI, Barabashko MS, Dolbin AV, Sumarokov VV (2012) The specific heat and the radial thermal expansion of bundles of single-walled carbon nanotubes. Fiz Nizk Temp 38 (6):667–673

Laser Optogalvanic Spectroscopy and Collisional State Dynamics Associated with Hollow Cathode Discharge Plasmas

Michael Blosser, Xianming L. Han, Raul F. Garcia-Sanchez and Prabhakar Misra

Abstract In this chapter we will discuss the laser optogalvanic effect in a discharge plasma environment, specifically associated with an iron-neon (Fe–Ne) hollow cathode lamp. The history of the optogalvanic effect will serve as an introduction to the importance of the phenomena. The theoretical model behind the optogalvanic effect will provide insight into the importance of laser optogalvanic spectroscopy as a tool for spectral characterization of the plasma processes and enhanced understanding of the collisional state dynamics associated with the discharge species in hollow cathode lamps. The present chapter will focus on transition states of neon in the Fe–Ne hollow cathode lamp. The results presented here will use, for illustrative purposes, the waveforms associated with the laser-excited optogalvanic transitions of neon: $1s_4$–$2p_3$ (607.4 nm), $1s_5$–$2p_7$ (621.7 nm), $1s_3$–$2p_5$ (626.6 nm), $1s_5$–$2p_8$ (633.4 nm) and $1s_5$–$2p_9$ (640.2 nm). A comparison between the experimentally recorded optogalvanic signal waveforms and the Monte Carlo fitting routine, along with a discussion related to the variation of the (a_i and b_j) fitting coefficients as a function of the discharge current, will illustrate the success of our theoretical model. We will also briefly touch upon the potential applications of the optogalvanic effect at the nanoscale in fields such as graphene-based nanoelectronics and nanoplasmonics.

Keywords Laser optogalvanic spectroscopy · Collisional state dynamics · Hollow cathodes · Iron-neon hollow cathode lamp · Discharge plasmas · Neon · Transitional states · Optogalvanic signal waveforms · Monte Carlo fitting routine · Least-squares analysis · Graphene-based nanoelectronics · Nanoplasmonics

---

M. Blosser
Department of Physics, University of South Florida, Tampa, FL 33620, USA

X.L. Han
Department of Physics and Astronomy, Butler University, Indianapolis, IN 46206, USA

R.F. Garcia-Sanchez · P. Misra (✉)
Department of Physics and Astronomy, Howard University, Washington DC 20059, USA
e-mail: pmisra@howard.edu

P. Misra (ed.), *Applied Spectroscopy and the Science of Nanomaterials*,
Progress in Optical Science and Photonics 2, DOI 10.1007/978-981-287-242-5_2

1 Introduction

Laser optogalvanic spectroscopy is the study of the change in impedance that occurs as a result of irradiating the atoms in a gas discharge environment, such as a hollow cathode, with a laser source; the concomitant variation in impedance is often referred to as the OptoGalvanic Effect (OGE). OGE is a direct consequence of the laser excitation of the atoms/molecules populating the various energy levels of the species present in the discharge plasma, followed by collisions that result in redistribution of the plasma species in energetically favorable states. With advances in laser technology, such as the widespread availability of tunable and short-pulsed lasers, optogalvanic spectroscopy has gained a foothold in the study of plasma species and their atomic/molecular transitions by tuning the laser precisely to specific wavelengths that resonate with the plasma species of interest. In addition, specific collisional state dynamics of a particular species can be investigated because its rate of ionization undergoes changes due to collisions within the discharge plasma. The rate of ionization depends on the probability of atomic ionization occurring after the atom is hit by an electron and also on the duration of the resulting excited state [1], whereby, specifically, an atom has a higher rate of ionization if it is in a metastable state.

At the heart of the laser optogalvanic spectroscopy investigations is a Hollow Cathode Lamp (HCL), which is typically composed of an anode, a cathode and a buffer gas (e.g. neon) encased in a glass enclosure, and is exposed to lasers and other spectral line sources. The use of the HCL allows us to obtain OG signals for the cathode species in addition to the buffer gas. The experiments presented here make use of an Fe–Ne (the notation refers to the composition of the electrode and the buffer gas used) HCL, which allows us to not only determine the OG signals that result from the iron, but also determine the neon transitions at very precise wavelengths. Thus, laser optogalvanic spectroscopy is an ideal tool also for spectral calibration in atomic and molecular spectroscopy.

With the advent of nanotechnology, the study of nanoplasmonics is becoming increasingly important. Laser optogalvanic spectroscopy can play a significant role in understanding the behavior of nanoplasmas [2–4]. In the same fashion that laser optogalvanic spectroscopy has been employed to determine the characteristics of hollow cathode tubes involving metals such as iron and noble gases such as neon [5], this technique has potential applications in the determination of collisional rates and population densities for electrodes made of nanomaterials, such as graphene, which has found extensive use in research involving Li-ion batteries [6–10]. Currently, while research on graphene-based nanoelectronics and nanoplasmas is being carried out, to our knowledge there are no extensive studies using laser optogalvanic spectroscopy and graphene electrodes available in the refereed literature.

2 Literature Review

In this section we will review the advances in the optogalvanic effect and the evolution of the laser optogalvanic spectroscopy research from a historical perspective and delve into more detail into the possible applications of its use in the field of nanoscience.

2.1 The Optogalvanic Effect

The first known publications relating to the optogalvanic effect are seen as early as the 1920s [11, 12]. Foote and Mohler [11] observed ionization of caesium (Cs) vapor occurring when the sample was irradiated with light of wavelength longer than 318 nm, which corresponded to wavelengths at which Cs emits light. The authors deduced that the reason for the ionization was due to the fact that the excited atoms collided with others and caused the ionization of the sample. In 1928, Penning observed this same phenomenon, to which he referred to as a new photoelectric effect [12]. Penning detected a variation in the impedance associated with the discharge of an element when irradiated with the emission of another identical discharge, which later came to be known as the OptoGalvanic Effect (OGE). Other similar effects were observed through a wide variety of experiments during the subsequent decades. In 1930, Terenin [13] observed the OGE in molecules while studying the photoionization of salt vapors. Joshi [14–16] independently discovered the same effect (and referred to it as the Joshi Effect), while studying chlorine in a discharge tube, and noted that the current through the discharge would change when the sample was irradiated with energies equal to the difference in energy between the ground and excited states. However, some scientists argued that the Joshi Effect was different from the Optogalvanic Effect on the grounds that one referred to a voltage rise and the other to a current rise [17]. Kenty [18] and Meissner and Miller [19] studied the effect in mercury (Hg) and helium (He) discharges, respectively. In the 1960s, Garscadden et al. [20] also observed the OGE when a change in the discharge current manifested itself as moving striations in a He–Ne discharge.

While the OGE was observed through multiple experiments over half a century, it was not until the development of the tunable dye laser that such investigations took center-stage and became mainstream. Green et al. [21] showed that high-sensitivity spectra could be obtained using tunable lasers using galvanic detection of optical absorptions in a gas discharge [21] and also via detection of species in flames [22]. Green et al. measured a change in the voltage across the discharge tube while irradiating the sample at a laser wavelength corresponding to a transition of the sample under investigation. The authors determined that such a change in voltage could be used as a way to characterize materials in the discharge without the use of an optical detection device, unlike techniques such as fluorescence. This type of research approach also illustrated the usefulness and applications of using a hollow cathode lamp as the gas discharge environment, which allowed not only excited state transitions but resonance transitions as well to be observed. A year later, in 1977, King et al. [23] showed how direct calibration of laser wavelength and bandwidth using the OG signal provided a simple and yet highly accurate calibration tool for tunable dye lasers. Due to the continued rapid development of tunable dye lasers and Green's pioneering research, laser optogalvanic spectroscopy became a widespread spectroscopic technique used in the study of atoms and molecules in a discharge environment, especially as it relates to their ionization, transition states and collisional dynamics in response to laser excitation.

2.2 Optogalvanic Spectroscopy Techniques

In 1979, Lawler et al. [24] reported Doppler-Free Intermodulated Optogalvanic Spectroscopy, a single-photon spectroscopy technique similar to intermodulated fluorescence spectroscopy that utilizes the detection of the impedance change in gaseous discharges. The primary concept of Doppler-free spectroscopy revolved around the circumvention of the Doppler Effect in spectral data. The Doppler Effect in spectroscopy refers to the fact that, if an atom is moving towards a laser source at a certain velocity v then the laser frequency will be upshifted relative to the motion of the atom. Unlike typical optogalvanic techniques at the time, which were Doppler-limited, the laser beam was split into two equal parts, but with different frequencies, which were then sent through the discharge in opposite directions. Due to the fact that the split laser beams propagate in dissimilar directions, they interact with different velocity groups, those with equal but opposite velocities. However, when the laser frequency matches the atomic/molecular absorption frequency, the two laser beams interact with the same atoms/molecules, and in this case with the ones with zero velocity in the direction of the absorption, resulting in a Doppler-free spectrum. Lawler et al. [24] used this technique to study the ^{3}He transitions 2^3P–3^3D and found that this technique compared favorably with other Doppler-free techniques of the time, such as High-Resolution Saturation Spectroscopy [25], Saturated-Interference Spectroscopy [26] and Doppler-Free Laser Polarization Spectroscopy [27]. The Optogalvanic Double-Resonance Spectroscopy [28] technique utilized the same concepts as Doppler-Free Intermodulated Optogalvanic Spectroscopy [24]. In this case, the first laser would be tuned to a specific wavelength that would cause precise transitions between upper and lower energy levels, whereas the second laser would excite the molecules in the levels excited by the first laser. Such a twin laser approach resulted in increased sensitivity over other double resonance techniques that did not utilize the concept of optogalvanic spectroscopy.

Two-photon techniques serve as one of the standard tools for exciting atoms to states that are not possible by using single-photon processes. Goldsmith et al. [29] used a continuous wave (CW) dye laser in a He–Ne discharge to observe transitions in ^{20}Ne using Doppler-Free Two-Photon Optogalvanic Spectroscopy. The use of two-photon optogalvanic spectroscopy allowed the study of non-metastable transitions in addition to metastable and ground state transitions. As a result, a ground-breaking study by Shawlow et al. [29] reported the first optogalvanic spectroscopy observations of non-metastable state transitions of atoms/molecules in a discharge [29].

2.3 Optogalvanic Spectroscopy Studies in a Hollow Cathode Lamp Discharge

Misra et al. [30] reported a comprehensive list of over 350 optogalvanic transitions in the 337–598 nm region using a pulsed tunable dye laser and an iron-neon (Fe–Ne) hollow cathode discharge. In conjunction with simultaneously recorded interference fringes, a subset (223) of these OG transitions arising from neon permitted the calibration of dye laser frequencies in the visible and near ultraviolet to 0.3 cm^{-1} accuracy. The authors also reported in the same paper [30] that the waveform shapes of the distinct OG transitions of neon at wavelengths (in air) 3510.721, 3515.190, and 3520.471 Å, respectively, recorded with a steady discharge current of 0.5 mA and a laser pulse energy of 20 μJ, were quite different. The same authors investigated the polarity of OG transitions of neon further [31] by focusing on a total of twenty nine OG transitions in the near UV and visible that could be explained in terms of processes that affect the population of neon atoms in metastable states. A close inspection of the oscilloscope traces showed conclusively that the OG signals arising from metastable neon energy levels exhibit waveforms that are negative in character (corresponding to increased ionization current) and subsequently became positive (due to decreased ionization current). In contrast, the OG signals arising from transient non-metastable levels were either completely negative or inverted in shape as compared to those from longer-lived metastable levels. More recently, experimental investigations and mathematical modeling of the OG waveforms of laser-excited Fe–Ne hollow cathode lamp transitions have been carried out in great detail jointly by Han and Misra [5, 32, 33]. A mathematical rate equation model has been developed that incorporates the various processes contributing to the generation of the OG signal waveforms in a laser-stimulated hollow cathode discharge plasma. The experimentally observed waveform was fitted employing a nonlinear least-squares Monte Carlo technique to determine the pertinent amplitude coefficients, decay rates, and the instrumental time constant, in order to better characterize the collisional processes taking place in a hollow cathode discharge plasma.

2.4 Applications Involving Nanomaterials and Nanoplasmas

As mentioned in the introduction, laser optogalvanic spectroscopy has potential prospects in applications utilizing nanomaterials and nanoplasmas. In particular, the development of graphene-based electrodes for use in hollow cathode lamps holds promise for intriguing discharge plasma experiments using laser optogalvanic spectroscopy. Additionally, investigations of nanoplasmas, plasmas with a lifetime of the order of femtoseconds [2–4], holds potential for ultrashort timescale optogalvanic studies that will shed light on fast processes occurring in discharge plasmas that hitherto have escaped detection and study.

3 Theoretical Model

The laser-induced optogalvanic effect in a discharge environment, such as a hollow cathode lamp, can be summarized in the sequence of events shown in Fig. 1.

First, the relevant energy levels of the atom (e.g. neon) can be identified as follows: the ground energy level (not shown in Fig. 1), a group of relatively closely-spaced energy levels (two are shown and labeled |1> and |2> in Fig. 1, which will be referred to as the "first group"), and another group of relatively closely spaced energy levels above the first group (two are shown and are labelled as $|U_1>$ and $|U_2>$ in Fig. 1, which will be called the "second group").

Collisions with the electrons in the discharge current excite a large number of atoms into the first group of energy levels, but relatively few of these are excited into the second group. Some of the energy levels in the first group are metastable states, which causes the population in these states to be large. Even if an energy level is not a metastable state and can radiatively decay down to the ground state, radiation trapping by the ground state can significantly lengthen its effective lifetime. The electron collisional ionizations from the first group of energy levels, in turn, contribute electrons to the discharge current. If the atomic population in the first group of energy levels is altered, then the contribution to the total number of free electrons will change due to the different ionization rates of each energy level, resulting in changes in the discharge current.

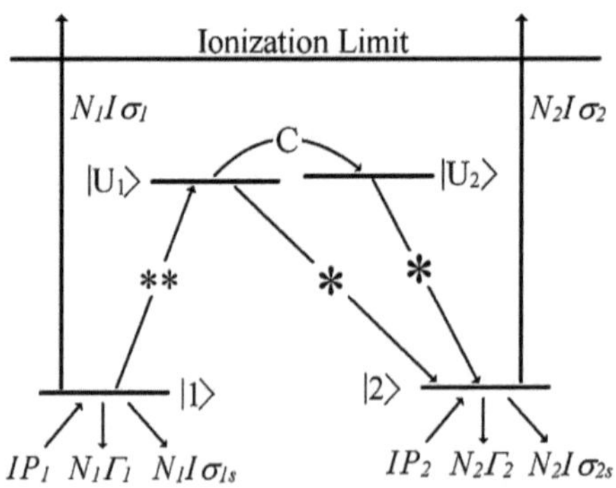

Fig. 1 An energy-level schematic illustrating the collisional dynamics and population distribution giving rise to the Optogalvanic Effect in a Hollow Cathode discharge plasma

A pulsed laser can be used to alter the atomic population in the first group of energy levels. First, a laser can be tuned between |1> and |U_1>, exciting the atom from |1> to |U_1>. Subsequently, the atom can radiatively decay down to |2> in the first group of energy levels. Alternatively, a collision may transfer the atom from |U_1> to |U_2> and then radiatively decay down to |2>. In the end, the net effect of the laser pulse is to transfer some atoms from |1> to |2>. This process happens very quickly, on the order of tens of nanoseconds (lifetime of the second group of energy levels). On the time scale of the optogalvanic effect, this transfer may be said to occur instantaneously.

The different population distributions will result in a dissimilar ionization rate, which in turn causes a change in the discharge current. Specifically, the contribution to the discharge current from |1> is reduced because some of the atoms in that energy level are transferred to other energy levels (|2> in the above discussion), while contributions from other energy levels on the receiving end of the population transfer (such as |2>) will increase. As a result, the discharge current will undergo a modification, which in turn is captured in the varying OG waveform.

Finally, the ensuing collisions will eventually fill back the population hole in |1> and return its population back to its steady-state value; and this in turn will also occur for the populations in the other energy levels. As a result, the discharge current in the hollow cathode lamp will eventually return to its steady-state value as well.

In order to describe the evolution of the atomic populations in |1> and |2>, following the population transfer mentioned above, we write down the following differential equation:

$$\frac{dN_i}{dt} = IP_i - N_i(\Gamma_i + I\sigma_i + I\sigma_{is}) \qquad i = 1, 2 \tag{1}$$

where P_i is related to the electron collisional excitation probability from the ground state (not shown in the figure) to |*i*>, in this case, |1> or |2>, I is the discharge current, N_i is the atomic population for state |*i*>, Γ_i is the effective spontaneous decay rate for state |*i*>, σ_i is related to the electron collisional ionization probability of state |*i*>, and σ_{is} is related to the electron collisional de-excitation probability of state |*i*>.

Prior to laser excitation, by setting $\frac{dN_i}{dt} = 0$, we obtain the steady-state populations n_i for state |*i*>:

$$n_i = \frac{IP_i}{\Gamma_i + I\sigma_i + I\sigma_{is}} \qquad i = 1, 2 \tag{2}$$

To solve Eq. (1), we assume that the ground state holds the vast majority of the atoms so that its population can be treated as constant. As a consequence, we can treat P_i as constant. Taking into account that Eq. (2) gives the populations for $t = \infty$, we obtain the solutions for Eq. (1):

$$N_i(t) = n_i + c_i \exp[-(\Gamma_i + I\sigma_i + I\sigma_{is})t] \qquad i = 1, 2 \tag{3}$$

where $\{c_i\}$ are constants determined by initial conditions due to the pulsed laser excitation to $|U_1>$ and the quick relaxation down to $|2>$. The population changes caused by the laser excitation are:

$$\Delta N_i(t) = N_i(t) - n_i = c_i \exp[-(\Gamma_i + I\sigma_i + I\sigma_{is})t \quad i = 1, 2 \tag{4}$$

After laser excitation, the atomic population in $|1>$ and $|2>$ can be written as

$$\Delta N_1(0) = -\Delta N_0 \text{ and } \Delta N_2(0) = +\Delta N_0. \tag{5}$$

Inserting the two equations Eq. (5) into Eq. (4) and writing the two equations separately, we obtain:

$$\Delta N_1(t) = -\Delta N_0 \exp[-(\Gamma_1 + I\sigma_1 + I\sigma_{1s})t] \tag{4a}$$

$$\Delta N_2(t) = +\Delta N_0 \exp[-(\Gamma_2 + I\sigma_2 + I\sigma_{2s})t] \tag{4b}$$

Therefore, the Optogalvanic (OG) signal due to ionizations from $|1>$ and $|2>$ is

$$\begin{aligned} S(t) &= \Delta N_1 I\sigma_1 + \Delta N_2 I\sigma_2 \\ &= \Delta N_0 \{-I\sigma_1 \exp[-(\Gamma_1 + I\sigma_1 + I\sigma_{1s})t] + I\sigma_2 \exp[-(\Gamma_2 + I\sigma_2 + I\sigma_{2s})t]\} \end{aligned} \tag{6}$$

Normalizing it to ΔN_0, we obtain the experimental time-resolved OG signal as

$$s(t) = \frac{S(t)}{\Delta N_0} = a_1 e^{-b_1 t} + a_2 e^{-b_2 t} \tag{7}$$

where

$$a_1 = -I\sigma_1 \tag{8}$$

$$b_1 = \Gamma_1 + I(\sigma_1 + \sigma_{1s}) \tag{9}$$

$$a_2 = I\sigma_2 \tag{10}$$

$$b_2 = \Gamma_2 + I(\sigma_2 + \sigma_{2s}) \tag{11}$$

3.1 The Effect of Plasma/Instrument Relaxation Time on the OG Signal

Neither the discharge plasma nor the detection instrument can respond to changes instantaneously. Let us assume the relaxation time constant to be τ. The actual OG signal observed in an experiment is a convolution of Eq. (7) with a time constant τ:

$$s(t) = \frac{a_1}{1 - b_1\tau}\left(e^{-b_1 t} - e^{-t/\tau}\right) + \frac{a_2}{1 - b_2\tau}\left(e^{-b_2 t} - e^{-t/\tau}\right) \tag{12}$$

We want to point out that since the relaxation time constant τ depends on the discharge conditions, it is not a constant if the discharge current changes.

3.2 Physical Significance of Fitting Coefficients

As seen from Eqs. (8) and (9), the a_1 term in Eq. (12) is negative and the a_2 term is positive. This is because the effect of the pulsed laser is to transfer some population from |1> to |2>, resulting in a drop in the contribution to the total ionization from | 1>, and hence a negative OG signal results due to energy level |1>; whereas the reverse holds true for the positive OG signal from energy level |2>.

If the first group of energy levels consists of more than 2 individual energy levels, then the OG signal will be an extension of Eq. (12):

$$s(t) = \sum_{i=1}^{n} \frac{a_i}{1 - b_i\tau}\left(e^{-b_i t} - e^{-t/\tau}\right) \tag{13}$$

where n is the total number of energy levels in the first group. One term in the summation has negative amplitude, which corresponds to the contribution from the initial energy level |1> from which atomic population is transferred to other energy levels initiated by the laser pulse, and the rest of the terms are positive because their corresponding energy levels are on the receiving end of the population transfer.

4 Experimental Set-up for Laser Optogalvanic Spectroscopy

The experimental set-up used is shown in Fig. 2 [5]. A pulsed laser is tuned to each of the studied neon transitions (607.4, 621.7, 626.6, 633.4, and 640.2 nm) and directed to enter an iron-neon (Fe–Ne) Hollow Cathode Lamp (HCL). The discharge current in the HCL is limited by the RC circuit and the current is adjusted to

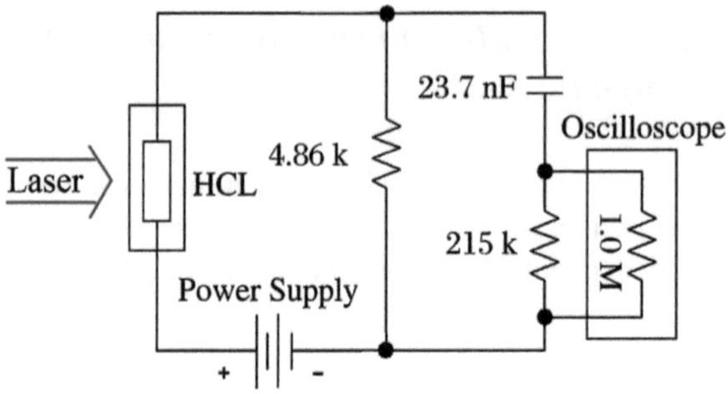

Fig. 2 Schematic experimental arrangement for laser optogalvanic spectroscopy using a Hollow Cathode Lamp (HCL)

be between 2–19 mA via the power supply voltage. A Tektronix TSD 224 digital oscilloscope displays and stores the OG signal. Subsequently, the OG signal waveforms from the oscilloscope are used in the Monte Carlo fitting routine [5].

5 Neon Transitions

Unlike other noble gases, neon can discharge at low currents, such as those used in the present investigation, making it an ideal gas for the study of discharge plasma environments in a hollow cathode lamp. For the neon 1s–2p transitions, there are four 1s states and ten 2p states, which yields a total of 30 radiative transitions [32]. Figure 3 shows these transitions with their corresponding wavelengths in nanometers. Laser optogalvanic spectroscopy, along with the above mathematical formulation, makes it possible to determine the population density distribution in the discharge. We will illustrate the theoretical model by applying it specifically to the following OG transitions of neon in the visible: $1s_4$–$2p_3$ (607.4 nm), $1s_5$–$2p_7$ (621.7 nm), $1s_3$–$2p_5$ (626.6 nm), $1s_5$–$2p_8$ (633.4 nm) and $1s_5$–$2p_9$ (640.2 nm). As seen on the arrow connections in Fig. 3, the main idea is that when neon atoms are excited from their original state, $1s_5$ for example, to a state such as $2p_7$, they can subsequently relax back from $2p_7$ to $1s_2$, $1s_3$, $1s_4$ and $1s_5$. Table 1 below summarizes the possible relaxation states for the investigated OG transitions.

5.1 $1s_4$–$2p_3$ Transition (607.4 Nm)

This particular transition for atomic neon occurs at 607.4 nm, where the neon atoms after excitation can relax down to another state; additionally, it can also relax back to where it started, the $1s_4$ state. The Monte Carlo approach cited above is used to fit the theoretical Eq. (13) to the observed waveform recorded at 607.4 nm. A fuller discussion regarding the fitting process is given below in the Results and Discussion section. The fitted values for the $1s_4$–$2p_3$ waveform are summarized in Table 2.

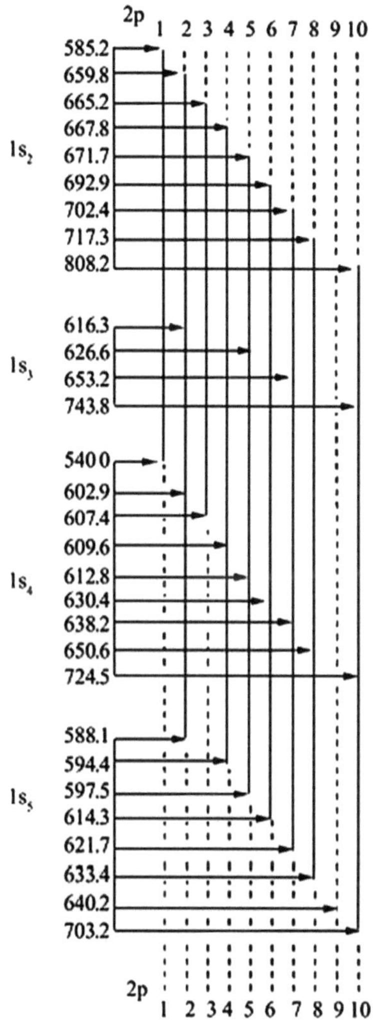

Fig. 3 Neon atom transitions between $1s_i$–$2p_j$ states with corresponding wavelengths (nm) (adapted from Ref. [32] with permission)

Table 1 Investigated OG transitions of neon and associated relaxation states

Transition	Wavelength (nm)	Relaxation states
$1s_4 \rightarrow 2p_3$	607.4	$1s_2$ | $1s_4$
$1s_5 \rightarrow 2p_7$	621.7	$1s_2$ | $1s_3$ | $1s_4$ | $1s_5$
$1s_3 \rightarrow 2p_5$	626.6	$1s_2$ | $1s_3$ | $1s_4$ | $1s_5$
$1s_5 \rightarrow 2p_8$	633.4	$1s_2$ | $1s_4$ | $1s_5$
$1s_5 \rightarrow 2p_9$	640.2	$1s_5$ | $2p_j \rightarrow 1s_i$

Table 2 Fitted decay rates obtained from a non-linear least-squares fit of the observed OG waveforms for neon at 607.4 nm for 2–19 mA currents. The fitted a-coefficients are in arbitrary units, while the b-coefficients are in units of $(\mu s)^{-1}$

I (mA)	a_1	b_1	a_2	b_2	a_3	b_3
2	18.258	0.349	−14.491	0.252	0.182	0.005
3	17.672	0.255	−16.919	0.159	3.011	0.075
4	15.130	0.229	−15.984	0.129	4.516	0.074
5	18.754	0.165	−19.918	0.113	3.961	0.061
6	19.813	0.134	−19.922	0.105	2.128	0.050
7	3.280	0.270	−1.345	0.089	–	–
8	2.114	0.672	−0.87164	0.103213	3.165	0.137
9	2.047	0.642	−3.34231	0.100771	3.540	0.124
10	1.950	0.613	−3.67748	0.102807	3.834	0.121
11	1.687	0.704	−5.05266	0.116876	5.427	0.135
12	1.452	0.944	−2.08021	0.120003	2.687	0.171
13	1.316	1.065	−2.06138	0.127232	2.739	0.181
14	1.261	1.172	−2.58873	0.136241	3.341	0.184
15	1.251	1.416	−2.32504	0.142215	3.158	0.198
16	1.194	1.441	−2.40965	0.144766	3.272	0.201
17	1.153	1.452	−3.08785	0.150483	3.913	0.195
18	1.223	1.450	−3.24175	0.148668	4.054	0.192
19	1.165	1.517	−3.62614	0.158177	4.498	0.199

5.2 *$1s_5$–$2p_7$ Transition (621.7 Nm)*

The $1s_5$–$2p_7$ transition is observed around 621.7 nm, where the neon atoms after excitation can relax down to three additional states, namely $1s_2$, $1s_3$, and $1s_4$, plus of course the original $1s_5$ state. The fitted values for the $1s_5$–$2p_7$ waveform are given in Table 3.

5.3 *$1s_3$–$2p_5$ Transition (626.6 Nm)*

This particular transition at exists around 626.6 nm, where the neon atoms after excitation can relax back to three state, namely the $1s_2$, $1s_4$ and $1s_5$ in addition to relaxing back to $1s_3$. According to the work put forward, the observed waveform can be fitted to the expression in Eq. (13). The values fitted are given in Table 4.

Table 3 Fitted decay rates obtained from a non-linear least-squares fit of the observed OG waveforms for neon at 621.7 nm for 2–19 mA currents. The fitted a-coefficients are in arbitrary units, while the b-coefficients are in units of $(\mu s)^{-1}$

I (mA)	a_1	b_1	a_2	b_2
2	1.467	0.164	−0.625	0.019
3	2.202	0.144	−1.167	0.031
4	3.419	0.105	−2.366	0.040
5	8.575	0.054	−7.866	0.041
6	2.875	0.028	−2.564	0.016
7	0.758	0.107	−0.453	0.011
8	0.776	0.200	−0.411	0.038
9	1.298	0.145	−0.978	0.070
10	1.551	0.147	−1.230	0.082
11	1.698	0.150	−1.381	0.090
12	1.868	0.153	−1.554	0.097
13	1.956	0.157	−1.641	0.103
14	1.951	0.163	−1.632	0.107
15	1.862	0.170	−1.535	0.109
16	1.780	0.176	−1.445	0.110
17	1.799	0.179	−1.457	0.112
18	1.750	0.187	−1.402	0.116
19	1.781	0.190	−1.433	0.120

Table 4 Fitted decay rates obtained from a non-linear least-squares fit of the observed OG waveforms for neon at 626.6 nm for 2–19 mA currents. The fitted a-coefficients are in arbitrary units, while the b-coefficients are in units of $(\mu s)^{-1}$

I (mA)	a_1	b_1	a_2	b_2
2	14.037	0.237	−4.392	0.034
3	14.195	0.180	−6.314	0.041
4	27.220	0.092	−21.431	0.056
5	20.285	0.069	−16.302	0.041
6	4.849	0.093	−2.320	0.009
7	4.935	0.454	−0.993	0.006
8	5.131	0.353	−1.906	0.036
9	5.368	0.323	−2.402	0.051
10	5.290	0.335	−2.426	0.056
11	5.165	0.356	−2.361	0.059
12	5.914	0.306	−3.218	0.088
13	5.859	0.314	−3.232	0.093
14	5.773	0.324	−3.188	0.098
15	5.764	0.332	−3.167	0.101
16	5.827	0.336	−3.191	0.105
17	5.909	0.343	−3.159	0.105
18	5.765	0.354	−3.036	0.106
19	5.711	0.360	−3.019	0.110

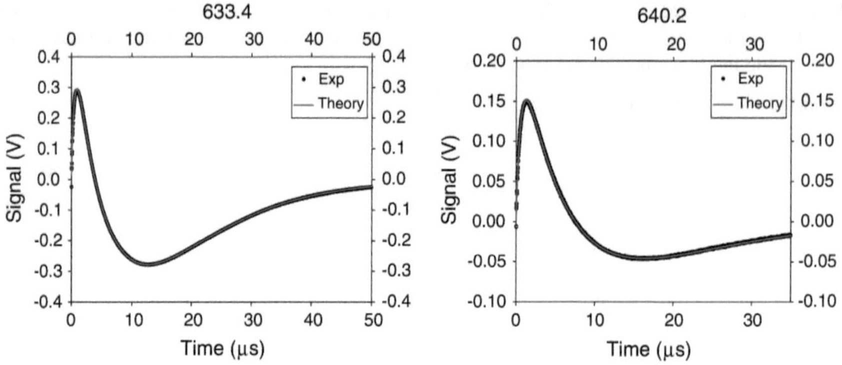

Fig. 4 Time variation of the OG signal at 2.0 mA fitted to the mathematical model for the OG waveforms at 633.4 and 640.2 nm illustrating the similarity between experimental and theoretical fits for each neon transition studied

5.4 $1s_5$–$2p_8$ Transition (633.4 Nm)

This particular OG transition is recorded around 633.4 nm, where the neon atoms after excitation can relax to three states, namely $1s_2$, $1s_4$, and $1s_5$. The observed waveform can then be fitted to the expression in Eq. (13). The fitted coefficients for the $1s_5$–$2p_8$ waveform (shown in Fig. 4) are summarized in Table 5.

Table 5 Fitted decay rates obtained from a non-linear least-squares fit of the observed OG waveforms for neon at 633.4 nm for 2–19 mA currents. The fitted a-coefficients are in arbitrary units, while the b-coefficients are in units of $(\mu s)^{-1}$

I (mA)	a_1	b_1	a_2	b_2
2	2.983	0.169	−1.317	0.028
3	14.192	0.179	−6.330	0.041
4	30.874	0.089	−25.157	0.057
5	22.296	0.068	−18.324	0.042
6	4.854	0.092	−2.334	0.009
7	5.004	0.464	−0.989	0.006
8	5.132	0.352	−1.915	0.036
9	5.375	0.322	−2.408	0.052
10	3.221	0.137	−2.776	0.094
11	3.221	0.144	−2.774	0.099
12	3.096	0.152	−2.642	0.103
13	2.999	0.159	−2.520	0.105
14	3.080	0.165	−2.591	0.111
15	2.852	0.172	−2.356	0.113
16	3.221	0.173	−2.720	0.120
17	3.571	0.176	−3.070	0.128
18	2.130	0.205	−1.622	0.117
19	2.153	0.210	−1.644	0.121

Table 6 Fitted decay rates obtained from a non-linear least-squares fit of the observed OG waveforms for neon at 640.2 nm for 2–19 mA currents. The fitted a-coefficients are in arbitrary units, while the b-coefficients are in units of $(\mu s)^{-1}$

I (mA)	a_1	b_1	a_2	b_2
2	4.113	0.068	−3.875	0.059
3	4.280	0.069	−4.036	0.061
4	4.384	0.064	−4.134	0.055
5	9.890	0.063	−7.829	0.033
6	1.035	0.017	−0.957	0.010
7	0.289	0.168	−0.114	0.033
8	0.554	0.131	−0.371	0.073
9	0.945	0.126	−0.746	0.092
10	1.007	0.132	−0.792	0.097
11	0.616	0.156	−0.382	0.087
12	0.772	0.153	−0.522	0.097
13	0.789	0.159	−0.522	0.101
14	1.018	0.156	−0.736	0.111
15	1.125	0.159	−0.826	0.117
16	1.139	0.163	−0.826	0.120
17	1.047	0.171	−0.718	0.121
18	1.003	0.178	−0.657	0.122
19	1.254	0.175	−0.893	0.131

5.5 *$1s_5$–$2p_9$ Transition (640.2 Nm)*

At first glance, the $2p_9$ population can only radiatively relax back to the starting state $1s_5$, resulting in no changes in the population distribution among the 1s states, and thus no OG transitions should be observed. However, we did observe OG transitions following laser excitation. Collisions transfer population from $2p_9$ to other 2p state(s), which in turn relax down to 1s state(s), resulting in population distribution changes in the 1s states. The fitted parameters for the $1s_5$–$2p_9$ waveform (shown in Fig. 4) are summarized in Table 6.

6 Results and Discussion

The stored data on the digital oscilloscope were fitted to the mathematical model and analyzed in this section of the work. Similar to previous experiments [5], Eq. (14) was used to fit the OG waveforms (index j replacing i in Eq. (13)):

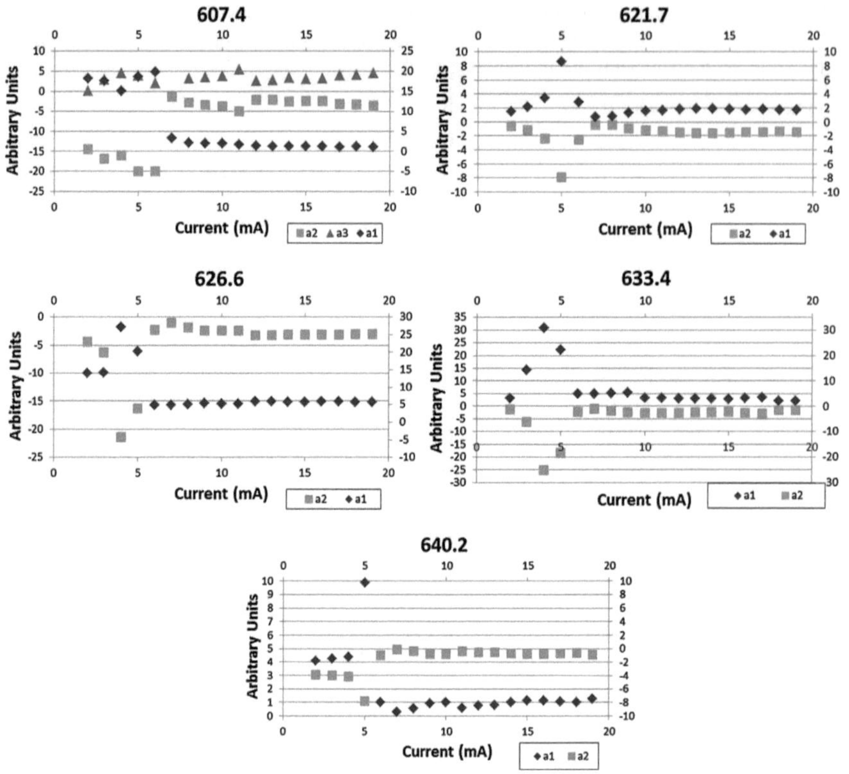

Fig. 5 Variation of the a-coefficients as a function of discharge current for the five OG transitions (607.4, 621.7, 626.6, 633.4, and 640.2 nm) studied for neon

$$S(t) = \sum_{j=1}^{j_{max}} \frac{a_j}{1 - b_j} \left[e^{-b_j t} - e^{t/\tau} \right] \quad (14)$$

Furthermore, the parameters a_j, b_j and τ were obtained following least-squares fitting of each of the transitions $1s_i$–$2p_k$ using the above equation. The number of terms in the equation that were used to fit the waveform depended on the wavelength and the current. However, in most cases, not all of the four states were equally important and involved in a particular optogalvanic transition. As a result, we can fit the experimental signal using less than four terms in Eq. (14) for a particular optogalvanic signal, and the minimum case involved two states.

Of the several OG waveforms recorded in the present investigation: $1s_4$–$2p_3$ (607.4 nm), $1s_5$–$2p_7$ (621.7 nm), $1s_3$–$2p_5$ (626.6 nm), $1s_5$–$2p_8$ (633.4 nm) and $1s_5$–$2p_9$ (640.2 nm), two are shown for illustrative purposes in Fig. 4.

Figure 4 serves to illustrate the high quality of the Monte Carlo fits for the OG transitions studied for fixed circuit currents in the range 2.0–19 mA and also demonstrates the variations in the amplitude and shape of the optogalvanic waveforms as a function of current and wavelength change. As a result of the interaction between the laser photons and the population in a specific 1 s_i level, the neon atoms are excited to the $2p_k$ level, which in turn decays via spontaneous emission: $2p_k \rightarrow 1s_i$ (i = 2, 3, 4, 5,…), whereby the population of the initial 1s level decreases and one or more of different 1s levels increases. Depending on the ionization rates from these 1s levels, one observes an increase or decrease in ionization (with a corresponding increase or decrease in the OG signal) that lasts for varying lengths of time. As can be seen from Fig. 4, the values from the theoretical model are essentially identical to the experimental values. In Tables 2, 3, 4, 5 and 6, the fitted a-coefficients are in arbitrary units, while the b-coefficients are in units of $(\mu s)^{-1}$.

As seen from the above tables, the exponential rates, b_j, do not change significantly as the discharge current decreases from 19 to 8 mA. However, for 6 and 7 mA the exponential rates suddenly become very small. In addition, the tabulation shows a similar behavior for the a-coefficient as a function of the discharge current in the 19–8 mA range. Figures 5 and 6 help illustrate the rapid change in waveform behavior as the discharge current reaches 6 and 7 mA. It appears that there are

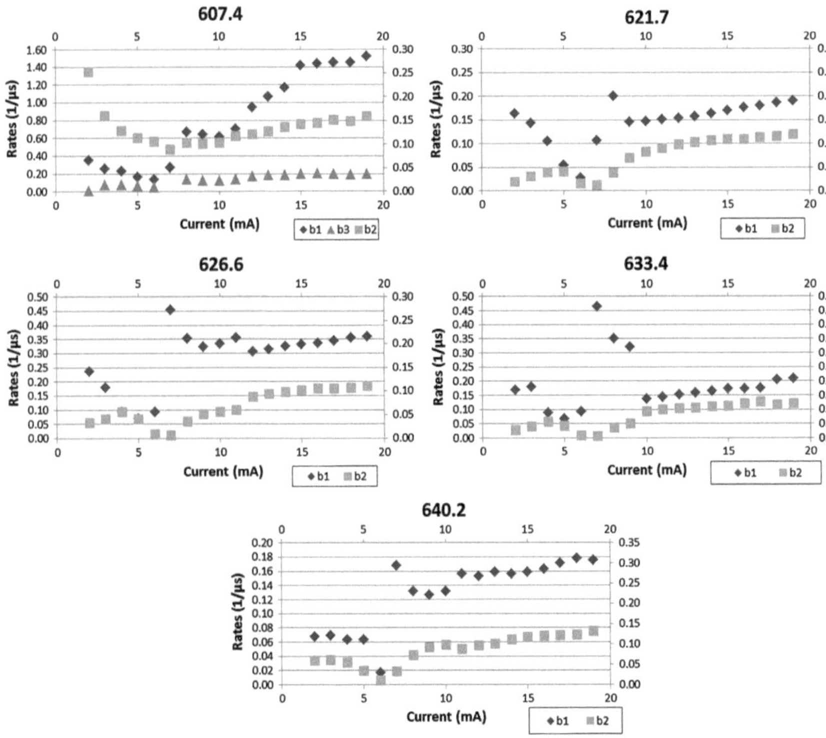

Fig. 6 Variation of the b-coefficients as a function of discharge current for the five OG transitions (607.4, 621.7, 626.6, 633.4, and 640.2 nm) studied for neon

noticeable plasma resonance effects around 7 mA, especially for the 607.4 nm transition, whereby our model is able to fit the experimental optogalvanic waveform with two terms only, as reflected in Table 2.

7 Concluding Remarks

In this chapter we have provided a historical perspective regarding the evolution of the optogalvanic effect and discussed both the experimental investigations and the mathematical formulation that helps explain the optogalvanic signal waveforms obtained by laser-exciting the discharge plasma in a hollow cathode lamp. In order to illustrate laser optogalvanic spectroscopy and collisional state dynamics associated with discharge plasmas, we have used a tunable laser to excite the plasma in an iron-neon (Fe–Ne) hollow cathode lamp and studied in great detail the neon transitions $1s_i$–$2p_k$ in the 590–670 nm region (for the current range 2–19 mA). Specifically, we focused on the following transitions: $1s_4$–$2p_3$ (607.4 nm), $1s_5$–$2p_7$ (621.7 nm), $1s_3$–$2p_5$ (626.6 nm), $1s_5$–$2p_8$ (633.4 nm) and $1s_5$–$2p_9$ (640.2 nm), to illustrate the applicability of the rate equation model developed and the excellent quality of the fit obtained employing the Monte Carlo fitting routine. The variations in the a- and b-coefficients as a function of current in the range 2–19 mA have also been determined and we noted a significant change in these parameters around 6–7 mA current values, while relatively small changes in the 8–19 mA range.

Current optogalvanic research continues to focus on the studies involving atomic states of various buffer gases, including metastable states, in a hollow cathode lamp employing a variety of electrode materials and gas combinations. Since the optogalvanic effect is driven by the change in conductivity of the plasma under the effects of a light source, studies such as the present one can provide insight into the use of laser optogalvanic spectroscopy in newly emerging fields, such as graphene-based nanoelectronics, nanoplasmonics and nanoplasma dynamics.

References

1. Rai SB, Rai DK (1996) Review article: optogalvanic spectroscopy. Proc Indian Nat Sci Acad 62(6):475–512
2. Morsch S, Brown PS, Badyal JPS (2012) Nanoplasma surface electrification. J Mater Chem 22:3922–3929. doi:10.1039/C2JM16044G
3. Heidenreich A, Last I, Jortner J (2008) Nanoplasma dynamics in Xe clusters driven by ultraintense laser fields. Eur Phys J D 46(1):195–202. doi:10.1140/epjd/e200,-00288-0
4. Heidenreich A, Jortner J (2010) Effects of the nanoplasma electrons on Coulomb explosion of xenon clusters. J Phys Chem C 114(48):20636–20647. doi:10.1021/jp105291u
5. Han XL, Chandran H, Misra P (2010) Collisional rate parameters for the $1s_4$ energy level of neon 638.3 nm and 650.7 nm transitions from the analyses of the time-dependent optogalvanic signals. J At Mol Sci 1, 2:118–125. doi:10.4208/jams.051509.071209a

6. Chen L, Zhang M, Wei W (2013) Graphene-based composites as cathode materials for lithium ion batteries. J Nanomaterials 2013(2013), Article ID 940389. doi:10.1155/2013/940389
7. Yang J et al (2013) LiFePO4–graphene as a superior cathode material for rechargeable lithium batteries: impact of stacked graphene and unfolded graphene. Energy Environ Sci 6:1521–1528. doi:10.1039/C3EE24163G
8. Pitchai R, Thavasi V, Mhaisalkarb SG, Ramakrishna S (2011) Nanostructured cathode materials: a key for better performance in Li-ion batteries. J Mater Chem 21:11040–11051. doi:10.1039/C1JM10857C
9. Daniov V, Rudnev V, Kretov V (2012) Simulation of the heat transfer in the nanocathode. Open J Appl Sci 2(4B):78–81. doi:10.4236/ojapps.2012.24B019
10. Shi Y et al (2012) Graphene wrapped LiFePO4/C composites as cathode materials for Li-ion batteries with enhanced rate capability. J Mater Chem 22:16465–16470. doi:10.1039/C2JM32649C
11. Foote PD, Moehler FL (1925) Photo-electric ionization of caesium vapor. Phys Rev 26:195
12. Penning FM (1928) Demonstração de um novo efeito fotoelétrico. Physica 8:137
13. Terenin A (1930) Photoionization of salt vapors. Phys Rev 36:147–148. doi:10.1103/PhysRev.36.147.2
14. Joshi SS (1939) Threshold potentials and reactivity under electrical discharge. Curr Sci 8:548
15. Joshi SS, Narasimham V (1940) A light effect in chlorine under electrical discharge. Curr Sci 9:535
16. Joshi SS, Deshmukh GS (1945) Interaction of nitrous oxide and hydrogen in the silent discharge. Nature 155:483
17. Wagh SM, Deshpande DA (2011) Revisiting the joshi effect. Curr Sci 101(9):1182–1190
18. Kenty C (1950) Role of metastable (3P2) Hg atoms in low current discharges in Hg rare gas mixtures. Phys Rev 80(1):95–96
19. Meissner KW, Miller WF (1953) Influence of irradiation on the characteristic of a glow discharge in pure rare gases. Phys Rev 92:896
20. Garscadden A, Bletzinger P, Friar EM (1964) Moving striations in a He-Ne laser. J Appl Phys 35:3432. doi:10.1063/1.1713244
21. Green RB, Keller RA, Luther GG, Schenck PK, Travis JC (1976) Galvanic detection of optical absorptions in a gas discharge. Appl Phys Lett 29:727–729
22. Green RB, Keller RA, Schenck PK, Travis JC, Luther GG (1976) Opto-galvanic detection of species in flames. J Am Chem Soc 98(26):8517–8518. doi:10.1021/ja00442a035
23. King DS, Schenck PK, Smyth KC, Travis JC (1977) Direct calibration of laser wavelength and bandwidth using the optogalvanic effect in hollow cathode lamps. Appl Opt 16 (10):2617–2619
24. Lawler JE, Ferguson AI, Goldsmith JEM, Jackson DJ, Schawlow AL (1979) Doppler-free intermodulated optogalvanic spectroscopy. Phys Rev Lett 42:1046. doi:http://dx.doi.org/10.1103/PhysRevLett.42.1046
25. Hänsch TW, Shahin IS, Schawlow AL (1971) High-resolution saturation spectroscopy of the sodium D lines with a pulsed tunable dye laser. Phys Rev Lett 27:707. doi:http://dx.doi.org/10.1103/PhysRevLett.27.707
26. Kowalski FV, Hill WT, Schawlow AL (1978) Saturated-interference spectroscopy. Opt Lett 2:112
27. Wieman C, Hänsch TW (1976) Doppler-free laser polarization spectroscopy. Phys Rev Lett 36:1170. doi:http://dx.doi.org/10.1103/PhysRevLett.36.1170
28. Vidal CR (1980) Optogalvanic double-resonance spectroscopy. Opt Lett 5(4):158
29. Goldsmith JEM, Ferguson AI, Lawler JE, Schawlow AL (1979) Doppler-free two-photon optogalvanic spectroscopy. Opt Lett 4(8):230–232
30. Zhu X, Nur AH, Misra P (1994) Laser optogalvanic wavelength calibration with a commercial hollow cathode iron-neon discharge lamp. J Quant Spectrosc Radiat Transfer 52:167–177

31. Nur AH, Zhu X, Misra P (1995) Polarity of laser excited optogalvanic transitions in neon. Spectrosc Lett 28:367–377
32. Han XL, Su MC, Haridass C, Misra P (2004) Collisional dynamics of the first excited states of neon in the 590–670 nm region using laser optogalvanic spectroscopy. J Mol Struct 695–696:155–162
33. Han XL, Blosser MC, Misra P, Chandran H (2012) Abrupt changes in neon discharge plasma detected via the optogalvanic effect. Thin Solid Films 521:155–157. doi:10.1016/j.tsf.2011.12.088

Applications of Fluorescence Anisotropy in Understanding Protein Conformational Disorder and Aggregation

Neha Jain and Samrat Mukhopadhyay

Abstract Fluorescence spectroscopy is an ultra-sensitive multiparametric technique that provides key insights into protein conformational dynamics and size changes simultaneously. Fluorescence polarization (anisotropy) is one of the parameters related to the rotational dynamics of a fluorophore either intrinsic to the molecule or attached to a biomolecule. The anisotropy measurements can be utilized to unravel the structural and dynamical properties of biomolecules. The advantage of fluorescence anisotropy measurements is that it is a concentration-independent parameter; it can be measured either in the steady-state or in the time-resolved format. Steady-state fluorescence anisotropy provides important information about the overall size/dynamics of biomolecules, whereas the time-resolved fluorescence anisotropy can distinguish between the local and the global dynamics of a fluorophore. Therefore, the time-resolved anisotropy measurements allow one to determine the conformational flexibility as well as the size of biomolecules and assemblies. In recent years, it has been demonstrated that fluorescence anisotropy can be effectively utilized to obtain structural and dynamical information of protein-based assemblies such as aggregates, protein–lipid complexes etc. This chapter provides an overview of the applications of fluorescence anisotropy to study protein conformational disorder, misfolding and aggregation, leading to the formation of nanoscopic amyloid fibrils that are implicated in a range of human diseases.

N. Jain · S. Mukhopadhyay (✉)
Department of Biological Sciences and Centre for Protein Science, Design and Engineering, Indian Institute of Science Education and Research (IISER), Mohali, India
e-mail: mukhopadhyay@iisermohali.ac.in

S. Mukhopadhyay
Department of Chemical Sciences, Indian Institute of Science Education and Research (IISER), Mohali, India

Present Address:
N. Jain
Department of Molecular, Cellular and Developmental Biology, University of Michigan, Ann Arbor, MI 48109, USA

P. Misra (ed.), *Applied Spectroscopy and the Science of Nanomaterials*,
Progress in Optical Science and Photonics 2, DOI 10.1007/978-981-287-242-5_3

Keywords Fluorescence spectroscopy · Fluorescence polarization · Rotational dynamics · Fluorescence anisotropy · Time-resolved measurements · Protein conformation · Dynamics · Disorder · Protein misfolding · Aggregation · Protein–lipid complexes · Amyloid fibrils

1 Principles of Fluorescence Anisotropy

Fluorescence spectroscopy is an ultra-sensitive tool to investigate many dynamic processes including conformational and size changes of proteins in response to the changes in the surrounding environment. Among the fluorescence readouts, the fluorescence anisotropy is often the method of choice both in vitro as well as in vivo since the changes in the anisotropy are highly sensitive. Additionally, being a ratiometric readout, the anisotropy is independent of the concentration of fluorophores. The theory of fluorescence polarization was first introduced by Perrin in 1926. It was introduced in the field of biochemistry many decades ago by the pioneering work of Weber and has gained enormous importance in the decades that have followed [1]. It provides unique insights into different molecular parameters including the changes in conformation, orientation, fluidity of the medium and biomolecular size. Therefore, it has been successfully utilized in the investigations of complex mechanisms of biological folding and assembly [2–6].

Fluorescence anisotropy is defined as the ratio of the difference between the intensities of vertically and horizontally polarized emissions to the total fluorescence intensity (Eq. 1). Being a ratiometric measurement, fluorescence anisotropy is a dimensionless parameter. Two fundamental properties need to be considered while measuring the fluorescence anisotropy: (i) the fluorescent probe possesses both excitation and emission dipoles and (ii) there is always a time-lag between the absorption and emission of a photon (lifetime of the fluorophore). The depolarization of fluorescence anisotropy occurs due to the angular displacement of the fluorophore which takes place in between the absorption and emission of photons and therefore, is a measure of the rate of rotational diffusion during the lifetime of the excited state of a fluorophore (nanoseconds). The rate of fluorescence depolarization, as a result of rotational diffusion, is dictated by the viscosity of the medium and the size (and shape) of the fluorophore [7–9]. The size of a rotating unit increases upon aggregation and interaction with the other molecules. Therefore, fluorescence anisotropy offers a reliable mean to probe the increase in the size. The larger the size is, the slower the fluorescence depolarization is and larger is the observed fluorescence anisotropy (Fig. 1).

Amyloid aggregation, mediated by protein conformational disorder/misfolding, accounts for a variety of human diseases such as Parkinson's, Alzheimer's and Huntington's diseases [10–12]. These aggregates are composed of cross β-sheet structure and are highly ordered. Since most proteins are naturally fluorescent due to the presence of fluorescent amino acid (such as tryptophan), or they can be

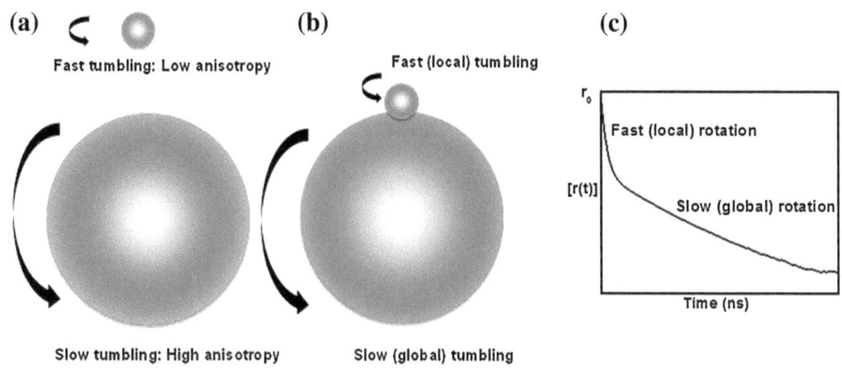

Fig. 1 Schematic of fluorescence anisotropy: **a** Fast depolarization arising out of rapid tumbling of a monomer or a small molecule. **b** Dampening of rotational dynamics due to binding/association. **c** Time-resolved fluorescence anisotropy showing local and global rotational dynamics

chemically labeled by fluorescent dyes of choice, measuring the fluorescence anisotropy provides a convenient approach to probe the aggregation process. The anisotropy measurements can also be extended to the microscopy format to carry out cellular anisotropy imaging [13–15].

The anisotropy measurements are carried out in two broadly classified formats, namely, steady-state and time-resolved measurements. The steady-state fluorescence anisotropy (r_{ss}) is measured under continuous excitation source. To measure the steady-state anisotropy, the sample is excited with a vertically polarized light and the emission is measured using polarizers in two orthogonal orientations $\left(I_{\parallel} \text{ and } I_{\perp}\right)$. These polarized intensities are used to obtain the steady-state anisotropy as follows [7, 8]:

$$r_{ss} = (I_{\parallel} - GI_{\perp})/(I_{\parallel} + 2GI_{\perp}) \tag{1}$$

The perpendicular component is always corrected by a geometry factor called G-factor. The changes in the steady-state anisotropy (r_{ss}) indicate the overall changes in the protein conformation (flexibility) and/or size of the protein. These measurements are well suited for monitoring self-association of proteins and protein–lipid interactions since these interactions result in larger size and hence slower rotation leading to the increase in the fluorescence anisotropy (Fig. 1). Steady-state anisotropy measurements provide a convenient and quick approach of detecting conformational change, flexibility and aggregation of biomolecules but are not adequate to describe these changes in a quantitative fashion. This is due to the fact that the anisotropy has contributions from both local/internal motions of the fluorophore and the global tumbling of the entire molecule/assembly that cannot be distinguished by steady-state measurements alone. This limitation can be overcome by fluorescence anisotropy measurements carried out in the time-resolved format in which the time-dependent decays of polarized emission intensities (parallel and perpendicular) are acquired using a polarized and pulsed excitation source [16–19]. The fluorescence anisotropy decay monitors the depolarization events that occur during the lifetime of the excited state of a fluorophore. The time-resolved anisotropy measurements enable one to distinguish between the local and the global dynamics of a fluorophore present in the system under investigation. Typically, the time-resolved decay of fluorescence anisotropy can be expressed as sum of exponentials as follows:

$$r(t) = r_0[\alpha_1 \exp(-t/\phi_{fast}) + \alpha_2 \exp(-t/\phi_{slow})] \quad (2)$$

where r_0 is the intrinsic fluorescence anisotropy, ϕ_{fast} and ϕ_{slow} are fast and slow rotational correlation times, respectively and α_1 and α_2 are amplitudes associated with fast and slow rotational times, respectively.

The two different rotational correlation times (ϕ_{fast} and ϕ_{slow}) describe the local and the global motion, respectively. The fast rotational correlation time represents the local tumbling of the fluorophore whereas the slow rotational correlation time attributes to the global dynamics. The slow rotational correlation time (ϕ_{slow}) is often related to the volume (V) and hence the size of the biomolecules, as well as the viscosity (η) of the medium by the Stokes-Einstein-Debye equation as follows:

$$\phi_{slow} = \eta V/kT \quad (3)$$

$$V = {}^4\!/_3 \pi R^3 \quad (4)$$

where R is the mean hydrodynamic radius of the molecule or aggregate.

In the subsequent sections, we describe the applications of fluorescence anisotropy in studying the protein misfolding and aggregation. For a more comprehensive description on the theory and practice of fluorescence anisotropy, readers are referred to Refs. [7, 8].

2 Fluorescence Anisotropy in the Description of Protein Misfolding and Aggregation

2.1 Probing Protein Oligomerization and Amyloid Assembly

A number of studies have reported that the fluorescence anisotropy (steady-state and time-resolved) technique can be utilized efficiently to decipher the aggregation mechanisms of various proteins [20–25]. For example, measuring the fluorescence anisotropy of Thioflavin-T (ThT), an amyloid indicator, has been established as a simple tool to monitor the amyloid aggregation of HET-s prion protein [20]. A steady increase in ThT anisotropy was observed as a function of time following which the nucleation (k_n) and elongation (k_e) of prion amyloid assembly were determined. In a recent report, the oligomer formation of amyloid-β (Aβ) peptide, known to play a key role in the pathogenesis of Alzheimer's disease, was established by monitoring the changes in the fluorescence anisotropy of pyrene-labeled pronucleon peptide [21]. A decrease in the pyrene anisotropy was observed when the peptide was bound to oligomers, which could be attributed to the long lifetime of pyrene. Similarly, time-resolved fluorescence anisotropy technique was utilized to investigate the oligomerization and self-assembly of HIV-1 integrase, as well as to characterize the oligomeric aggregates of the prion protein [22, 23]. In another study, the protein–protein interactions in a crowded milieu were elucidated using time-resolved anisotropy decays where fluorophores such as 1,8-anilino naphthalene sulfonic acid (ANS) and fluorescein isothiocyanate were employed to observe the interactions between apomyoglobin (apoMb) and ribonuclease A (RNase A), as well as human serum albumin (HSA) at high concentrations [24]. The anisotropy decay profiles suggested that apoMb forms dimers with RNase A, whereas, no dimerization was observed upon interaction with HSA. The time-resolved anisotropy experiments were also utilized to reveal different conformational states of HSA at different pH [25]. In this chapter, we have compiled a few selected studies wherein steady-state and time-resolved fluorescence anisotropy techniques were efficaciously used to understand the key molecular events pertaining to protein conformational change and misfolding leading to amyloid aggregation.

2.2 Chain Collapse of an Intrinsically Disordered Protein

As mentioned previously, fluorescence anisotropy is widely used to monitor the protein conformational and size changes since it is related to the rotational motion of a fluorophore. The faster the rotation, the lower the anisotropy is and the slower the rotation, the higher the anisotropy. Recently, this technique was elegantly implemented in a study to distinguish between two distinct, predominantly populated states of a natively-unfolded protein namely, κ-casein, under two different solution conditions [26]. κ-casein, containing a single tryptophan residue (intrinsic fluorophore), is an intrinsically disordered protein (IDP) which can adopt different conformations such as a collapsed, compact state or an extended state depending upon the solution conditions. The structural and dynamical properties of these two conformers were distinguished using both steady-state and time-resolved fluorescence anisotropy in addition to other fluorescence observables. Under native (physiological) condition, the steady-state fluorescence anisotropy of tryptophan was higher compared to that in the presence of a denaturant, suggesting that the rotational mobility of tryptophan was restricted under native condition owing to the compaction of κ-casein, whereas, the polypeptide chain was expanded in the presence of a denaturant [26]. In order to obtain insights into the rotational dynamics of these two conformers, time-resolved fluorescence anisotropy measurements were performed and following the decay analyses using a biexponential function, two rotational correlation times (fast: local and slow: global) were recovered (Fig. 2). The slow rotational correlation time ($\sim$7 ns) was used to

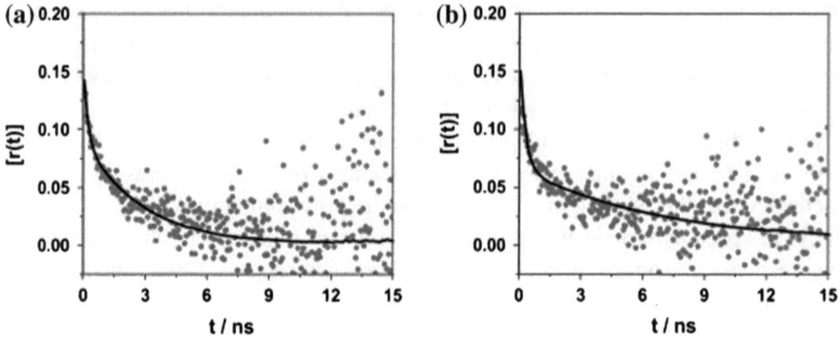

Fig. 2 Time-resolved anisotropy decays of single tryptophan in κ-casein depicting fast (local) and slow (global) motion under native (**a**) and denatured (**b**) condition. The *grey filled circles* represent the observed anisotropy decays, and the *black lines* represent the fits obtained using a biexponential function. Reproduced with permission (Ref. [26])

estimate the mean hydrodynamic radius of the protein which corresponded to ~1.9 nm for the collapsed globule. However, for the expanded conformer, the slow rotational correlation time is independent of the size of the protein because segmental mobility/fluctuations of an extended chain can depolarize the fluorescence much faster compared to the global rotation of the chain. Therefore, dynamic light scattering (DLS) technique was utilized for size estimation which indicated an average hydrodynamic radius of ~4.6 nm (for the expanded conformer) that decreased to ~1.9 nm (for the collapsed conformer) upon transferring κ-casein from the denaturant-containing solution into the native buffer. Thus, fluorescence anisotropy was able to distinguish between denatured (expanded) and collapsed (compact) forms of κ-casein [26]. The collapsed state can further initiate and promote the formation of protein oligomers and amyloid aggregate at higher protein concentrations as observed by much longer rotational correlation times (Unpublished results from the Mukhopadhyay laboratory).

2.3 Characterization of Molten-Globule State of Ovalbumin

Steady-state and time-resolved fluorescence anisotropy experiments were performed to illustrate the structure and dynamics of a low pH-induced partially-unfolded, molten-globule state of ovalbumin [27]. Partially-unfolded conformers are conjectured to be amyloidogenic, which play a pivotal role in protein aggregation and amyloid assembly. In these experiments, the fluorescence anisotropy of the intrinsic tryptophans was measured. Results from the steady-state anisotropy measurements revealed a conformational expansion at low pH (pH 2) because the low pH form had higher anisotropy compared to the native form at pH 7. Investigation into the dynamical properties of the native and the low pH conformers by

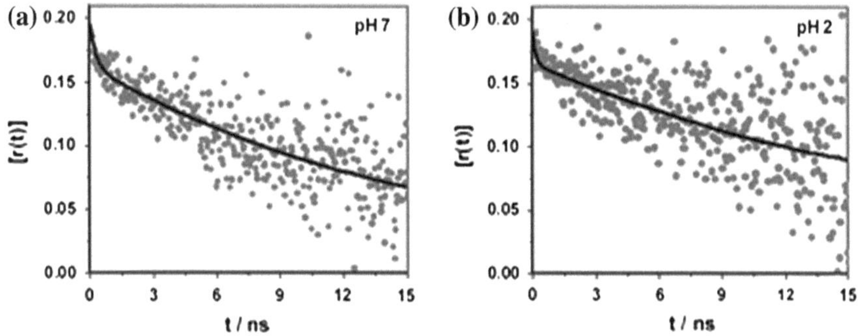

Fig. 3 Time-resolved fluorescence anisotropy of tryptophan in ovalbumin in the native form at pH 7 (**a**) and in the molten-globule form at pH 2 (**b**) forms. The *grey filled circles* represent the observed anisotropy decays, and the *black lines* represent the fits obtained using a biexponential function. Reprinted (adapted) with permission from Ref. [27]; Copyright (2012) American Chemical Society

time-resolved fluorescence anisotropy technique yielded two rotational correlation times. The fast rotational correlation time was found to be the same ($\sim$0.2 ns) for both the conformers, whereas, the slow rotational correlation time increased from $\sim$14 ns (native form) to $\sim$21 ns (low pH form) (Fig. 3). These results indicated that the protein undergoes a conformational expansion at the acidic pH, thus leading to slow global tumbling of the fluorophore. However, the fluorescence lifetime of tryptophan is short, which can reduce the precision in the estimation of the slow rotational correlation time. To circumvent this issue, an extrinsic fluorophore, namely, 1,8-anilino naphthalene sulfonic acid (ANS) was employed since it has a longer mean lifetime (>10 ns) when bound to the hydrophobic pockets inside the protein. The fluorescence anisotropy decay measurements of ANS, non-covalently bound to the low pH-induced conformer of ovalbumin, revealed a long correlation time of $\sim$24 ns, further supporting the conformational expansion. An overall increase of $\sim$40 % in the mean hydrodynamic volume of the molten globule state of ovalbumin was observed which is in line with the size expansion for a typical molten globule [27]. Upon the formation of amyloid, ovalbumin demonstrated a very long rotational correlation time (>100 ns) suggesting higher order protein assembly (Unpublished results: Bhattacharya).

2.4 *Conformational Dynamics of Amyloidogenic Compact Disordered State of β_2-Microglobulin*

A similar study was recently carried out to characterize the disordered, yet compact, state of β_2-microglobulin, a protein involved in dialysis-related amyloidosis [28]. In these experiments, the tryptophan fluorescence anisotropy decays were obtained under both native and low pH (pH 2.5) conditions. Two rotational correlation times, fast and slow, were recovered upon decay analyses. Comparison between the fast

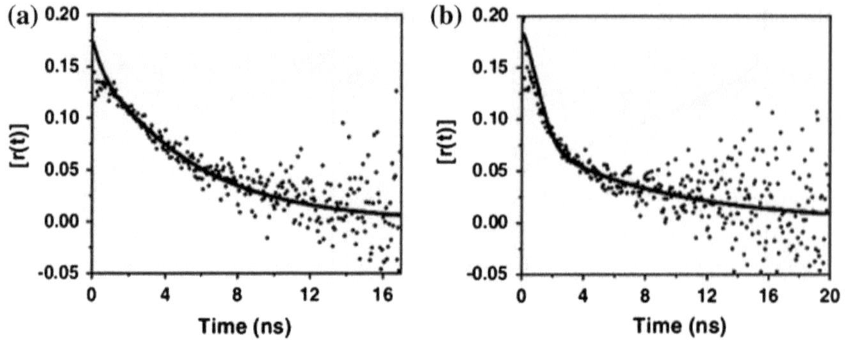

Fig. 4 Tryptophan fluorescence anisotropy decays of β_2-microglobulin at pH 7 (**a**) and at pH 2.5 (**b**). The *black filled circles* represent the observed anisotropy decays, and the *black lines* represent the fits obtained using a biexponential function. Reproduced with permission from Springer (Ref. [28])

rotational correlation times, an indicator of local tumbling of the fluorophore, of the native protein and the low pH conformer revealed no significant differences. However, a significant increase in the amplitude (α_1; Eq. 2) of the fast component, from ~37 % (native) to ~61 % (low pH form), was observed, indicating that the low pH form of β_2-microglobulin possesses higher conformational flexibility. Similarly, an increase in the slow rotational correlation time (from ~5.9 to ~8.8 ns) was observed when β_2-microglobulin was transferred from the native buffer into the low pH buffer suggesting an increase in the mean hydrodynamic volume of the protein at low pH which is accompanied by a volume expansion of ~70 % (Fig. 4). Overall, the observations obtained from time-resolved anisotropy measurements suggested that the low pH-induced conformer of β_2-microglobulin exists in a 'collapsed disordered' form accompanied by an increase in conformational flexibility and swelling of the polypeptide chain [28].

2.5 Detection of Oligomers and Fibrils of Amyloid-β Peptide

In one of the early findings, it was demonstrated that the time-resolved fluorescence anisotropy technique could be successfully utilized to recover the early and late events of protein aggregation [29]. In this work, the aggregation of the amyloid-β peptide (Aβ), known to play a key role in the pathogenesis of Alzheimer's disease, was investigated using fluorescence anisotropy decays whereby pre-fibrillar and fibrillar stages could be detected during Aβ aggregation. The fluorescence anisotropy experiments were carried out using a fluorescently labeled synthetic Aβ(1–40) peptide. The peptide, denoted as 'fluo-peptide', contained a fluorescein marker that was covalently attached to a cysteine residue. At a very low concentration of the fluo-peptide, the fluorescein anisotropy decayed very rapidly to zero indicating that the fluo-peptide tumbled very rapidly. Moreover, when the fluo-peptide was incubated for

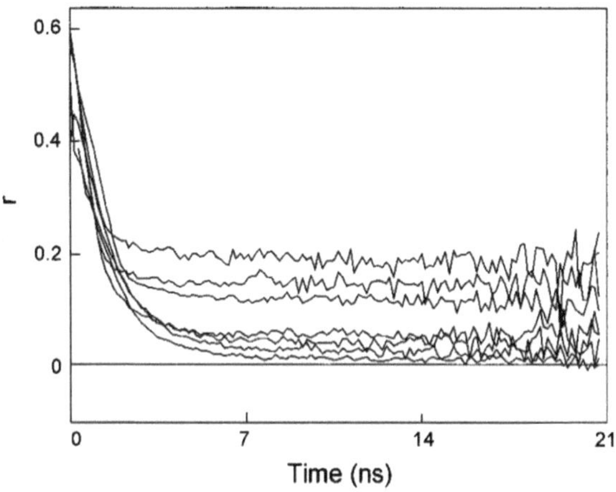

Fig. 5 Time resolved anisotropy decays of Aβ 1–40 (normal peptide with fluo-peptide). The traces in order of increased 'r' values represents anisotropy decay due to fluo-peptide only followed by addition of normal peptide after incubation for 0, 1, 24, 72, 96, and 120 h. Reproduced with permission from Elsevier (Ref. [29])

6 days under native conditions, the fluorescein anisotropy decays remained almost the same, suggesting that the fluo-peptide did not undergo any self-association at low concentrations. Rather, the peptide existed as a dimer because its estimated mean hydrodynamic radius (~ 1.5 nm) is comparable to that of the Aβ dimers (~ 1.8 nm) that were measured by dynamic light scattering (DLS) technique. In order to monitor Aβ aggregation at higher concentration, the unlabeled and fluorescein-labeled Aβ peptides were mixed in an appropriate ratio (100:1) and the fluorescence anisotropy decays were measured immediately after mixing. It was observed that the fluorescein anisotropy did not decay completely to zero. This non-zero anisotropy or "residual anisotropy" in the decay profile was attributed to the formation of a complex consisting of the fluo-peptide and the preformed aggregates (from unlabeled Aβ peptide), that were already present in the solution, prior to mixing. Upon further incubation, there was a progressive increase in the relative contribution of the residual anisotropy to the overall anisotropy decay due to the formation of large aggregates. After a certain period of time, the relative contribution of residual anisotropy dropped which was presumably caused by the precipitation of the large aggregates (Fig. 5). The above observations suggested that only two types of fluorescent species were detectable in the anisotropy data, the "uncomplexed" Aβ dimer and large aggregates. Thus, the study suggested that the aggregation of Aβ is initiated by a nucleation event which is followed by the formation of larger aggregates. This technique can be of great use for diagnostic purposes, as the different species formed during aggregation could be detected in the early stages. On the other side, labeling the protein with a fluorophore having long lifetime would facilitate capturing the larger aggregates which would have long correlation times. This study provides a new strategy to detect early oligomers and to unravel the underlying molecular mechanism.

2.6 Characterization of Protofibrils in Barstar

It was demonstrated that the time-resolved fluorescence anisotropy technique could be successfully employed to gain insights into the structural reorganization of barstar during various stages of aggregation at a residue-specific resolution [30]. For these experiments, the authors generated seven different single-cysteine mutants of barstar, spanning along the entire polypeptide chain, whereby the thiol group of each of the cysteine residue/mutant was chemically labeled by 5-((((2-iodoacetyl) amino)ethyl)amino)naphthalene-1-sulfonic acid (IAEDANS). In this study, the amplitude (α_1; Eq. 2) of the fast rotational component (ϕ_{fast}; Eq. 2), which is associated with the flexibility of the fluorescent probe (AEDANS), was efficiently utilized to decipher the structural details of barstar oligomers and protofibrils en-route to barstar fibrillation (Fig. 6). Briefly, analyses of each of the residue-specific AEDANS anisotropy decay at different stages of barstar aggregation enabled the recovery of both the local (fast) and global (slow) rotational correlation times and the respective amplitudes. Since the amplitudes of the fast components showed a large variability among the residue positions, a "dynamic amplitude map" was created that delineated the following salient features of barstar aggregates: (i) during barstar oligomerization at low pH, the C-terminal end becomes more structured whereas the N-terminus does not participate in the oligomer formation and the middle region also remains flexible, (ii) as the aggregation proceeds, and protofibrils are formed, all the amino acid residues of barstar except one become more ordered inside the aggregate core without any drastic structural reorganization and finally, (iii) the fibrils share a similar "internal structure" as that of the protofibrils because the residue-specific AEDANS fluorescence dynamics remained almost constant after the protofibrils were formed [30]. Additionally, it was reported that the low pH-induced oligomers are the "on-pathway" species to protofibril and fibril formation owing to the similarity of the dynamic amplitude maps at each stage (Fig. 6).

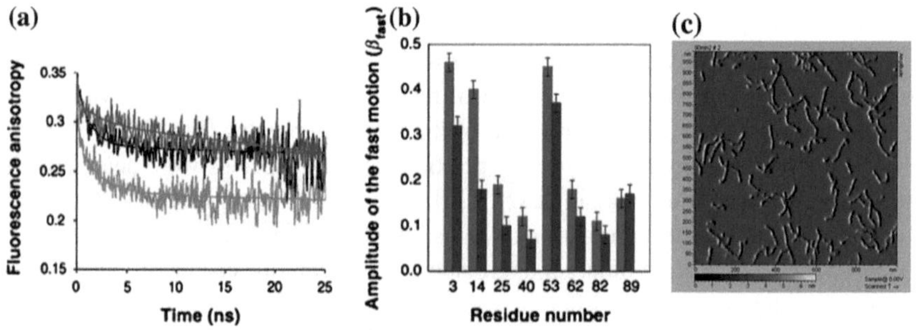

Fig. 6 **a** Time-resolved fluorescence anisotropy decays of barstar in protofibrils with (AEDANS labeled at Cys3, *pink*; at Cys14, *blue*; and at Cys40, *red*). **b** Dynamic amplitude map (β_{fast}) of barstar in the oligomeric-form (*red*) and in protofibrils (*blue*). **c** Atomic force microscopy *image* of protifibrils. Reproduced with permission from Elsevier (Ref. [30])

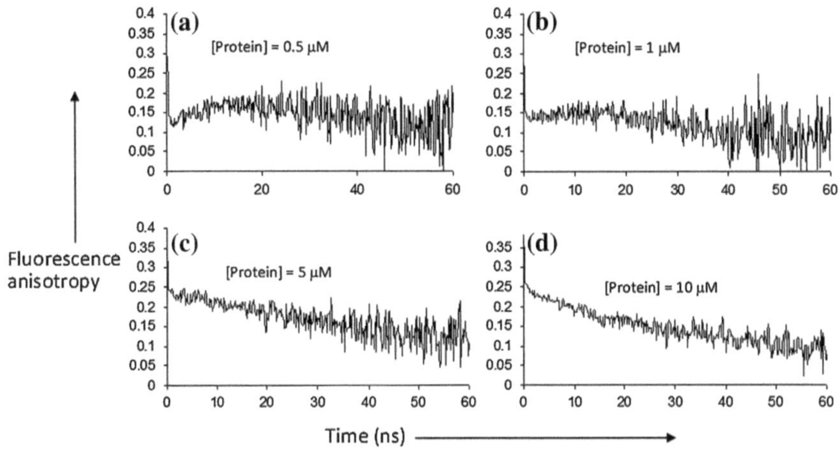

Fig. 7 Protein concentration dependence of time-resolved anisotropy decays of barstar (AEDANS labeled at Cys82) 5 min after a pH-jump to 2.7 at room temperature. **a** 0.5 μM; **b** 1 μM, **c** 5 μM and **d** 10 μM. Analysis using a two-population model revealed that the fraction of aggregates approaches 1.0 beyond 5 μM of barstar concentration (Unpublished results by S. Mukhopadhyay and G. Krishnamoorthy)

Studies on protein concentration dependence revealed that barstar aggregates even at sub-micromolar concentrations. At low concentrations (≤1 μM), monomers and the aggregates appear to coexist. This is evident from a *dip-and-rise* behavior in the time-resolved fluorescence anisotropy decay (Fig. 7). It has been reported that a dip-and-rise behavior in the time-dependent anisotropy is a definitive indication of the presence of at least two populations of molecules consisting of a species having a short fluorescence lifetime and fast rotational motion, along with another species having a long fluorescence lifetime and slow rotational motion in the timescale of fluorescence [31]. This is due to the fact that photons emitted at early time are highly depolarized due to the fast rotational motion and/or extensive angular diffusion, whereas photons emitted at late times are predominantly from the large aggregate, which has a slow rotational motion and/or more restricted angular diffusion. The two-population analysis of time-resolved anisotropy revealed that the fraction of the aggregate is altered from ~0.4 (0.5 μM of protein) to ~1.0 (10 μM of protein) at pH 2.7 at room temperature. The unusual dip-and-rise behavior is not seen at protein concentrations higher than 5 μM, suggesting that nearly all the molecules are transformed into aggregates at high protein concentrations. It should be pointed out, however, that the absence of dip-and-rise behavior in the fluorescence anisotropy decay would not have ruled out the presence of two (or more) populations (Unpublished results: S. Mukhopadhyay and G. Krishnamoorthy). This technique will be useful to identify the population distributions of monomers and aggregates as a function of protein concentration for many other pathologically-relevant proteins.

2.7 Mechanistic Insights into Aggregation of Serum Albumin

Recently, it has been demonstrated that steady-state fluorescence anisotropy could be efficiently utilized to monitor the aggregation kinetics of serum albumin [32]. It was reported that at low pH (pH 3), bovine serum albumin (BSA) forms a partially-unfolded amyloidogenic precursor at room temperature that formed amyloid fibrils when incubated at high temperature and in the presence of salt [32]. In order to delineate the aggregation mechanism of BSA, a variety of fluorophores were used independently. At high protein concentrations and before heating, the authors observed a small increase in the anisotropy of tryptophan (intrinsic probe) and a significant increase in the ANS (extrinsic probe; non-covalently bound to BSA) anisotropy which was ascribed to the formation of soluble oligomers at room temperature. Upon heating, a progressive increase in both the tryptophan and ANS anisotropy was observed that reached to saturation quickly without any lag phase. This observation suggested that at an elevated temperature, the large-sized aggregates are formed from pre-formed oligomers. The BSA aggregation kinetics was also monitored by labeling the single free cysteine (located in domain I of BSA) with IAEDANS and changes in AEDANS anisotropy were recorded as a function of aggregation. Due to its location the change in AEDANS anisotropy reflected the changes in the domain I of BSA. At low protein concentration the steady-state anisotropy of AEDANS-labeled BSA was low ($\sim$0.17) compared to that at higher protein concentration and in the presence of salt at room temperature ($\sim$0.19) indicating the formation of soluble oligomers. A sharp increase in AEDANS anisotropy was observed upon heating and saturated at $\sim$0.26. Therefore, combining all the fluorophore anisotropy data collected as a function of time, the study on BSA aggregation suggested that, at higher protein concentration and in the presence of salt, collapse of the extended hydrophobic patches occurs at room temperature leading to oligomerization that eventually drives the formation of amyloid-like fibrils accompanied by a size increase. Thus the anisotropy measurements of different fluorophores provided detailed information on the mechanism of aggregation in BSA [32].

2.8 Detergent-Induced Oligomerization of Lysozyme

In another study, the steady-state fluorescence anisotropy technique was used to decipher conformational and size changes during the lysozyme aggregation in the presence of an anionic surfactant namely, sodium dodecyl sulphate (SDS) [33]. A variety of fluorophores (intrinsic and extrinsic) were used to gain insights into the mechanism. Under native conditions all the probes, including tryptophan, ANS and dansyl, showed a high degree of rotational motion and thus low fluorescence anisotropy. When the protein was incubated at pH 9.2 with SDS at a sub-micellar concentration and at room temperature, there was a sharp increase in the steady-state anisotropy of all the fluorophores. This observation suggested that the positively charged protein at pH 9.2 interacts with negatively charged SDS and undergoes conformational changes, leading to the exposure of hydrophobic patches. These hydrophobic patches promote the formation of aggregates, wherein the probes get buried inside, resulting in a highly restricted mobility and a significant increase in the anisotropy due to the formation of large-sized aggregates.

2.9 Membrane-Induced Folding of Amyloidogenic α-Synuclein

A very recent report has demonstrated the utilization of steady-state fluorescence anisotropy technique, in addition to other fluorescence methodologies, in discerning the precise location of various residues of α-synuclein on a membrane surface. α-Synuclein, a protein implicated in Parkinson's disease, belongs to a class of Intrinsically Disordered Proteins (IDPs) that are devoid of stable structures and can adopt different forms depending upon the binding partners [34]. The authors took the advantage of non-availability of tryptophan and created 10 single tryptophan mutants (F4W, A27W, Y39W, A56W, A69W, A78W, A90W, A107W, A124W, and A140W) of α-synuclein. In order to determine the rotational mobility of α-synuclein at a residue-specific resolution upon binding to a lipid membrane, the tryptophan fluorescence anisotropies of all the locations were recorded. It was conjectured that, due to binding-induced folding, the interaction between the membrane and few specific residues of α-synuclein would dampen the rotational flexibility of tryptophan that could be captured by the increase in the fluorescence anisotropy. The changes in the anisotropy would also be affected by the exact positioning of the fluorophore on the membrane surface. In the native/free (disordered) form, the anisotropies of all the locations were low (~ 0.05) but demonstrated marked location-dependent variation upon membrane binding-induced folding to a helical state. A fluorescence anisotropy map was constructed based on the tryptophan steady-state anisotropy data that were recorded at various residue positions under membrane-bound conditions. A substantial increase in the fluorescence anisotropy ($r_{ss} > 0.15$) for a few residues in the N-terminus and the middle (NAC) domain (positions 4, 56 and 78) revealed that the residues are tightly held with the membrane and therefore experience low rotational flexibility. No significant changes in the tryptophan anisotropy ($r_{ss} < 0.10$) were observed for the residues located in the C-terminus (positions 107, 124 and 140) suggesting that the C-terminal region is not involved in the binding-coupled folding event. A slightly lower anisotropy was observed for the residue positions 27, 39, 69 and 90 as compared to the positions 4, 56 and 78 indicating that the former residues are located on the outer surface of the membrane and hence, demonstrate higher degree of flexibility (Fig. 8). Taken together, these anisotropy data allowed us to delineate residue-specific structural details of membrane-bound α-synuclein [34]. The time-resolved fluorescence anisotropy measurements on this lipid–protein system is currently underway in our laboratory.

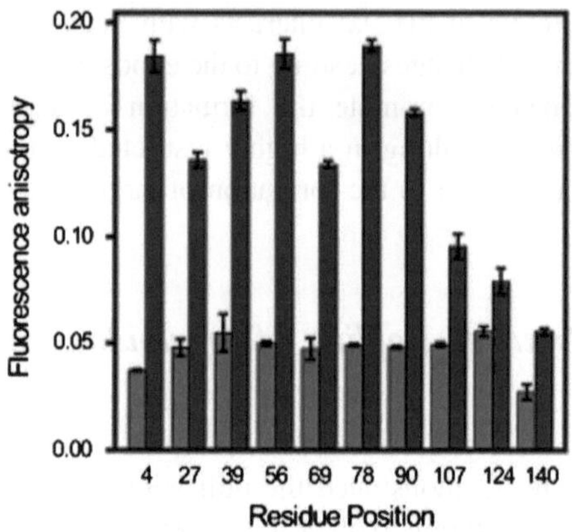

Fig. 8 Steady-state fluorescence anisotropy of single tryptophan located at different residue positions in α-synuclein; free IDP state (*light grey*) and membrane-bound form (*dark grey*) (Adapted from Ref. [34])

2.10 Watching Protein Association in the Cell by Anisotropy Imaging

Fluorescence anisotropy has been extended in the field of imaging to monitor the spatial localization of proteins in the living cells [13, 35–38]. This has given a new direction to the study of assembly, arrangement of cellular constituents and membrane–protein interaction at the micro- and nano-scale resolutions. Confocal fluorescence anisotropy imaging is a highly sensitive tool to elucidate both the dynamics and the structural features of protein aggregates in the living system, which can be further extended to study amyloid systems and drug screening procedures (Fig. 9). α-Synuclein aggregation was directly monitored in the cell lines by

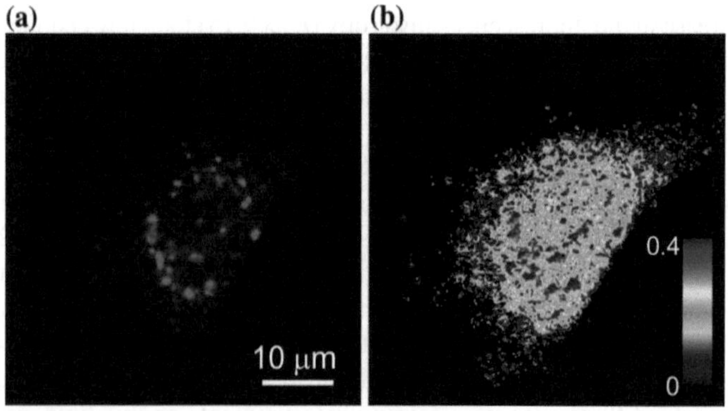

Fig. 9 Confocal fluorescence anisotropy (CFA) microscopy of α-synuclein-C4-FlAsH aggregates in SH-SY5Y cells. **a** CFA *image* and **b** associated fluorescence anisotropy *image* (Adapted from Ref. [36])

employing the fluorescence anisotropy imaging technique [36]. It was observed that α-synuclein amyloidogenic aggregates were composed of multimeric and immobile structures. Fluorescence anisotropy imaging can be coupled with homo-fluorescence resonance energy transfer (FRET) to understand the molecular details of protein oligomerization since homo-FRET results in the fast depolarization and thus leads to a decrease in the anisotropy even without a significant rotational mobility [35]. This approach has provided valuable insights into the organization of serotonin$_{1A}$ receptor assembly in the membrane [37] and homo-dimerization of amyloid precursor protein at the plasma membrane [38] in live cells.

3 Conclusion and Future Directions

The application of fluorescence spectroscopy and imaging in biological systems has expanded tremendously over the past decades and a number of biological processes can now be studied at the single-molecule level. Fluorescence anisotropy measurements have emerged as a very attractive technique to interrogate various steps involved in the complex process of protein folding, misfolding, aggregation and amyloid formation. The approach is not limited to detection of amyloids but can potentially be extended to study the inhibition and/or reversal of aggregation that is of paramount importance for designing the therapeutics against pathological consequences of amyloids. We envision that the fluorescence anisotropy techniques discussed in this article will find broad applications in the structural, molecular and cellular understanding of various biologically important protein association, interactions and aggregation.

Acknowledgments We thank Dr. Mily Bhattacharya for critically reviewing and for making extremely valuable suggestions on this manuscript and the members of the Mukhopadhyay laboratory of Amyloid Biology for the research contributions described in this book chapter. Research grant from CSIR (to SM) and financial support from IISER Mohali is gratefully acknowledged.

References

1. Steiner RF (1991) Fluorescence anisotropy: theory and applications. In: Lakowicz JR (ed) Topics in fluorescence spectroscopy, vol 2: principles. Plenum Press, New York
2. Kawski A (1993) Fluorescence anisotropy: theory and applications of rotational depolarization. Crit Rev Anal Chem 6:459–529
3. Munishkina LA, Fink AL (2007) Fluorescence as a method to reveal structures and membrane-interactions of amyloidogenic proteins. Biochim Biophys Acta 1768:1862–1885
4. Jha A, Narayan S, Udgaonkar JB, Krishnamoorthy G (2012) Solvent-induced tuning of internal structure in a protein amyloid protofibril. Biophys J 103:797–806
5. Sheynis T, Friediger A, Xue W-F, Hellewell AL, Tipping KW, Hewitt EW, Radford S, Jelinek R (2013) Aggregation modulators interfere with membrane interactions of β_2-microglobulin fibrils. Biophys J 105:745–755

6. Yan Y, Marriott G (2003) Analysis of protein interactions using fluorescence technologies. Curr Opin Chem Biol 7:635–640
7. Lakowicz JR (1999) Principles of fluorescence spectroscopy, 2nd edn. Kluwer Academic, New York
8. Valeur B (2001) Molecular fluorescence principles and applications. WILEY-VCH Verlag GmBH, Weinheim
9. Brown MP, Royer C (1997) Fluorescence spectroscopy as a tool to investigate protein interactions. Curr Opin Biotechnol 8:45–49
10. Dobson CM (1999) Protein misfolding, evolution and disease. Trends Biochem Sci 24:329–332
11. Luheshi LM, Crowther DC, Dobson CM (2008) Protein misfolding and disease: from the test tube to the organism. Curr Opin Chem Biol 12:25–31
12. Kumar S, Udgaonkar JB (2010) Mechanisms of amyloid fibril formation by proteins. Curr Sci 98:639–656
13. Gradinaru CC, Marushchak DO, Samim M, Krull UJ (2010) Fluorescence anisotropy: from single molecules to live cells. Analyst 135:452–459
14. Jameson DM, Ross JA (2010) Fluorescence polarization/anisotropy in diagnostics and imaging. Chem Rev 110:2685–2708
15. Jameson DM, Sawyer WH (1995) Fluorescence anisotropy applied to biomolecular interactions. Methods Enzymol 246:283–300
16. Saxena A, Udgaonkar JB, Krishnamoorthy G (2005) In applications of fluorescence spectroscopy. In: Hof M, Hutterer R, Fidler V (eds). Springer, New York
17. Bright FV, Munson CA (2003) Time-resolved fluorescence spectroscopy for illuminating complex systems. Anal Chim Acta 500:71–104
18. Krishnamoorthy G (2012) Motional dynamics in proteins and nucleic acids control their function: revelation by time-domain fluorescence. Curr Sci 102:266–276
19. Vogel SS, Thaler C, Blank PS, Koushik SV (2009) Time-resolved fluorescence Anisotropy. In: Periasamy A, Clegg RM (eds) FLIM microscopy in biology and medicine. Taylor & Francis, Boca Raton
20. Sabaté R, Saupe SJ (2007) Thioflavin T fluorescence anisotropy: an alternative technique for the study of amyloid aggregation. Biochem Biophys Res Comm 360:135–138
21. Matveeva EG, Rudolph A, Moll JR, Thompson RB (2012) Structure-selective anisotropy assay for amyloid beta oligomers. ACS Chem Neurosci 3:982–987
22. Deprez E, Tauc P, Leh H, Mouscadet J-F, Auclair C, Brochon J-C (2000) Oligomeric states of the HIV-1 integrase as measured by time-resolved fluorescence anisotropy. Biochemistry 39:9275–9284
23. Wang Y, Goodson T III (2007) Early aggregation in prion peptide nanostructures investigated by nonlinear and ultrafast time-resolved fluorescence spectroscopy. J Phys Chem B 111:327–330
24. Zorrilla S, Rivas G, Lillo MP (2004) Fluorescence anisotropy as a probe to study tracer proteins in crowded solutions. J Mol Recognit 17:408–416
25. Otosu T, Nishimoto E, Yamashita S (2010) Multiple conformational state of human serum albumin around single tryptophan residue at various pH revealed by time-resolved fluorescence spectroscopy. J Biochem 14:191–200
26. Jain N, Bhattacharya M, Mukhopadhyay S (2011) Chain collapse of an amyloidogenic intrinsically disordered protein. Biophys J 101:1720–1729
27. Bhattacharya M, Mukhopadhyay S (2012) Structural and dynamical insights into the molten-globule form of ovalbumin. J Phys Chem B 116:520–531
28. Narang D, Sharma PK, Mukhopadhyay S (2013) Dynamics and dimension of an amyloidogenic disordered state of human β_2-microglobulin. Eur Biophys J 42:767–776
29. Allsop D, Swanson L, Moore S, Davies Y, York A, El-Agnaf OMA, Soutar I (2001) Fluorescence anisotropy: a method for early detection of Alzheimer and β-peptide (Aβ) aggregation. Biochem Biophys Res Comm 285:58–63

30. Mukhopadhyay S, Nayak PK, Udgaonkar JB, Krishnamoorthy G (2006) Characterization of the formation of amyloid protofibrils from barstar by mapping residue-specific fluorescence dynamics. J Mol Biol 358:935–942
31. Ludescher RD, Peting L, Hudson S, Hudson B (1987) Time-resolved fluorescence anisotropy for systems with lifetime and dynamic heterogeneity. Biophys Chem 28:59–75
32. Bhattacharya M, Jain N, Mukhopadhyay S (2011) Insights into the mechanism of aggregation and fibril formation from bovine serum albumin. J Phys Chem B 115:4195–4205
33. Jain N, Bhattacharya M, Mukhopadhyay S (2011) Kinetics of surfactant-induced aggregation of lysozyme studied by fluorescence spectroscopy. J Fluoresc 21:615–625
34. Jain N, Bhasne K, Hemaswasthi M, Mukhopadhyay S (2013) Structural and dynamical insights into the membrane-bound α-synuclein. PLoS ONE 8(12):e83752. doi:10.1371/journal.pone.0083752
35. Ghosh S, Saha S, Goswami D, Bilgrami S, Mayor S (2012) Dynamic imaging of homo-FRET in live cells by fluorescence anisotropy microscopy. Methods Enzymol 505:291–327
36. Roberti MJ, Jovin TM, Jares-Erijman E (2011) Confocal fluorescence anisotropy and FRAP imaging of α-synuclein amyloid aggregates in living cells. PLoS ONE 6(8):e23338. doi:10.1371/journal.pone.0023338
37. Ganguly S, Clayton AHA, Chattopadhyay A (2011) Organization of higher-order oligomers of the serotonin$_{1A}$ receptor explored utilizing homo-FRET in live cells. Biophys J 100:361–368
38. Devauges V, Marquer C, Lécart S, Cossec J-C, Potier M-C, Fort E, Suhling K, Lévêque-Fort S (2012) Homodimerization of amyloid precursor protein at the plasma membrane: a homoFRET study by time-resolved fluorescence anisotropy imaging. PLoS ONE 7(9):e44434. doi:10.1371/journal.pone.0044434

30. Mukhopadhyay S, Nayak PK, Udgaonkar JB, Krishnamoorthy G (2006) Characterization of the formation of amyloid protofibrils from barstar by mapping residue-specific fluorescence dynamics. J Mol Biol 358:935–942
31. Ludescher RD, Peting L, Hudson S, Hudson B (1987) Time-resolved fluorescence anisotropy for systems with lower symmetry [illegible] Biophys Chem [illegible]
[illegible]

Nuclear Magnetic Resonance Spectroscopy in Nanomedicine

Ping-Chang Lin

Abstract Retaining the essentials of noninvasive measurement and deep penetration through living bodies, nuclear magnetic resonance (NMR) spectroscopy presents as a useful adjunct to the *in vitro* and in vivo studies of nanomedicine. This chapter is aimed at introducing basic NMR principles and certain NMR techniques relevant to the field of nanomedicine, followed by exploration of physicochemical characterization and metabolic profiles responding to treatment and development in nanomedicine.

Keywords Noninvasive measurements · Nuclear magnetic resonance · NMR spectroscopy · Physicochemical characterization · Metabolic profiles · Nanomedicine

1 Introduction to Nanomedicine

Nanomedicine, emerging from nanotechnology that usually generates and implements materials at less than 100 nm scale, is a discipline of "*science and technology of diagnosing, treating, and preventing disease and traumatic injury, of relieving pain, and of preserving and improving human health, using molecular tools and molecular knowledge of the human body*" [1]. According to the definition proposed by the European Medical Research Councils of the European Science Foundation in 2004, nanomedicine was categorized into analytical tools, nanoimaging agents, nanomaterials, nanodevices, novel therapeutics and drug delivery systems [1]. On the other hand, the US National Institutes of Health Roadmap for Medical Research in 2004 stated that nanomedicine "*refers to highly specific medical interventions at the molecular scale for curing disease or repairing damaged tissues, such as bone, muscle, or nerve*" (http://commonfund.nih.gov/nanomedicine/overview).

P.-C. Lin (✉)
Department of Radiology, Howard University, 2041 Georgia Ave, N.W, Washington, DC 20060, USA
e-mail: pingchang.lin@howard.edu

P. Misra (ed.), *Applied Spectroscopy and the Science of Nanomaterials*, Progress in Optical Science and Photonics 2, DOI 10.1007/978-981-287-242-5_4

Typically, the initiative structures of nanomedicine fabrication contain a length scale between 1 and 100 nm and at least one dimension in the range of sub-nm to 100 nm [2]. Emphasis on structures controlled at the nanometer scale has gathered numerous new properties that do not exhibit at atomic or molecular levels, but instead, permit a potential for a diverse array of applications, including those in biomedical disciplines [3].

Conventional medicine generally relies on certain expression of disease symptoms (e.g. metabolic responses, infections and cancers) detected or diagnosed through use of sophisticated, high-cost medical instruments, devices or implants, while the therapy is then developed upon the corresponding clinical trials. On the contrary, nanomedicine, which keeps more attention on nanoscale interactions within biomolecules, cell organelles and cells, is aimed at achieving early diagnosis and ensuring accurate dosage for treatment at the molecular or cellular level prior to the appearance of traditional symptoms. At the present time, two major challenges remain at the forefront of nanomedicine development: (1) how reliably to regulate the assembly of more than one active component in a nanoscale vector, and (2) how well to control the forms and dosages of the active agents spatially and temporally released [4].

Despite being in the early stages of development, the realm of nanomedicine embraces a myriad of nanomaterials retaining potential for positive impacts on healthcare applications and revolution in medical research in the areas of drug delivery, therapy, diagnostic and therapeutic multi-functional systems, imaging agents and sensing/monitoring techniques for disease diagnosis or response to disease pathogenesis, and nanotechnologies for facilitation of biomedical discovery and research [5, 6]. Once the intrinsic characteristics of nanomedicine, such as pharmacokinetics and cytotoxicity, are fully comprehended, the progress of nanomedicine development can be expected to improve the delivery, efficacy, bioavailability, correct dosage, targeting ability and personalization of nanopharmaceuticals associated with disease diagnosis and treatment [7–9].

Overall, engineered materials with dimensions similar to biological moieties at the nanoscale can possibly be implemented as diagnostic nanomedicine, therapeutic nanomedicine, or both [10–12]. Available nanomaterials that can be considered nanomedicines include various polymer, protein or drug conjugates/complexes, liposomes, niosomes, micelles, dendrimers, emulsions, nanocrystals, nanoparticles, nanocapsules, carbon nano-platforms and quantum dots. Certain types of nanomaterials are shown in Fig. 1 [5, 13–16].

The physiological behavior associated with nanomedicines can influence the therapeutic efficacy, diagnostic accuracy, or both; therefore, in this context, it is necessary to understand the physicochemical characteristics of nanomedicines affecting their distribution and behavior in vivo [18]. Because the physiological interactions in the body may be highly impacted by the in vivo biodistribution, passage, phagocytosis and endocytosis of nanomedicines, which are essentially different from those of conventional medicines, it is important to develop robust

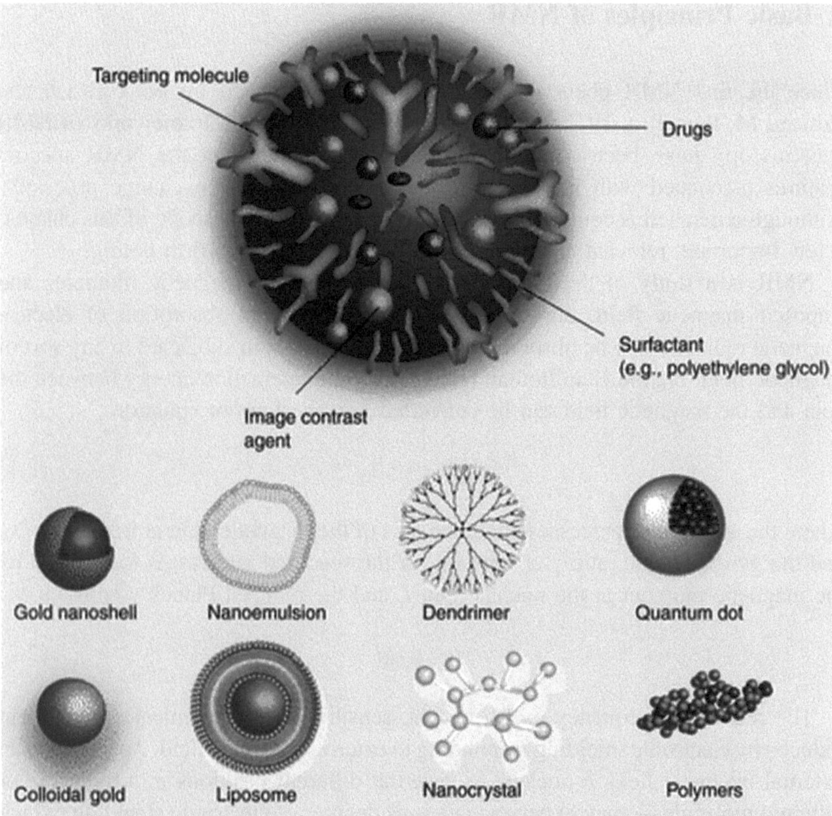

Fig. 1 Nanomaterials used in nanomedicine. The surface of a nanomedicine is possibly covered with hydrophilic surfactants, such as polyethylene glycol, and its interior core can be formed by solid materials, liquid materials or an encapsulated drug. Reproduced with permission from Ref. [17], Future Medicine Ltd

standards to directly respond to assessment of the potential of fabricated nanomaterials and to provide guidance for ensuring the quality control, safety and toxicity of nanomedicines [19, 20]. For evaluation of the physicochemical properties of manufactured nanomedicines, many techniques routinely applied to conventional pharmaceuticals can also be used in the same manner [21]. Indeed, several nanomaterial-related characteristics are crucially important and require investigation through commonly applied modalities or new specific technologies to comprehend the in vivo behavior of the nanomaterials. In this chapter, we mainly address the applications of NMR spectroscopy to nanomedicine, after briefly describing the NMR principle and relevant NMR techniques applicable to physicochemical characterization of nanomedicines.

2 Basic Principles of NMR

Since the first NMR phenomenon experimentally detected by Felix Bloch and Edward M. Purcell in 1946, the theoretical and experimental frameworks of NMR spectroscopy have been continuously developed to interpret the NMR spectral features associated with molecular structures and chemical exchange processes. Although a detailed account of NMR theory is not within the scope of this chapter, a few important, relevant equations and milestones are still worth noting.

NMR is a study of the interactions between nuclear magnetic moments and imposed magnetic field, during which the emission and absorption of electromagnetic radiation can be observed [22]. For a nuclear spin subjected to any sort of magnetic field, B_0, the Hamiltonian representing the interaction energy between the spin and the magnetic field can be converted into the Larmor equation:

$$\omega_0 = -\gamma B_0 \tag{2.1}$$

where the spin vector precesses about the axis of the magnetic field at frequency ω_0, and the gyromagnetic ratio γ, a constant for the specified nucleus, is formulated by the magnetic moment μ, the nuclear spin I, and the reduced Planck's constant $\hbar$:

$$\gamma = \mu/\hbar I \tag{2.2}$$

The resonance frequency ω_0 is, in fact, sensitive to the chemical environment induced by electronic shielding responding to external magnetic field. As placed in an external magnetic field, a nucleus—located at different positions in a molecule or different molecules—may experience various degrees of electronic shielding, which are caused by the opposite rotation of the surrounding electrons to the nuclear spin precession and are referred to as the chemical shift. Therefore, the Larmor equation is then modified by introducing the electron-generated magnetic field into the effective magnetic field B at the nucleus:

$$\omega_0 = -\gamma B = -\gamma B_0(1 - \sigma) \tag{2.3}$$

where σ is the shielding constant.

To detect the precessional motion of nuclei from the viewpoint of classical mechanics, the net longitudinal magnetization, which is the ensemble of nuclear magnetic moments constructed by the population difference between the nuclear spin states along the direction of the external magnetic field B_0, must be tipped from its equilibrium position towards the transverse plane by applying a second magnetic field B_1 in the transverse plane [23]. In modern NMR techniques, a RF pulse or a RF pulse sequence, i.e. the B_1 field, is applied to induce the magnetization precession at the Larmor frequency, which can be expressed by the Bloch equations in the presence of relaxations in the laboratory frame [24]:

$$\frac{dM_x(t)}{dt} = \gamma(M(t) \times B(t))_x - \frac{M_x(t)}{T_2} \tag{2.4}$$

$$\frac{dM_y(t)}{dt} = \gamma(M(t) \times B(t))_y - \frac{M_y(t)}{T_2} \tag{2.5}$$

$$\frac{dM_z(t)}{dt} = \gamma(M(t) \times B(t))_z - \frac{M_z(t) - M_0}{T_1}, \tag{2.6}$$

where the magnetization $M(t) = M_x(t)\hat{i} + M_y(t)\hat{j} + M_z(t)\hat{k}$, the external magnetic field $B(t)$ consists of the static magnetic field B_0 and the time-dependent B_1, and the magnetization relaxation times include T_1, the spin-lattice relaxation time, and T_2, the spin-spin relaxation time. The RF signal induced from magnetization precession is then obtained by a receiving coil placed in the transverse plane.

Free induction decay (FID), which is the time-dependent signal generated from nuclear magnetization precession, can be processed into the frequency-dependent spectrum through Fourier transformation (FT). This process was first introduced as pulsed FT NMR by Ernst and Anderson in 1966 [23, 25]. Due to the capability of rapid data accumulation, the FT NMR method since then has been applied in different aspects: collection of all spectral information in an FID, increase of sensitivity and improvement of signal-to-noise ratio in a rather short period of time, and dynamic studies of rapid processes including chemical reactions and relaxations [22]. In addition, studies of biomolecules, cells, tissues and in vivo applications have been fostered through many modern NMR techniques, including relaxation measurements and two-dimensional (2D) FT NMR spectroscopy.

Magnetization relaxation is the process of the spin system regaining its equilibrium state after the RF pulse perturbation. The two relaxation components, T_1 and T_2, are depicted in the Block equations (Eqs. 2.4–2.6). T_1 relaxation is associated with the restoration of spin populations to the Boltzmann distribution by transferring the excited-state energy of the spins to the surrounding lattice, while T_2 relaxation is related to the entropy process of transverse magnetization disappearance caused by elimination of phase coherence through exchanging rather than losing energy between spins [23, 26]. The mechanism of longitudinal T_1 relaxation, also referred to as spin-lattice relaxation, is caused by the ensemble of changes in individual magnetic moments interacting with the local magnetic field oscillating at the Larmor frequency, i.e., the magnitudes and orientations of the magnetic moments are altered due to the thermal motion of associated atoms or molecules [27]. The process of transverse T_2 relaxation, also named spin-spin relaxation, results from the interaction of individual magnetic moments with the local magnetic field oscillating at the Larmor frequency and from a slightly different precession frequency caused by the distributed local magnetic field at every individual magnetic moment [27].

2D FT NMR spectroscopy is a set of NMR methods that process acquired data into a two-dimensional frequency space and are particularly useful for determination of molecular structures. The first 2D NMR experiment, which is known as

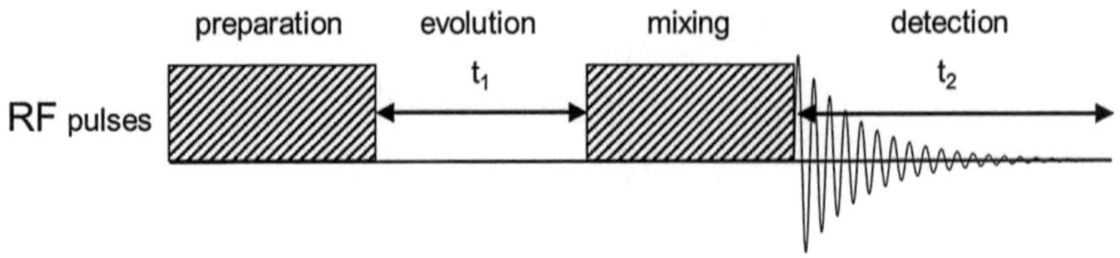

Fig. 2 Schematic diagram of a generic 2D NMR experiment

homonuclear correlation spectroscopy (COSY), was proposed by Jean Jeener in 1971, followed by a theoretical description of COSY and a thorough statement of 2D NMR spectroscopy reported by Richard Ernst and coworkers in 1976 [23, 28]. This contribution opened a gate to the development of modern multidimensional NMR techniques that play an important role in various disciplines such as organic chemistry, structural biology, and molecular and cellular pharmacology. Although 3D or higher dimensional spectra can be achieved in high-resolution liquid state NMR spectroscopy, this chapter rather provides a brief description of 2D methodology as frequently encountered for in vivo NMR spectroscopy [23].

Most of the 2D NMR experiments consist of four time periods: preparation, evolution, mixing and detection (Fig. 2) [23]. The duration of preparation time is typically designed long enough to permit the spin system to achieve the equilibrium state, and the spin system is generally perturbed into some coherence by applying certain types of pulses and delays at the end of this period. Evolution time t_1 is referred to as the indirectly detected domain because the spin system is not detected by the receiver coil but would rather freely precess. For example, the spins precess and dephase at their Larmor frequency according to the spin-spin relaxation T_2, if a 90° pulse is applied prior to the evolution time. The length of evolution time t_1 is set to zero in the beginning and then incremented by a constant Δt_1, with an individual FID stored for each increment of t_1. In mixing time, the coherence present at the end of t_1 is manipulated into an observable signal by another set of pulses and delays. Alternatively, magnetization associated with one spin may be transferred to another spin with a sinusoidal amplitude modulation during this period. During detection time t_2, the magnetization in the transverse plane is observed as an FID in the same fashion used in the one-dimensional FT NMR spectroscopy.

In practice, a time-domain 2D NMR experiment of variables t_1 and t_2 is Fourier transformed with respect to both the time variables into the 2D frequency-domain spectrum, in which the contour plot possesses an advantage in identifying and studying coupled resonances embedded in more intense resonances shown in the regular 1D NMR spectra. Among thousands of 2D NMR pulse schemes proposed, certain pulse sequence techniques have been representatively used in the widespread applications across various research areas.

COSY, the first introduced 2D NMR technique, has been used extensively to identify nuclei sharing a scalar coupling [22]. The loci of cross-peaks (i.e. off-diagonal peaks) in a COSY spectrum correlate the scalar-coupled nuclei at the corresponding chemical shifts. To study the secondary and tertiary molecular structures, 2D nuclear

Överhauser enhancement spectroscopy (NOESY) is performed to examine the occurrence of magnetization transfer between nuclei, which is through the mechanism of dipole-dipole interaction during the mixing time. The appearance of cross-peaks in a NOESY spectrum indicates the spatial closeness of the corresponding nuclei, and the cross-peak intensities can be used for conformational analysis [22, 27]. In addition, the exchange spectroscopy, used for the study of chemical exchange processes, adopts the same NOESY pulse sequence, in which the mixing time is designed to observe chemical exchanges. Further, multiple quantum spectroscopy has been increasingly implemented in 2D experiments, through which the evolution of unobservable multiple-quantum coherence during time t_1 can be detected indirectly [27]. Also, the idea of multiple quantum coherence can be very useful in in vivo applications via applying a spectral editing technique selectively to exclude undesired spin signals in order to simplify an NMR spectrum [23].

3 Physicochemical Characterization of Nanomedicines

3.1 *Overview of Physicochemical Characteristics of Nanomedicines*

Compared with the bulk material counterparts, nanoscale materials, including what are considered as nanomedicines, possess distinct physicochemical properties such as size, size distribution, shape, composition, molecular weight, purity, surface properties, stability, solubility and aggregation state [29]. Most of these properties are highly associated with particular physiological interactions and may provide benefits to medical applications, e.g. improvements in efficacy, reduction of side effects, prevention and treatment [6, 17, 21].

Among the physicochemical properties of nanomedicine, size plays a critical role in regulating circulation in the bloodstream, penetration across the physiological drug barriers, site- and cell-specific localization as well as induction of cellular responses [30–32]. The shape of nanomedicine affects efficiency of drug delivery, flow and adhesion of drug delivery carriers throughout the circulatory system, in vivo circulation time, cellular uptake, degradation, biocompatibility and retention in tissues and organs [33–42]. Furthermore, the disposition and translocation of a nanomedicine in an organism may be influenced by the shape, together with the size and agglomeration state, of the nanomedicine [43]. Chemical composition is an important factor that determines toxicity of nanomaterials and also influences the transport, delivery and biodistribution of nanomedicines [44, 45]. Purity determination is crucial because the impurities of nanomedicine, which typically result from the side products of nanomedicine formulation, can not only connect with nanomedicine reproducibility but significantly impact drug efficacy as well as introduce unfavorable side effects [18, 21]. Surface properties are of intrinsic importance due to typical occurring interactions of nanomedicines with biological organisms at the surfaces [43]. Among the surface properties of nanomedicines, surface composition is associated with the superficial layers but not with the bulk materials, while surface energy is related to the dissolution, aggregation and accumulation of nanomedicines [18]. Surface charge, which is typically estimated by zeta potential, controls the dispersion stability or aggregation of nanomaterials and may affect physiological barrier penetration and receptor binding. Beyond the surface properties, species absorbance/adhesion potentially changes the activity and conformation of the attached species and leads to alteration in the surface composition of the nanomedicine [18]. The stability of a nanomedicine may impact its corresponding toxicity, which is possibly induced by several factors, such as temperature, pH, particle size and the presence of other excipients or impurities, during synthesis and storage, or even in vivo [21, 45, 46].

3.2 NMR-Accessible Physicochemical Characteristics

NMR spectroscopy is a nondestructive/noninvasive technique that can be utilized to detect the composition of the sample examined [18]. In addition to requiring limited sample preparation to reduce artefacts or structural deterioration, it possesses a high sensitivity to changes in the local environment for resolving the structures of amorphous materials, polymers and biomolecules with apparent lack of long-range order [18, 47, 48]. Moreover, compared with the other spectroscopic approaches applied to nanomedicine characterization, NMR spectroscopy provides an advantage in simultaneously, unambiguously identifying the chemical composition and structure in an inhomogeneous system (Table 1) [49, 50]. In many circumstances the molecular mobility, intermolecular distances and diffusion properties of individual components can be answered via NMR spectroscopy as well [49, 50]. Through specific designation of RF and gradient pulse sequences for NMR measurements, the dynamics of species investigated, including relaxation, molecular conformation and molecular mobility interaction, can be evaluated in different conditions [47, 51, 52]. Although the main drawback of low detection sensitivity of NMR spectroscopy indicates the demands of increasing sample amount and

Table 1 NMR spectroscopy for evaluation of the physicochemical characteristics of nanomedicines

Physicochemical characteristics analyzed	Strengths	Limitations	Refs
Size (indirect analysis) Structure Composition Purity Conformational change	Non-destructive/non-invasive method Unambiguous identification of molecules and compounds Limited sample preparation requirement	Low sensitivity Time consuming tatic Relatively large sample size requirement Scarce NMR active nuclei	[21, 50, 52–60]

acquisition time for measurement to reach a certain level of signal-to-noise ratio, the increasing sample amount can be of benefit to exclusion of the influence of sample-sample holder interference (Table 1) [18, 48, 52].

Several physiochemical characteristics belonging to nanomedicine candidates can be determined or investigated by NMR spectroscopy, for example, the structure, purity and functionality in dendrimers, polymers and fullerene derivatives, and the conformational changes occurring in the interactions between ligands and nanomaterials [21, 53–56]. The diffusivity of nanomedicine can be evaluated using pulsed field gradient NMR technique, through which the sizes and interactions of investigated species were calculated [57]. The subsections below briefly address the physicochemical characteristics of nanomedicine generally accessible to NMR spectroscopy.

3.2.1 Size

The size of nanomedicine or nanomaterial is generally considered as the diameter of a sphere whose selected physical property is equivalent to that of the examined nanomaterial in the same condition [61, 62]. A frequently adopted model is the volume diameter of a nanomedicine, meaning the nanomedicine of equal volume to the modeled sphere [61]. Another model is the hydrodynamic diameter of a nanomaterial, referred to as the effective size calculated from the diffusion coefficient using the Stokes-Einstein relationship [61].

The size of a material governs its total surface area, which reflects the interference probability of the reactive components. For a sphere of radius r, the volume V of the sphere is linearly proportional to r^3 while the surface area of the sphere scales with r^2. The fraction of reactive components on the surface of a spherical material is associated with the ratio of the surface area to the volume of the material, which is proportional to the reciprocal of radius r and is depicted by dispersion F (Eq. 3.1 and Fig. 3) [63]. For cubic materials of edge length d and for cylindrical materials of cross-sectional radius r and fixed length L, the dispersions retain in the very similar manner (Eqs. 3.2 and 3.3) and are also shown in Fig. 3:

$$F = \frac{4\pi r^2}{4\pi r^3/3} = 3\sqrt{[3]}\frac{4\pi}{V} \tag{3.1}$$

$$F = \frac{6d^2 - 12d + 8}{d^3} = \frac{6}{\sqrt{[3]}V} - \frac{12}{\sqrt{[3]}V^2} + \frac{8}{V} \tag{3.2}$$

$$F = \frac{2\pi r^2 + 2\pi rL - 4\pi r}{\pi r^2 L} = 2.2\sqrt{[3]}\frac{10\pi}{V} - 0.4\sqrt{[3]}\left(\frac{10\pi}{V}\right)^2 \quad \text{for} \quad L = 10r \tag{3.3}$$

Overall, simply reducing the size of nanomedicine can dramatically increase its surface area per unit mass, which enhances the interference probability of the

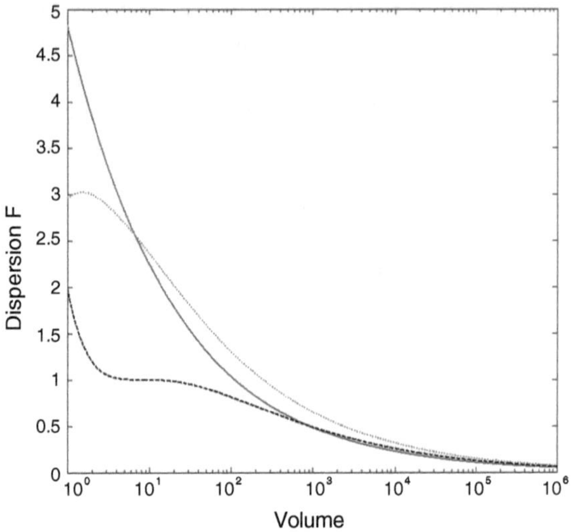

Fig. 3 Relationship between volume (or size) and dispersion F for materials of spherical (*solid curve*), cubic (*dashed curve*) and cylindrical shapes (*dotted curve*). The volume of a material is defined as the number of atoms or reactive component units in the material

reactive surfactants, possibly causing alterations in the physicochemical and physiological properties of the nanomedicine [21].

3.2.2 Composition and Purity

In the process of fabricating nanomedicine, it usually comprises two or more types of nanomaterials to form a complex, e.g. a chelate, a conjugant or a capsule; therefore, even engineered nanomedicine formulation without contamination may exhibit notable heterogeneity and polydispersity [21, 64]. Further, the complexity of engineered nanomedicine causes the distinct difference between physicochemical characterization of nanomedicine and that of traditional pharmaceuticals [64]. Because numerous types of nanomaterials can be implemented in the production of nanomedicine, consequent composition analysis of the nanomaterial complex is more complicated than that for a single entity [18, 21, 65].

Due to potential heterogeneity and polydispersity in an engineered nanomedicine, the purity of the nanomedicine must be determined in the complex of different components by defining a rational range rather than absolute purity standard [21, 64]. Indeed, the acceptable range of purity must meet necessary safety regulation and efficacy profile for the formulation through testing a large number of the randomly sampled component materials and batch-to-batch synthetic reproducibility [64]. In general, determination of nanomedicine purity can be accomplished through analysis of the chemical composition. Prior to finalizing the formulation of a nanomedicine and proceeding with the composition analysis, proper purification processes are required to remove any residual manufacturing components or side products to ensure the absence of endotoxin contamination [66].

3.2.3 Structure and Conformational Alteration

Available in a regular crystalline structure, an amorphous form or a non-crystallographically symmetric pack is the structure of a fabricated nanomedicine, which is generally governed by the size, chemical composition, number of components or atoms, and chemical interactions between the components/atoms of the nanomedicine [67]. In addition, covalent bonding, ionic bonding, hydrogen bonding, van der Waals forces and dipole-dipole interaction can engage in the settlement of a nanomedicine structure [67]. For instance, covalent bonding conducts the development of rigid bond angles, and metallic bonding, which is referred to as the constitution of electrostatic forces between delocalized conducting electrons from electron clouds and positively-charged metallic ions, performs a close-packed arrangement for metal atoms [68].

While engineered nanomedicines such as nanoparticles enter living environments, proteins or enzymes presenting in a biological fluid adhere to the surface of nanomedicines to form a nanomaterial-biological component interface that is the primary structure of a protein corona in a dynamic, competitive process [69–71]. The nano-bio interface is not at steady state and is likely formed in an inhomogeneous and dynamic/transient environment that may be affected by the nature of binding ligands, hydrophobic and charged regions, as well as free-surface-energy-induced conformational changes or reactive oxygen species [70]. These adhered proteins or enzymes undergoing conformational alteration or aggregation in different manners can result in exposure to cryptic epitopes, potentially trigger inappropriate cellular responses and induce unknown consequences [71–73].

3.3 Physicochemical Characterization of Nanomedicines Using NMR Spectroscopy

For characterizing the physicochemical properties of nanomedicines addressed in Sect. 3.2, several NMR spectroscopic techniques have been applied in a frequent manner, such as the pulsed gradient spin-echo (PGSE) technique for size measurement, the liquid- and solid-state NMR techniques, as well as the 2D NMR spectroscopy, for structural examination, and the High-Resolution Magic Angle Spinning (HR-MAS) NMR for composition confirmation. The following examples illustrate the applications of NMR spectroscopy to nanomedicine characterization.

The PGSE NMR spectroscopy has been implemented to measure the size and size distribution of colloidal materials fabricated from macromolecular amphiphiles and nanoparticles by analyzing the objects' diffusivities in solution or dispersion [57]. Given the self-diffusion coefficient D of a nanomaterial, determined in the PGSE NMR measurement, the hydrodynamic radius r_H of the nanomaterial can be calculated from the Stokes-Einstein equation, on the assumption of spherical objects:

$$r_H = \frac{k_B T}{6\pi\eta D} \quad (3.4)$$

where k_B stands for Boltzmann constant, T is absolute temperature and η represents solution viscosity. The capability of the PGSE NMR method has been demonstrated in a study that the accurate size distributions of polysulfide nanoparticles and polymersome vesicles were performed by examining the diffusion coefficients of the poly-ethylene glycolylated surfaces in both the colloidal systems [57]. In the analysis of these two colloidal systems, the PGSE NMR measurements were also used to detect the population differences among the nanoparticles, vesicles and colloidal components such as micelles, and to investigate the purification efficiency [57].

Similar to the PGSE method, the pulsed gradient stimulated echo NMR measurement has the ability of measuring the apparent diffusion coefficients of polystyrene latex nanoparticles [74]. As considering the long-range electrostatic interaction between polystyrene latex nanoparticles, the hydrodynamic sizes of the nanoparticles in pure and sodium dodecyl sulfáte aqueous solutions were accurately determined through use of the Stokes-Einstein equation after extrapolative determination of the true diffusion coefficients [75].

In a structural study of butyl-passivated silicon nanocrystals nc-Si/Bu, the multinuclear liquid- and solid-state NMR spectroscopies, which are critical to characterize passivated reaction products, were used to observe the butyl group bonded to the silicon nanoparticle surface [76]. Upon analysis of the scalar J couplings of ^{29}S-^{13}C in the liquid-state NMR spectra and the through-space dipolar couplings in the ^{1}H-^{29}Si solid-state NMR spectra, the formation of butyl-capped nanocrystalline silicon was verified [76]. Similar NMR approaches can be applied to a variety of nanoparticle systems of different isotopes including ^{111}Cd and ^{197}Au [76].

A ligand shell coating on a metal core-shell nanoparticle is the key of response to the particle's solubility and interfacial properties, while the structure of the ligand shell can be determined by implementing 1D and 2D NMR techniques. In a study of nanoparticles consisting of a gold core and a self-assembled thiolated monolayer constructed by a binary mixture of aliphatic and aromatic ligand molecules at varying composition, the structure of a thiolated monolayer shell was investigated through using 1D NMR and 2D NOESY [77].

Solution NMR can also be used to assess aqueous solubility enhancement and cytotoxicity reduction through, for example, characterizing the structure of hydrophilic polymer-capped nanoparticles [78]. By directly heating carboxylated multiple-walled carbon nanotubes (MWCNT) in poly(propionylethylenimine-*co*-ethylenimine) (PPEI–EI) random amino-polymer melt, the proton signals at ~2.0 and ~2.8 ppm were absent in the ^{1}H NMR spectra of the PPEI–EI polymer-functionalized MWCNTs, in comparison of the spectra of PPEI–EI [79]. The signal absence was responsible for the amidation and esterification of ethylenimine units with the nanotube surface-bound carboxylic groups in the functionalization of MWCNT samples [78, 79].

In the last few years, HR–MAS NMR using magic angle spinning for non-solid materials has been extensively applied to the biological and biomedical fields because it can generate high-resolution NMR spectra for tissue and cellular samples measured in heterogeneous nature [80]. Through utilizing the advantage of HR-MAS NMR accurately to characterize surface-attached ligands and modified surfaces, a study has successfully investigated individual synthetic steps of the cyclopeptide immobilized on the surface of poly(vinylidene fluoride) based nanoparticles, while another study examined the process of thermolytically produced thiol-derivatized silver clusters [80–82].

4 NMR-Based Metabonomics

4.1 What Is Metabonomics?

Several developing disciplines are currently applied thoroughly to understand the nature of many normal physiological and disease processes, including (i) genomics, which focuses on the systematic study of the whole genome of a cell or organism for investigation of a genetic complement, (ii) transcriptomics, which primarily concerns the complete set of RNAs encoded by the genome of a specified cell or organism for examination of gene expression, (iii) proteomics, which aims at the large-scale study of proteins within a cell or organism for analysis of protein synthesis and cell signaling, (iv) metabolomics, which emphasizes on metabolic profiling and fluxes, particularly with normal endogenous metabolism, at a cellular or organ level, and (v) metabonomics, which mainly examines perturbations of metabolic profiling and regulation of function caused by diseases, drugs, environmental factors and extragenomic influences [83, 84]. In these 'omics' techniques, alterations in gene expression and proteomic data do not always act as the surrogate endpoints in diagnosis of disease or evaluation of drug efficacy, but merely reflect the potential for the endpoint changes [84]. On the contrary, changes in metabolic profiles caused by metabolism disruption after a drug or noxious agent/stressor exposure are the endpoint observations, revealing the connection of the metabonomics measurements with cellular function and the overall physiological state in a cell or organism [84, 85].

Metabonomics, referring to the quantitative assessment of the dynamic responses of metabolites in a multicellular system to xenobiotic exposures such as pathophysiological intervention and genetic modification, provides a platform for investigating the complex consequences of toxic reactions, disease processes and genetic manipulation [84, 86]. A myriad of sample types in various biological systems at different degrees of complexity from biofluids, cells and tissues to animal models and human subjects can be analyzed using metabonomics [85]. Metabonomics is a powerful tool consisting specific analytical techniques and pattern recognition methods. The analytical techniques are implemented to explore metabolic signatures, followed by applying pattern recognition methods, such as

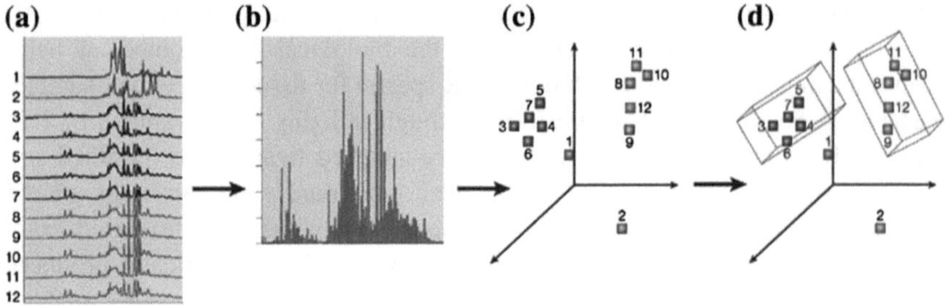

Fig. 4 Procedure of NMR spectroscopy based metabonomics. Reproduced with permission from Ref. [84], Nature Publishing Group. **a** NMR spectra, **b** Primary data reduction, **c** Unsupervised mapping of data in three-dimensional space, **d** Supervised classification and calculation of confidence limits

multivariate statistics, to latent-information extraction, primary pattern discrimination and identification/prediction of objects unknown to the implemented analytical techniques in complex datasets (Fig. 4) [84]. Among a number of spectroscopic techniques employed to produce the metabolic profiles of biomaterials for a metabonomic study, NMR spectroscopy possesses advantages in sample handling, non-destructivity, analytical reproducibility, molecular identification and instrumental robustness, accompanied with the capabilities of molecular dynamics analysis and tissue sample measurement [84, 87].

4.2 Applications of NMR-Based Metabonomics

Over the last years, NMR-based metabonomics has been increasingly utilized in disease and disorder diagnoses, drug discovery and development, and nutrition and toxicology researches through studying different biofluids, e.g. urine, serum and cerebrospinal fluid, on a metabolic basis [85, 88, 89]. In the aspect of drug development, for instance, environmental complications may modulate drug metabolism and toxicity, and may further restrain the predictions of drug-induced responses through pharmacogenomics [90]. Thus, a metabonomics-based approach named pharmacometabonomics has been proposed to predict the beneficial or adverse effects of potential drug intervention through analyzing biofluid metabolite profiles obtained from NMR spectroscopy [91]. An example of pharmacometabonomic study exhibited that subjects at high predose urinary levels of *p*-cresol sulfate reduced the capacity for acetaminophen sulfonation, and also predicted the connection of an individual's predose urinary metabolite profile with the postdose urinary fate of a normal dose of analgesic and antipyretic drug acetaminophen [90].

In a nutrition study, ^{1}H NMR spectroscopy was incorporated into a metabonomic approach to the biochemical effects of dietary phytoestrogens [92]. The 1D and 2D NMR spectral profiles were acquired in urine collected from premenopausal women before and after soy or miso consumption in the controlled dietary intervention. The chemometric investigation of subtle changes in the NMR spectral profiles exhibited the impact of soy isoflavone-rich diets, particularly the unconjugated soy isofavone form such as miso, on balance perturbation in a few osmolytes such as betaine, choline, creatinine and creatine [92].

Instead of examination of collected urine, another NMR-based metabonomic study was to investigate the serum metabonome in the obese and lean strains of swine [93]. Multivariate analysis of the NMR spectra of serum collected in these two strains of swine revealed distinct metabonome profiles in which certain metabolites' levels in the obese pigs were significantly higher than those in the lean ones, indicating different metabolisms existing between these two strains [93].

As an approach of toxicological screening, the high-resolution ^{1}H NMR spectroscopy was applied to the analysis of biological fluids to illustrate metabolic response with respect to the mechanism or the texture of toxic processes in an organism [94]. Examples include a study of toxicological responses to acetaminophen overdose through analysis of urine and plasma in human hepatocyte transplanted chimeric mice [95]. The NMR-based metabonomic analysis showed a decrease in urinary endogenous 2-oxoglutarate, an intermediate metabolite in the TCA cycle, and an increase in β-hydroxybutyrate, a ketone body that is the final product of the acetyl-CoA metabolic pathway, in plasma after acetaminophen administration [95]. The alterations in the observed endogenous metabolites may be caused by suppression of energy metabolism and TCA cycle, referring to the effect of acetaminophen-induced hepatotoxicity [95, 96].

Another toxicity study using ^{1}H NMR-based metabonomics was to examine the metabolic response after administration of a toxic organophosphorous compound tributyl phosphate [97]. The ^{1}H NMR spectra collected in the urine of the tributyl phosphate exposed rats revealed the disturbance of endogenous urinary metabolites, among which three tributyl phosphate-associated metabolites, namely, dibutyl phosphate, N-acetyl-(S-3-hydroxybutyl)-L-cysteine and N-acetyl-(S-3-oxobutyl)-L-cysteine were the key factors in segregating the tributyl phosphate treated group from the control group in the metabonomic analysis [97].

4.3 NMR-Based Metabonomics in Nanomedicine

Metabonomics has potential to provide a thorough evaluation of the bioactivities of nanomedicines, although it is still relatively underexplored in characterizing or profiling nanomedicines [85]. However, from the aspect of investigation modern NMR technology possesses an advantage that various metabolites in biofluids can be detected simultaneously at high sensitivity, which is suitable for metabolomic studies. Therefore, the combination of NMR spectroscopy and metabonomic analysis techniques will most likely benefit the characterization and regulation of nanomedicines.

The concern of in vivo toxicity may overwhelm the application of a particular nanomedicine in clinical practice. Nanomedicine-induced metabolic changes are possibly associated with in vivo toxicity or toxic effect that can be examined through use of metabonomic analysis of the acquired NMR data. By adopting the NMR-based metabonomics, several studies have successfully demonstrated the toxic or biochemical impacts of extensively used nanoparticles including titanium dioxide (TiO_2), silicon dioxide (SiO_2) and ultra-small superparamagnetic particles of iron oxides (USPIO):

Toxicity of TiO_2 nanoparticles. Among a wide range of applications in manufactured TiO_2 nanoparticles, a recent study examined the toxicological effect of the administered nanoparticles on rats at three concentration levels of oral exposure [98]. The collected rat urine and serum were investigated by ^{1}H NMR spectroscopy, followed by a metabonomic approach using principal component analysis and partial least squares analysis. According to the NMR-based metabonomic analysis, the complex disturbance in endogenous metabolites of urine and serum exhibited the impact of TiO_2 nanoparticle treatment on energy and amino acid metabolisms as well as on the minor injuries in liver and heart [98].

Toxicity of SiO_2 nanoparticles. SiO_2 nanoparticles have been innovated as an emerging drug delivery system due to their possessing particular structures and functions. As no surprise, they also raise toxicological concerns in public. Recently, a study of SiO_2 nanoparticles showed that sub-acute hepatoxicity was induced in rats after intranasal exposure of SiO_2 nanoparticles, in addition to pulmonary inflammation, myocardial ischemic damage, atrioventricular blockage and neurotoxicity responding to SiO_2 inhalation reported in literature [99]. The NMR-based metabonomic approach to SiO_2-nanoparticle exposed rat serum showed the elevation of lactate, alanine, acetate and creatine levels, indicating hepatic dysfunction verified in the serum biochemistry test of significant increases in ALT and AST enzyme activity levels [99].

Influence of USPIO nanoparticles on a biochemical process. USPIO, typically composed of an iron oxide core and a hydrophilic coating, is an example of nanoparticles used to enhance magnetic resonance imaging contrast for detection of lymph node metastasis, carotid atherosclerotic lesions, stroke, brain tumor proliferation, multiple sclerosis and spinal cord injury based on their properties of "superparamagnetic" susceptibility, long circulating lifetime in the bloodstream and prevention of early uptake by monocytes or macrophages after intravenous administration [100, 101]. However, the influence of nanoparticles on the biochemistry of macrophage cells is nearly opaque, particularly in the intracellular biochemical processes [102]. Through the HR-MAS ^{1}H NMR spectroscopy and multivariate analysis, Feng et al. demonstrated that macrophage-like cells responded to USPIO exposure with changes in cellular metabonome linked to energy production and the TCA cycle [102]. According to the study, the metabolic properties of macrophage appeared to be the temporal index of phagocytosis and release of USPIO, as well as the information on cell membrane modification [102].

The universal, non-selective vision of metabolic profiles performed by the NMR-based metabonomics has demonstrated the capability of exploring unexpected but valuable assets, particularly for those nanomedicines' properties hardly characterized by conventional parameters [85]. However, the low sensitivity of NMR techniques may generate an obstacle of detecting the metabolites or compounds at low concentration levels, which can, fortunately, be complemented by high-sensitivity Mass Spectroscopy. Given the accessibility of metabonomics to cell culture, animal model and human subject researches, the method of NMR-based metabonomics practically integrates the *in vitro* and in vivo toxicological studies of nanomedicines through characterization of the cellular metabolic phenotypes *in vitro* [85]. Still, oversimplification of the *in vitro* model should be taken into account. In this manner, an adequately large sample size is required to minimize bias, variance, or both resulting from different undesired factors in individual samples and to develop appropriate multivariate statistics models. Having the strength of integrating data collected, for example, by NMR spectroscopy from the different biological matrices of the same organism, the metabonomics will likely offer a more complete and comprehensive picture of the effects of a particular nanomedicine.

5 Other NMR Spectroscopy Applications in Nanomedicine

NMR spectroscopy is one of the major tools for elucidation and confirmation of the structures of newly synthesized and natural compounds in pharmacology and the pharmaceutical industry [103]. It has also been widely used to determine the composition, impurities and complexity of conventional drugs, as well as the drug behaviors in acidic and basic solutions. In the area of nanomedicine, NMR spectroscopy has been frequently involved in physicochemical characterization of engineered nanomaterials and metabonomic analysis of organismic response (Sects. 3.3 and 4.3). In addition, NMR spectroscopy plays an important role in the development of nanomedicine, in investigating the toxicity of nanomedicines, in detecting inflammatory response to nanomedicine treatment in particular organs or tissues, in evaluating the in vivo targeting efficiency of nanomedicine, et cetera, which are exemplified in the studies summarized below.

NMR for nanomedicine development—study of dendrimer-heparin interactions. Dendrimers, the polymers of tree architecture, are considered a new episode of multifunctional drug delivery systems to improve therapeutic efficacy and reduce side effects. Retaining the flexible choice of terminal functionality and versatility in surface chemical modification, dendrimers were the preferred candidates of non-invasive vehicles for delivering heparin, which is a commonly used anticoagulant and employed in the treatment of thrombosis—typically administrated through intravenous or subcutaneous injections in clinical trials [104]. In a recent study of dendrimer-heparin interactions, the authors utilized NMR spectroscopy to investigate the binding activities and structures of the heparin-dendrimer aggregates [104, 105]. The complexes under investigation were made by mixing generation-5 (G-5) primary amine-terminated and ethylenediamine-cored poly(amidoamine) (PAMAM) dendrimer with heparin at different concentration ratios. The ^{1}H NMR measurement of the complexes revealed downshifts of methylene protons (b' and d') in the outermost layer of the G-5 dendrimer at high heparin/dendrimer ratios

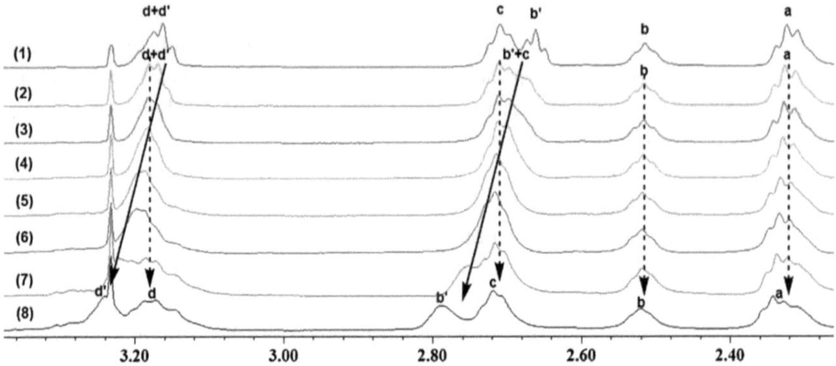

Fig. 5 ^{1}H NMR spectra of G-5 dendrimer with heparin at different concentration ratios. The concentration ratios of heparin/G5 dendrimer were from 0 to 20, as the concentration of G-5 dendrimer fixed at 2 mg/mL. Reproduced with permission from Ref. [104], the American Chemical Society

(Fig. 5), which were primarily caused by the electrostatic interaction between the cationic amine groups of G-5 dendrimer and the anionic sulfate/carboxylate groups of the heparin chain [104].

In addition, the size of the heparin-dendrimer aggregates was calculated from the Stokes-Einstein relationship using the apparent diffusion coefficients of the aggregates obtained from the PGSE ^{1}H NMR measurement. While raising the ratio of heparin to G-5 dendrimer from 0.7 to 20, a linear reduction in the diffusion coefficient of G-5 dendrimers was observed, which indicated the fraction of bound dendrimer increasing in the aggregate [104]. Furthermore, 2D ^{1}H -^{1}H NOESY was employed to examine the potential binding sites of heparin in the aggregate complex (Fig. 6). Accompanied with the reult of the 1D ^{1}H NMR measurement, the evidence of no peak shifts observed at the interior scaffold protons of the dendrimer and the absence of cross-peak signals between the heparin and dendrimer protons in the NOESY spectra pointed out that heparin binds to the surface of dendrimers through ionic interactions [104].

NMR for toxicity research—study of single-walled carbon nanotubes. Due to their novel physicochemical characteristics and ability to penetrate through particular biological barriers, single-walled carbon nanotubes (SWCNT) have been demonstrated as contrast agents and drug delivery vehicles [106]. To gain more insight into the in vivo biodistribution and pharmacological profile of SWCNT, several different types of SWCNT were engaged in a longitudinal study [106]. Given the capability of measuring the indexes of intrinsic hepatotoxicities—simultaneously, such as lipid acids and low molecular-weight metabolites—HR-MAS ^{1}H NMR spectroscopy was employed to assess any endogenous alterations in hepatic metabolism after SWCNT administration [106]. The spectra with a chemical shift range from 0.5 to 4.65 ppm, covering a myriad of metabolites, were acquired in the intact rat liver tissue under high-speed spinning at the magic angle.

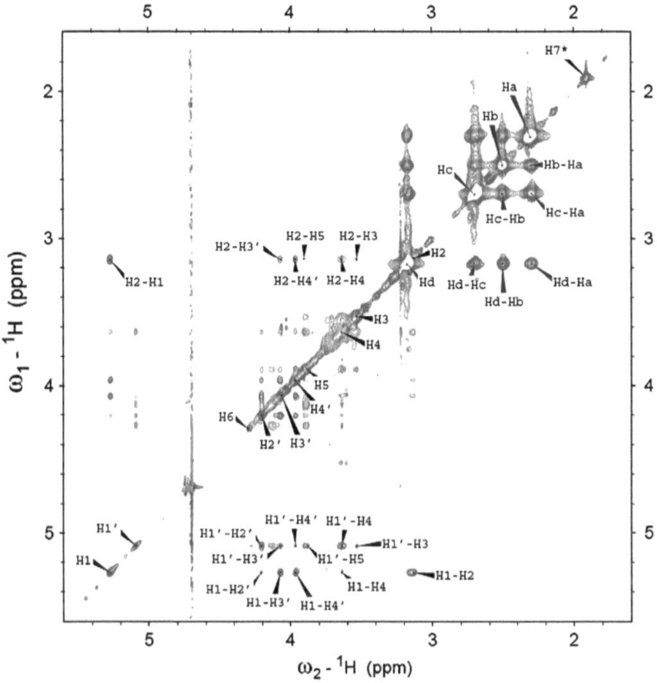

Fig. 6 1H–1H NOESY spectra (mixing time = 300 ms) of heparin/G5 dendrimer complexes at a concentration ratio of 2, when the G5 dendrimer concentration of 2 mg/mL. Reproduced with permission from Ref. [104], the American Chemical Society

The result showed no obvious variation across different experimental groups, indicating that hepatotoxicity was unlikely triggered by SWCNT injection; the analysis was consistent with other analytical outcomes [106]. Nevertheless, any concern related to organ/tissue toxicity caused by nanomedicine administration is possibly investigated by using the same or similar NMR approach.

NMR for inflammatory response—study of angiogenesis of aortic valve stenosis via *perfluorocarbon nanoparticles*. Similar to the early stage of atherosclerosis developing endothelial damage and dysfunction, the immune-inflammatory mechanisms are at the center of the pathogenesis of aortic valve stenosis, in which a pro-inflammatory agent, angiotensin II, performs upregulation of redox-sensitive cytokines, chemokines and growth factors [107–110]. Angiogenesis, the excrescence of new blood vessels from existing microvessels stimulated by local expression of cytokines and growth factors, is present in the pathogenesis of nonrheumatic aortic valve stenosis [111]. Angiogenic neovasculature can be markedly separated from mature vasculature by monitoring αvβ3 integrin, which is upregulated on the lumenal side of neovascular endothelium [109]. Upon a model of atherosclerosis, a recent study demonstrated the development of aortic valve sclerosis in the cholesterol-fed rabbits to imitate human aortic stenosis, while the synthesis of $\alpha_v\beta_3$ integrin-targeted perfluorocarbon (PFC) nanoparticles could be used in imaging contrast enhancement and drug delivery [109]. Two hours after the respective treatments of the $\alpha_v\beta_3$ integrin-targeted and integrin-nonspecific PFC nanoparticles, the aortic valve leaflets excised from the corresponding rabbits were measured using *ex vivo* ^{19}F magnetic resonance spectroscopy (MRS) to validate the binding ability of the targeted PFC nanoparticles. When administrating the targeted PFC nanoparticles to the rabbits, the intensities of the corresponding fluorine NMR signals collected at the rabbit aortic valves exhibited about three times stronger than those treated with the nonspecific PFC nanoparticles, making this modality a plausible assessment of preclinical aortic valve disease [109].

NMR for evaluation of nanomedicine targeting efficiency—study of ATP phosphorus signal enhancement via *galactose-chitosan oligosaccharide conjugated nanoparticles*. Although phosphorus-31 MRS retains potential in the study of energy metabolism in living systems, applying ^{31}P NMR to the in vivo investigation of hepatopathy is still very challenging due to no significant difference observed in the contents of the phosphorus compounds between the normal and pathological hepatic tissues. In a recent study, evaluations of targeting efficiency and ATP ^{31}P signal enhancement were proposed via employment of particular nanoparticles composed of chitosan oligosaccharide (CSO) and multiple galactose (Gal) residues, in which CSO and ATP form the CSO/ATP complex through electrostatic forces and Gal residues are specifically recognized by the asialoglycoprotein receptors on the surface of hepatocytes [112]. The ^{31}P NMR spectra showed the enhanced ATP signals of Gal-CSO/ATP nanoparticles in HepG-2 cells and the promising targeting efficiency *in vitro*, representing a strong likelihood of in vivo clinical applications [112].

6 Conclusion

Noninvasive NMR spectroscopy, which has been chosen in certain pre-clinical and clinical applications, expresses the capabilities of characterizing several physicochemical properties and bioactivities of nanomedicines, as well as for monitoring the dynamic responses of metabolites to nanomedicine administration. Many NMR techniques varying, for example, from one-dimensional spectroscopy to multi-dimensional spectroscopy and from pulsed field gradient to high-resolution magic angle spinning have been utilized to facilitate the nanomedicine studies. A recently emerging NMR-based metabonomics technique demonstrates the potential in evaluation of in vivo nanotoxicity through using *in vitro* models. Although, perhaps not as wide spread in every aspect of nanomedicine, NMR spectroscopy nonetheless will be more frequently applied when more comprehensive studies of nanomedicine are implemented and practiced in the near future.

References

1. European Science Foundation (2005) Nanomedicine: an ESF-European Medical Research Councils (EMRC) Forward Look report. In: ESF
2. McNeil SE (2005) Nanotechnology for the biologist. J Leukoc Biol 78(3):585–594
3. Webster TJ (2006) Nanomedicine: what's in a definition? Int J Nanomed 1(2):115
4. (CNRS) CNdlRS (2012) Nanomedicine : 'Russian doll' polymer vesicles mimic cell structure. ScienceDaily, Bordeaux
5. Duncan R, Gaspar R (2011) Nanomedicine(s) under the microscope. Mol Pharm 8 (6):2101–2141
6. Farokhzad OC, Langer R (2006) Nanomedicine: developing smarter therapeutic and diagnostic modalities. Adv Drug Deliv Rev 58(14):1456–1459
7. Galvin P, Thompson D, Ryan K, McCarthy A, Moore A, Burke C, Dyson M, MacCraith B, Gun'ko Y, Byrne M, Volkov Y, Keely C, Keehan E, Howe M, Duffy C, MacLoughlin R (2012) Nanoparticle-based drug delivery: case studies for cancer and cardiovascular applications. Cell Mol Life Sci 69(3):389–404
8. Bharali DJ, Mousa SA (2010) Emerging nanomedicines for early cancer detection and improved treatment: current perspective and future promise. Pharmacol Ther 128(2):324–335
9. Sajja HK, East MP, Mao H, Wang YA, Nie S, Yang L (2009) Development of multifunctional nanoparticles for targeted drug delivery and noninvasive imaging of therapeutic effect. Curr Drug Discov Technol 6(1):43–51
10. Kim BY, Rutka JT, Chan WC (2010) Nanomedicine. N Engl J Med 363(25):2434–2443
11. Del Burgo LS, Hernandez RM, Orive G, Pedraz JL (2013) Nanotherapeutic approaches for brain cancer management. Nanomedicine 10(5):905–919
12. Hachani R, Lowdell M, Birchall M, Thanh NT (2013) Tracking stem cells in tissue-engineered organs using magnetic nanoparticles. Nanoscale 5:11362–11373
13. Singh A, Sahoo SK (2013) Magnetic nanoparticles: a novel platform for cancer theranostics. Drug Discov Today 19(4):474–481
14. Mahajan KD, Fan Q, Dorcena J, Ruan G, Winter JO (2013) Magnetic quantum dots in biotechnology: synthesis and applications. Biotechnol J 8:1124–1134
15. Wong BS, Yoong SL, Jagusiak A, Panczyk T, Ho HK, Ang WH, Pastorin G (2013) Carbon nanotubes for delivery of small molecule drugs. Adv Drug Deliv Rev 65:1964–2015
16. Paszko E, Ehrhardt C, Senge MO, Kelleher DP, Reynolds JV (2011) Nanodrug applications in photodynamic therapy. Photodiagnosis Photodyn Ther 8(1):14–29
17. Hall JB, Dobrovolskaia MA, Patri AK, McNeil SE (2007) Characterization of nanoparticles for therapeutics. Nanomedicine (Lond) 2(6):789–803
18. Lin P-C, Lin S, Wang PC, Sridhar R (2013) Techniques for physicochemical characterization of nanomaterials. Biotechnol Adv 32(4):711–726
19. Pleus R (2012) Nanotechnologies: guidance on physicochemical characterization of engineered nanoscale materials for toxicologic assessment. ISO/TR 13014:2012
20. Gref R, Minamitake Y, Peracchia M, Trubetskoy V, Torchilin V, Langer R (1994) Biodegradable long-circulating polymeric nanospheres. Science 263(5153):1600–1603
21. Patri A, Dobrovolskaia M, Stern S, McNeil S, Amiji M (2006) Preclinical characterization of engineered nanoparticles intended for cancer therapeutics. In: Amiji M (ed) Nanotechnology for cancer therapy. CRC Press, Boca Raton, pp 105–138
22. Gupta RK (1987) NMR spectroscopy of cells and organisms, vol 1. CRC, Boca Raton
23. De Graaf RA (2008) in vivo NMR spectroscopy: principles and techniques. John Wiley and Sons, New York
24. Bloch F (1946) Nuclear induction. Phys Rev 70(7–8):460–474
25. Ernst RR, Anderson WA (1966) Application of Fourier Transform spectroscopy to magnetic resonance. Rev Sci Instrum 37(1):93–102
26. Levitt M (2001) Spin dynamics: basic principles of NMR spectroscopy. Wiley, Chichester
27. Keeler J (2011) Understanding NMR spectroscopy. Wiley. com, Berlin

28. Aue WP, Bartholdi E, Ernst RR (1976) Two-dimensional spectroscopy. Application to nuclear magnetic resonance. J Chem Phys 64(5):2229–2246
29. McNeil S (2011) Challenges for nanoparticle characterization. In: McNeil SE (ed) Characterization of nanoparticles intended for drug delivery, vol 697. Humana Press, Methods in Molecular Biology, Totowa, pp 9–15
30. Feng SS (2004) Nanoparticles of biodegradable polymers for new-concept chemotherapy. Expert Rev Med Devices 1(1):115–125
31. Jiang W, Kim BY, Rutka JT, Chan WC (2008) Nanoparticle-mediated cellular response is size-dependent. Nat Nanotechnol 3(3):145–150
32. Ferrari M (2008) Nanogeometry: beyond drug delivery. Nat Nanotechnol 3(3):131–132
33. Pal S, Tak YK, Song JM (2007) Does the antibacterial activity of silver nanoparticles depend on the shape of the nanoparticle? A study of the gram-negative bacterium escherichia coli. Appl Environ Microbiol 73(6):1712–1720
34. George S, Lin S, Ji Z, Thomas CR, Li L, Mecklenburg M, Meng H, Wang X, Zhang H, Xia T, Hohman JN, Lin S, Zink JI, Weiss PS, Nel AE (2012) Surface defects on plate-shaped silver nanoparticles contribute to its hazard potential in a Fish Gill Cell Line and Zebrafish Embryos. ACS Nano 6(5):3745–3759
35. Champion JA, Katare YK, Mitragotri S (2007) Particle shape: a new design parameter for micro- and nanoscale drug delivery carriers. J Control Release 121(1–2):3–9
36. Mitragotri S (2009) In drug delivery, shape does matter. Pharm Res 26(1):232–234
37. Euliss LE, DuPont JA, Gratton S, DeSimone J (2006) Imparting size, shape, and composition control of materials for nanomedicine. Chem Soc Rev 35(11):1095–1104
38. Geng Y, Dalhaimer P, Cai SS, Tsai R, Tewari M, Minko T, Discher DE (2007) Shape effects of filaments versus spherical particles in flow and drug delivery. Nat Nanotechnol 2 (4):249–255
39. Gratton SE, Ropp PA, Pohlhaus PD, Luft JC, Madden VJ, Napier ME, DeSimone JM (2008) The effect of particle design on cellular internalization pathways. Proc Natl Acad Sci U S A 105(33):11613–11618
40. Decuzzi P, Pasqualini R, Arap W, Ferrari M (2009) Intravascular delivery of particulate systems: does geometry really matter? Pharm Res 26(1):235–243
41. Jiang X, Qu W, Pan D, Ren Y, Williford JM, Cui H, Luijten E, Mao HQ (2013) Plasmid-templated shape control of condensed DNA-block copolymer nanoparticles. Adv Mater 25 (2):227–232
42. Doshi N, Prabhakarpandian B, Rea-Ramsey A, Pant K, Sundaram S, Mitragotri S (2010) Flow and adhesion of drug carriers in blood vessels depend on their shape: a study using model synthetic microvascular networks. J Control Release 146(2):196–200
43. Powers KW, Palazuelos M, Brown SC, Roberts SM (2009) Characterization of nanomaterials for toxicological evaluation. In: Sahu S, Casciano D (eds) Nanotoxicology from in vivo and in vitro models to health risks. John Wiley and Sons, New York, pp 1–27
44. Buzea C, Pacheco I, Robbie K (2007) Nanomaterials and nanoparticles: sources and toxicity. Biointerphases 2(4):MR17–MR71
45. Hardman R (2006) A toxicologic review of quantum dots: toxicity depends on physicochemical and environmental factors. Environ Health Perspect 114(2):165–172
46. Briscoe CJ, Hage DS (2009) Factors affecting the stability of drugs and drug metabolites in biological matrices. Bioanalysis 1(1):205–220
47. Wang L-Q, Exarhos GJ, Liu J (2001) Nuclear Magnetic resonance: characterization of self-assembled nanostructural materials. In: Wang ZL (ed) Characterization of nanophase materials, Wiley-VCH, Weinheim, pp 243–260
48. Furó I (2005) NMR spectroscopy of micelles and related systems. J Mol Liq 117 (1–3):117–137
49. Mayer C (2005) NMR studies of nanoparticles. In: Webb GA (ed) Annual reports on NMR spectroscopy, vol 55. Academic Press, Amsterdam, pp 205–258
50. Wissing S, Müller R, Manthei L, Mayer C (2004) Structural characterization of Q10-loaded solid lipid nanoparticles by NMR spectroscopy. Pharm Res 21(3):400–405

51. Tiede K, Boxall ABA, Tear SP, Lewis J, David H, Hassellöv M (2008) Detection and characterization of engineered nanoparticles in food and the environment. Food Addit Contam Part A 25(7):795–821
52. Sapsford KE, Tyner KM, Dair BJ, Deschamps JR, Medintz IL (2011) Analyzing nanomaterial bioconjugates: a review of current and emerging purification and characterization techniques. Anal Chem 83(12):4453–4488
53. Mullen DG, Fang M, Desai A, Baker JR, Orr BG, Banaszak Holl MM (2010) A Quantitative assessment of nanoparticle—ligand distributions: implications for targeted drug and imaging delivery in dendrimer conjugates. ACS Nano 4(2):657–670
54. Tomalia DA, Huang B, Swanson DR, Brothers HM, Klimash JW (2003) Structure control within poly(amidoamine) dendrimers: size, shape and regio-chemical mimicry of globular proteins. Tetrahedron 59(22):3799–3813
55. Pan BF, Cui DX, Gao F, He R (2006) Growth of multi-amine terminated poly(amidoamine) dendrimerson the surface of carbon nanotubes. Nanotechnology 17(10):2483–2489
56. Lundqvist M, Sethson I, Jonsson B-H (2005) Transient interaction with nanoparticles "freezes" a protein in an ensemble of metastable near-native conformations†. Biochemistry 44(30):10093–10099
57. Valentini M, Vaccaro A, Rehor A, Napoli A, Hubbell JA, Tirelli N (2004) Diffusion NMR spectroscopy for the characterization of the size and interactions of colloidal matter: the case of vesicles and nanoparticles. J Am Chem Soc 126(7):2142–2147
58. Mirau PA, Naik RR, Gehring P (2011) Structure of peptides on metal oxide surfaces probed by NMR. J Am Chem Soc 133(45):18243–18248
59. Tang Z, Xu B, Wu B, Germann MW, Wang G (2010) Synthesis and structural determination of multidentate 2,3-dithiol-stabilized Au clusters. J Am Chem Soc 132(10):3367–3374
60. Gun'ko V, Blitz J, Zarko V, Turov V, Pakhlov E, Oranska O, Goncharuk E, Gornikov Y, Sergeev V, Kulik T (2009) Structural and adsorption characteristics and catalytic activity of titania and titania-containing nanomaterials. J Colloid Interface Sci 330(1):125–137
61. Powers KW, Brown SC, Krishna VB, Wasdo SC, Moudgil BM, Roberts SM (2006) Research strategies for safety evaluation of nanomaterials. Part VI. Characterization of nanoscale particles for toxicological evaluation. Toxicol Sci 90(2):296–303
62. Shekunov BY, Chattopadhyay P, Tong HH, Chow AH (2007) Particle size analysis in pharmaceutics: principles, methods and applications. Pharm Res 24(2):203–227
63. Roduner E (2006) Size matters: why nanomaterials are different. Chem Soc Rev 35 (7):583–592
64. Grossman JH, McNeil SE (2011) Preclinical efficacy and toxicity testing of engineered nanomaterials. Nanomed Basic Clin Appl Diag Ther 2:71–79
65. Etheridge ML, Campbell SA, Erdman AG, Haynes CL, Wolf SM, McCullough J (2013) The big picture on nanomedicine: the state of investigational and approved nanomedicine products. Nanomed Nanotechnol Biol Med 9(1):1–14
66. Crist RM, Grossman JH, Patri AK, Stern ST, Dobrovolskaia MA, Adiseshaiah PP, Clogston JD, McNeil SE (2013) Common pitfalls in nanotechnology: lessons learned from NCI's nanotechnology characterization laboratory. Integr Biol (Camb) 5(1):66–73
67. Shevchenko VY, Madison AE (2002) Structure of nanoparticles: I. Generalized crystallography of nanoparticles and magic numbers. Glass Phys Chem 28(1):40–43
68. Rayner-Canham G (2003) Ch. 4 Metallic bonding. In: Descriptive inorganic chemistry. Macmillan, New York
69. Luyts K, Napierska D, Nemery B, Hoet PHM (2013) How physico-chemical characteristics of nanoparticles cause their toxicity: complex and unresolved interrelations. Environ Sci Process Impacts 15(1):23–38
70. Nel AE, Madler L, Velegol D, Xia T, Hoek EM, Somasundaran P, Klaessig F, Castranova V, Thompson M (2009) Understanding biophysicochemical interactions at the nano-bio interface. Nat Mater 8(7):543–557
71. Lynch I, Dawson KA, Linse S (2006) Detecting cryptic epitopes created by nanoparticles. Sci STKE 2006(327):pe14

72. Linse S, Cabaleiro-Lago C, Xue W-F, Lynch I, Lindman S, Thulin E, Radford SE, Dawson KA (2007) Nucleation of protein fibrillation by nanoparticles. Proc Natl Acad Sci 104 (21):8691–8696
73. Lundqvist M, Stigler J, Elia G, Lynch I, Cedervall T, Dawson KA (2008) Nanoparticle size and surface properties determine the protein corona with possible implications for biological impacts. Proc Natl Acad Sci 105(38):14265–14270
74. Callaghan PT, Pinder DN (1981) Self-diffusion of random-coil polystyrene determined by pulsed field gradient nuclear magnetic resonance: dependence on concentration and molar mass. Macromolecules 14(5):1334–1340
75. Kato H, Takahashi K, Saito T, Suzuki M, Kinugasa S (2008) Accurate size determination of PS latex nanoparticles in aqueous solution using pulsed field gradient nuclear magnetic resonance spectroscopy. Chem Phys Lett 463(1–3):150–154
76. Carter RS, Harley SJ, Power PP, Augustine MP (2005) Use of NMR spectroscopy in the synthesis and characterization of air- and water-stable silicon nanoparticles from porous silicon. Chem Mater 17(11):2932–2939
77. Liu X, Yu M, Kim H, Mameli M, Stellacci F (2012) Determination of monolayer-protected gold nanoparticle ligand–shell morphology using NMR. Nat Commun 3:1182
78. Du F, Zhang B, Zhou H, Yan B, Chen L (2009) Structure elucidation of nanoparticle-bound organic molecules by 1H NMR. TrAC Trends Anal Chem 28(1):88–95
79. Lin Y, Rao AM, Sadanadan B, Kenik EA, Sun Y-P (2002) Functionalizing multiple-walled carbon nanotubes with aminopolymers. J Phys Chem B 106(6):1294–1298
80. Alam T, Jenkins J (2012) HR-MAS NMR spectroscopy in material science. In: Farrukh M (ed) Advanced aspects of spectroscopy. Intech Open, Rijeka
81. Deshayes S, Maurizot V, Clochard M-C, Berthelot T, Baudin C, Déléris G (2010) Synthesis of specific nanoparticles for targeting tumor angiogenesis using electron-beam irradiation. Radiat Phys Chem 79(3):208–213
82. Conte P, Carotenuto G, Piccolo A, Perlo P, Nicolais L (2007) NMR-investigation of the mechanism of silver mercaptide thermolysis in amorphous polystyrene. J Mater Chem 17 (2):201–205
83. Horgan RP, Kenny LC (2011) 'Omic' technologies: genomics, transcriptomics, proteomics and metabolomics. Obstetrician Gynaecologist 13(3):189–195
84. Nicholson JK, Connelly J, Lindon JC, Holmes E (2002) Metabonomics: a platform for studying drug toxicity and gene function. Nat Rev Drug Discov 1(2):153–161
85. Duarte IF (2011) Following dynamic biological processes through NMR-based metabonomics: a new tool in nanomedicine? J Control Release 153(1):34–39
86. Nicholson JK, Wilson ID (2003) Understanding 'global' systems biology: metabonomics and the continuum of metabolism. Nat Rev Drug Discovery 2(8):668–676
87. Lindon JC, Nicholson JK (2008) Spectroscopic and statistical techniques for information recovery in metabonomics and metabolomics. Annu Rev Anal Chem (Palo Alto Calif) 1:45–69
88. Moolenaar SH, Engelke UF, Wevers RA (2003) Proton nuclear magnetic resonance spectroscopy of body fluids in the field of inborn errors of metabolism. Ann Clin Biochem 40 (Pt 1):16–24
89. Waters NJ (2013) Chpater 3 High resolution NMR based metabolic profiling in drug discovery and development. In: Garrido L, Beckmann N (eds) New applications of NMR in drug discovery and development. Royal Society of Chemistry, New Developments in NMR, Cambridge
90. Clayton TA, Baker D, Lindon JC, Everett JR, Nicholson JK (2009) Pharmacometabonomic identification of a significant host-microbiome metabolic interaction affecting human drug metabolism. Proc Natl Acad Sci U S A 106(34):14728–14733
91. Clayton TA, Lindon JC, Cloarec O, Antti H, Charuel C, Hanton G, Provost JP, Le Net JL, Baker D, Walley RJ, Everett JR, Nicholson JK (2006) Pharmaco-metabonomic phenotyping and personalized drug treatment. Nature 440(7087):1073–1077

92. Solanky KS, Bailey NJ, Beckwith-Hall BM, Bingham S, Davis A, Holmes E, Nicholson JK, Cassidy A (2005) Biofluid H-1 NMR-based metabonomic techniques in nutrition research metabolic effects of dietary isoflavones in humans. J Nutr Biochem 16(4):236–244
93. He QH, Ren PP, Kong XF, Wu YN, Wu GY, Li P, Hao FH, Tang HR, Blachier F, Yin YL (2012) Comparison of serum metabolite compositions between obese and lean growing pigs using an NMR-based metabonomic approach. J Nutr Biochem 23(2):133–139
94. Espina JR, Shockcor JP, Herron WJ, Car BD, Contel NR, Ciaccio PJ, Lindon JC, Holmes E, Nicholson JK (2001) Detection of in vivo biomarkers of phospholipidosis using NMR-based metabonomic approaches. Magn Reson Chem 39(9):559–565
95. Yamamoto T, Tomizawa K, Fujikawa M, Sato Y, Yamada H, Horii I (2007) Evaluation of human hepatocyte chimeric mice as a model for toxicological investigation using panomic approaches–effect of acetaminophen on the expression profiles of proteins and endogenous metabolites in liver, plasma and urine. J Toxicol Sci 32(3):205–215
96. Coen M, Lenz EM, Nicholson JK, Wilson ID, Pognan F, Lindon JC (2003) An integrated metabonomic investigation of acetaminophen toxicity in the mouse using NMR spectroscopy. Chem Res Toxicol 16(3):295–303
97. Neerathilingam M, Volk DE, Sarkar S, Alam TM, Alam MK, Ansari GA, Luxon BA (2010) 1H NMR-based metabonomic investigation of tributyl phosphate exposure in rats. Toxicol Lett 199(1):10–16
98. Bu Q, Yan GY, Deng PC, Peng F, Lin HJ, Xu YZ, Cao ZX, Zhou T, Xue AQ, Wang YL, Cen XB, Zhao YL (2010) NMR-based metabonomic study of the sub-acute toxicity of titanium dioxide nanoparticles in rats after oral administration. Nanotechnology 21 (12):125105
99. Parveen A, Rizvi SH, Gupta A, Singh R, Ahmad I, Mahdi F, Mahdi AA (2012) NMR-based metabonomics study of sub-acute hepatotoxicity induced by silica nanoparticles in rats after intranasal exposure. Cell Mol Biol (Noisy-le-grand) 58(1):196–203
100. Di Marco M, Sadun C, Port M, Guilbert I, Couvreur P, Dubernet C (2007) Physicochemical characterization of ultrasmall superparamagnetic iron oxide particles (USPIO) for biomedical application as MRI contrast agents. Int J Nanomed 2(4):609–622
101. Raynal I, Prigent P, Peyramaure S, Najid A, Rebuzzi C, Corot C (2004) Macrophage endocytosis of superparamagnetic iron oxide nanoparticles: mechanisms and comparison of Ferumoxides and Ferumoxtran-10. Invest Radiol 39(1):56–63
102. Feng JH, Zhao J, Hao FH, Chen C, Bhakoo K, Tang HR (2011) NMR-based metabonomic analyses of the effects of ultrasmall superparamagnetic particles of iron oxide (USPIO) on macrophage metabolism. J Nanopart Res 13(5):2049–2062
103. Holzgrabe U, Wawer I, Diehl B (1999) NMR Spectroscopy in drug development and analysis. Wiley. com., Weinheim
104. Feng X, Cheng Y, Yang K, Zhang J, Wu Q, Xu T (2010) Host-guest chemistry of dendrimer-drug complexes. 5. Insights into the design of formulations for noninvasive delivery of heparin revealed by isothermal titration calorimetry and NMR studies. J Phys Chem B 114 (34):11017–11026
105. Caminade AM, Laurent R, Majoral JP (2005) Characterization of dendrimers. Adv Drug Deliv Rev 57(15):2130–2146
106. Al Faraj A, Fauvelle F, Luciani N, Lacroix G, Levy M, Cremillieux Y, Canet-Soulas E (2011) In vivo biodistribution and biological impact of injected carbon nanotubes using magnetic resonance techniques. Int J Nanomed 6:351–361
107. Weiss D, Sorescu D, Taylor WR (2001) Angiotensin II and atherosclerosis. Am J Cardiol 87 (8A):25C–32C
108. Tedgui A, Mallat Z (2006) Cytokines in atherosclerosis: pathogenic and regulatory pathways. Physiol Rev 86(2):515–581
109. Waters EA, Chen JJ, Allen JS, Zhang HY, Lanza GM, Wickline SA (2008) Detection and quantification of angiogenesis in experimental valve disease with integrin-targeted nanoparticles and 19-fluorine MRI/MRS. J Cardiovasc Magn Reson 10:43

110. Sverdlov AL, Ngo DT, Chapman MJ, Ali OA, Chirkov YY, Horowitz JD (2011) Pathogenesis of aortic stenosis: not just a matter of wear and tear. Am J Cardiovasc Dis 1(2):185–199
111. Soini Y, Salo T, Satta J (2003) Angiogenesis is involved in the pathogenesis of nonrheumatic aortic valve stenosis. Hum Pathol 34(8):756–763
112. Yu R-S, Zhu X-L, Sun J-Z, Shi D, Chen Y, Wang Z-K, Tang K-Z, Du Y-Z (2013) Preliminary study on hepatocyte-targeted phosphorus-31 MRS using ATP-loaded galactosylated chitosan oligosaccharide nanoparticles. Gastroenterol Res Pract 2013:6

An Efficient Coupled Dipole Method for the Accurate Calculation of van der Waals Interactions at the Nanoscale

Hye-Young Kim

Abstract The van der Waals (VDW) force arises from purely quantum mechanical charge fluctuations and is variously called a dispersion or London or Casimir force. This often considered as weak, yet ubiquitous, attractive interaction is important in many nanoscale systems. This chapter provides an overview of the Coupled Dipole Method (CDM), an atomistic and accurate computational method widely adopted to predict the VDW forces between dielectric nanomaterials. There is a concern about the burden of memory and computing time needed to solve eigenvalue problems by either diagonalization or iteration, which have hindered the implementation of CDM for large systems. Here, an efficient way, named trace-CDM (TCDM), is presented. TCDM uses the simple fact that the trace of a square matrix is equal to the sum of its eigenvalues and thus calculates the accurate VDW energies without solving for the eigenvalues. Four examples are solved to demonstrate the advantages of the method.

Keywords van der Waals (VDW) forces and interactions · Dispersion force · London force · Casimir force · Coupled dipole method (CDM) · Trace-CDM (TCDM) · Nanoscale · Dielectric nanomaterials · Diagonalization · Iteration · VDW energies

1 Introduction

The van der Waals (VDW) interaction between atoms and molecules composing or being near a soft or hard condensed matter has always been one of the major research interests in diverse science and engineering fields. At separations large compared to atomic dimensions such that the overlap of electronic densities is negligible, atoms, molecules, self-assemblies, and surfaces interact via VDW dispersion forces giving arise to physical adsorption, atomistic friction, surface tension, and the aggregation

H.-Y. Kim (✉)
Department of Chemistry and Physics, Southeastern Louisiana University,
Hammond, LA 70402, USA
e-mail: hye-young.kim@selu.edu

P. Misra (ed.), *Applied Spectroscopy and the Science of Nanomaterials*,
Progress in Optical Science and Photonics 2, DOI 10.1007/978-981-287-242-5_5

and stability of nanomaterials. This interaction is often considered as a weak, and thus less significant, force in macroscopic systems. However, the degree of importance of VDW force varies for different systems and thus cannot be universally summarized. After all, the very existence of the attraction between any two neutral non-polar atoms or molecules was the mystery that made many scientists puzzle for a long time even after Johannes Diderik van der Waals empirically modified the ideal gas law and introduced the well-known van der Waals equation of state [1–5]

$$\left(P+\frac{a}{V^2}\right)(V-b)=Nk_BT \tag{1}$$

Here k_B, V, P, T, and N are the Boltzmann constant, total volume, pressure, temperature, and the number of gas atoms, respectively. The constants a and b in the correction terms can be empirically determined. The correction term in volume (b) represents the reduced accessible volume for each atom due to the repulsion from other atoms that already have occupied the volume b. The correction term in pressure (a/V^2) implies that the atoms at the boundary have less kinetic energy than those in the interior due to the attractive force exerted by other atoms. The attractive correction term in potential energy of atoms is proportional to $1/V^2$, i.e., proportional to $1/r^6$ if r is the mean separation of the gas atoms. The identity of this mysterious attractive force observed between neutral atoms was unsolved until the introduction of quantum mechanics. In 1930, London [6–10] solved the many-electron Schrödinger equation by second-order perturbation theory and represented the energy gain by the attractive force with an integral over the dynamic polarizaiblites of each atom. London's derivation of the VDW dispersion interaction is thus considered as one of the major early successes of quantum mechanics [3].

In short, the problem of calculating the interaction energy between two atoms reduces to a solution of the Schrödinger equation with a Hamiltonian:

$$\mathrm{H}=\mathrm{H}_1+\mathrm{H}_2+\mathrm{V} \tag{2}$$

where H_1 and H_2 are the Hamiltonians for the isolated atoms. The last term V is the Coulomb interactions between all the charges in atom 1 and those in atom 2 and is treated as a perturbation in the Shrödinger equation, which can then be solve by either perturbation theory or the variational method [2]. Eisenschitz and London [6] and London [7–10] made a dipole approximation which allows only one transition state per atom, and showed that the first term in the solution of the perturbation theory vanishes for nonpolar atoms. Therefore, from the second-order perturbation theory, London obtained the VDW interaction for two isotropic neutral atoms, 1 and 2, at separation r:

$$V=-C_6/r^6 \tag{3}$$

$$C_6=\left(\frac{3\hbar}{\pi}\right)\int d\omega\,\alpha_1(i\omega)\,\alpha_2(i\omega). \tag{4}$$

Here, α_1 (iω) and α_2 (iω) represents the dynamic polarizabilities of atoms 1 and 2, respectively.

For systems involving atoms or molecules with non-zero (or excess or non-vanishing or non-negligible) permanent electric multipoles (charge, dipole moment, quadrupole moment, etc.), the VDW interaction is definitely weaker than the electrostatic or induction contributions. However, for many systems composed of neutral non-polar particles, the VDW interaction becomes the dominant force and starts to play a crucial role in the function and stability of the system. This is mainly due to the fact that the dispersion interaction is ubiquitous like the gravitational interaction, and more importantly to the fact that it depends strongly on the detailed structure or shape of the nanoscale materials [11, 12]. The VDW interaction is ubiquitous since it is originated from the quantum mechanical fluctuation of electronic density (or dynamic polarization) of any atom or molecule in the universe. In fact, this is the force responsible for the condensation and crystallization of van der Waals crystals (such as solid helium), for the structures and energetics in colloid chemistry and biology, for the formation of liquid crystalline phases in solutions, for the anomalous swelling of clay soils on addition of water, for the formation of molecular crystals, and for viral self-assembly, to name a few [1–5, 13–20]. As the size of the system decreases into the region of micro and nanotechnology, the property of material in nanoscale changes from that of macro system [21]. The variety of modern technologies, especially electronic, mechanical, chemical, and bioengineering applications, indeed demand an accurate prediction that can lead to a delicate control over nano-sized small clusters of particles or devices [22, 23]. A specific example would be the micro-machine and nano-machine. Nanocolloids is also a research field in which the VDW forces play a key role in designing and manufacturing devices, such as nanoelectromechanical systems (NEMS), low-power circuits, miniature barcodes, etc. Another example among many others is the research field of physical adsorption [24–26] of atoms or molecules on various surfaces with possible applications to micro- and nano-sensors.

From a theoretical point of view in dealing with macro or even micro-sized systems, the common understanding until very recent years has been that the estimation of long-range van der Waals force is a well-defined problem and easily calculable. On the other hand, the estimation of short-range forces is considered as computationally more challenging because it involves the many-electron problem with overlapping electronic densities that requires inevitable approximations. It should be noted here that, in principle, the estimation of VDW dispersion forces also involves collective many-body problem: among atoms and molecules or between atoms and a surface or between clusters of atoms or molecules. However, this complication of atomistic many-body problem was often considered resolved by using the Clausius-Mossotti relation [27] to estimate the effective atomic properties (atomic polarizability) from the readily measurable macroscopic properties (permittivity function) of the corresponding condensed-matter bulk system. Using this effective atomic polarizability, the dispersion interaction of an atom and a cluster of atoms, for example, has been represented by a sum of London's two-body interactions between the atom and each constituent atoms of the cluster assigned with the effective atomic polarizabilities. This approach is commonly called the pair-sum or 2-body approximation [28].

The procedure of calculating the interaction energy of macroscopic bodies by summing the pair interactions between the individual atoms or molecules of these bodies, as mentioned above, is based on the assumption of additivity of the interatomic interaction energies. This assumption of additivity is justified only within the scheme of the second order perturbation theory [2, 29, 30] since the first nonadditive contribution occurs in the third-order perturbation theory. The nonadditive correction term due to three-body (or, triple-dipole) interactions was first obtained by Axilrod, Teller, and Muto (ATM) [31–33] in 1943 using the third order perturbation theory. Since then considerable efforts have been spent to evaluate the contribution of the nonadditive corrections to the pairwise 2-body interaction term, due to three-body and even four-body interactions derived from the third- and fourth-order perturbation theory, respectively [34–38]. The common reasoning behind the reluctance to explore further the higher-order many-body contributions was the assumption that the higher-order term in the perturbation theory would be smaller than the lower-order terms. However, at the same time, a concern recognized by some researchers was that the higher-order many-body contributions might not be so simply negligible in condensed matter systems. As will be see in the next section, the number of combinations of atoms that contribute to each n-th order many-body term in the perturbation theory increases significantly as the order n increases. Therefore, even though the VDW interaction of each combination of many-atoms might become weaker as n increases, the net contribution of the n-th order term after summing over the largely increased number of combinations might not necessarily be negligibly smaller than, for example, that of the (n − 1)-th order term. The main reason, at least to my opinion, that has kept many researchers away from further estimating the actual "quantitative" contribution of higher-order many-body interaction terms whose expression can be derived from the perturbation theory is the impractically cumbersome expressions of these many-body terms, as can be seen in Refs. [34–36]. Avoiding the use of these cumbersome expressions is one of the main advantages of using the trace-CDM (TCDM) which is presented in the next section.

One of the most studied VDW systems is the inert gases. The potential energy of like pairs of inert gas atoms is quite well known empirically. However, the total potential energy of a bulk system of inert gases is not precisely equal to the sum of the atomic pair potentials. There is an extensive body of work [39–41] which shows that if accurate pair potentials are used, the inclusion of the ATM three-body VDW interaction gives a good agreement with experiment for condensed rare gas systems. For example, the triple dipole dispersion energy for solid xenon is 10 % of the total cohesive energy of xenon, with similar fractional contributions for the lighter inert gases (Ne, Ar or Kr) [39]. More significantly found is that if the ATM interaction is omitted there is a definite disagreement with experiment, which cannot be remedied by simply modifying the pair potential. This is due to the strong dependence of the ATM interaction on the relative location of the three atoms involved, that is, on the type of triangle formed by these three atoms. As material becomes denser and as its shape becomes more anisotropic, the importance of the many-body interaction terms increases.

There is another extreme theoretical approach to obtain the dispersion interaction between an atom with a substrate or a large cluster of atoms, or between two large clusters of atoms. That is to consider the substrate or the large cluster of atoms as a continuum, characterized with a dielectric tensor $\varepsilon(\omega)$, instead of recognizing it as a cluster of many individual atoms characterized with atomic polarizabilities $\alpha(\omega)$. This continuum description of matter has an advantage over the prior-mentioned atomistic description of matter as all the many-body terms are intrinsically included. This continuum approximation method is, however, known to work well only when the distance between interacting particles is much larger than the interatomic distance of atoms within a cluster or a substrate [4, 5, 42–45]. In fact, the continuum methods are found to overestimate the many-body "screening" effect in nanoscale systems [28].

The significant improvement in accuracy achieved in recent experimental measurements of the force between nanoscale materials, especially in the force microscopy, such as atomic force microscopy (AFM), challenges for advanced calculations and thus more accurate predictions that can (1) incorporate the detailed atomistic nanostructures at the surface or edge of nanomaterials and (2) include all the nonadditive many-body interaction terms as well [46]. Because the aforementioned atomistic approach has the advantage of being sensitive to the detailed structure of the nanomaterials, many efforts have been made to find an atomistic approach that also can include "all" the many-body terms, beyond the three-body terms.

Since early 2005, the author was involved in an effort to explore better ways to estimate the van der Waals dispersion interaction accurately in nanoscale systems in collaboration with a group of enthusiastic theoretical and experimental experts. This collaborative effort led us to a series of studies using the coupled dipole method (CDM) [28, 47–52]. The focus of our original work was mostly on the accurate estimation of the van der Waals interaction between two nano-sized dielectric clusters of various shapes and relative orientations. However, the method is general and is readily applicable to other systems.

The aim of this chapter is to provide a self-contained overview of the subject of CDM. Also presented is an efficient way to execute CDM, henceforth called trace-CDM (TCDM), in order to reduce the computational burden of memory and run time, while maintaining accuracy. TCDM will allow cost-effective implementation of CDM for systems composed of large number of particles, such as those in computer simulation studies. The formulations and computational techniques are summarized in great details as some of these were omitted in the original publications due to space limit.

The outline of this chapter is as follows. The next section provides the formulation of CDM which calculates the nonretarded VDW interaction energies exactly within the dipole approximation. Also presented is the formulation of TCDM, an efficient computational way of solving CDM, which obtains accurate VDW energies without having to solve for all the eigenvalues. A few simple cases are solved analytically, as examples, to demonstrate the use of both CDM and TCDM. Section 3 concludes this chapter with a brief discussion on the importance of including all the many-body interaction terms (or at least up to the 10-th order term) and the practical advantages of using TCDM over two traditional ways of including many-body terms, the eigenvalue method of CDM and the perturbation theory.

2 Coupled Dipole Method (CDM) and Trace-CDM (TCDM) for Non-retarded VDW Interaction

The VDW attraction is truly a quantum mechanical phenomenon. That is why, as mentioned in the previous section, the mystery of VDW attractive dispersion interaction could only be unveiled by London adopting then newly established quantum mechanics theory. Quantum theory in its simplest form tells us that everywhere in space there are quantized electromagnetic radiation fields. In other words, there are photons everywhere in random motion. These photons get constantly absorbed and emitted by any particles in the system, so that induced dipole moments are formed instantaneously on the particles. The interaction energy between these instantaneously induced dipoles obtains the attractive VDW dispersion interaction energy which obeys the $1/r^6$ relationship for two atoms at separation r as shown in Eqs. (3) and (4). The final form of the VDW dispersion energy is given in terms of their atomic or molecular dynamic polarizabilities, and tends to increase rather than decrease with increasing temperature due to the increase of polarizabilites with increasing temperature [53].

Using the Schrödinger formalism, the correct free energy of the coupled electron-photon system is obtained only if both the electron and the photon remain essentially in their ground states. This implies the restriction to direct electron transitions to and from an excited state with no successive transitions through different excited states, and thereby implies that only zero or one photon of corresponding transition energy of each excited state is absorbed or emitted. Each transition may be represented by a single excitation of a quantum mechanical harmonic oscillator. London simplified the system by assuming that only one excited state is available for each harmonic oscillator, and thus the interacting ground-state atoms can be represented by equal number of interacting harmonic oscillators whose characteristic frequencies are equal to the electronic excitation energies. This is the point where the quantum mechanical formulation of London makes a connection to the classical Lorentz harmonic oscillator model of atoms (frequently also called Drude model) which has been used to explain the interaction between electromagnetic fields and matter since 1900 [54, 55].

It should be noted that the Schrödinger formalism adopted by London is non-relativistic, and the many-electron Schrödinger equation assumes static electromagnetic interaction potentials between all electrons. There is an exact quantum mechanical treatment, the quantum electrodynamics (QED) procedure, of the dispersion interaction which involves quantization of both matter and electromagnetic fields (or photons). The full quantum mechanical analysis is far too complicated and would not add much to our current purpose. The conclusion in non-relativistic regime and thus in the non-retarded VDW interaction, which is the subject of interest in this chapter, remains unaltered.

In practice, to calculate the VDW interaction energies, certainly among condensed matter scientists and engineers, the oscillator model has been strongly favored due to its simplicity and to the fact that they already have a readily available tool to estimate the atomistic polarizability from the macroscopic dielectric constant using the Clausius-Mossotti relation. In 1963, McLachlan [56, 57] derived the VDW dispersion interaction starting from the Drude model of atoms. In CDM, following McLachlan's approach, the oscillator model is employed to derive the nonretarded VDW interaction energy of a cluster of atoms or between clusters of atoms. The general form of the dynamic atomic polarizability of a Drude harmonic oscillator is:

$$\alpha(\omega) = \left(\frac{e^2}{m}\right)\sum_k \frac{1}{\left(\omega_{ok}^2 - \omega^2\right)} = \sum_k \frac{\alpha_{ok}}{\left(1 - \frac{\omega^2}{\omega_{ok}^2}\right)} \tag{5}$$

Here ω_{0k} and α_{0k} is the characteristic frequency and static polarizability of the k-th electron transition in the atomic states. For the simplest Drude atomic model, where an atom is represented with one characteristic frequency (ω_0) and the isotropic static polarizability (α_0) as assumed by London, the dynamic atomic polarizability becomes:

$$\alpha(\omega) = \frac{\alpha_o}{\left(1 - \frac{\omega^2}{\omega_o^2}\right)} \tag{6}$$

Substituting Eq. (6) to (4) yields, after the integration over frequency, another form of London's formula for the VDW interaction energy of two identical atoms at separation r:

$$V = -\frac{3}{4}\hbar\omega_0 \frac{\alpha_0^2}{r^6} \tag{7}$$

Before going any further, it should be noted here that the London's simple oscillator model has been generalized [2, 29, 30] by allowing more than one electronic transition states per atom. With this generalized oscillator model, there appear additional nonvanishing higher-order terms in the second order perturbation theory for two isolated atoms which correspond to the instantaneous dipole-quadrupole, quadrupole-quadrupole interaction, etc. (see Eqs. 36 and 37 in Ref. [2]) Although these higher-order multipole interaction terms are interesting on their own, they are out of scope of this chapter. Here, we will focus on the dispersion interaction due to the instantaneous fluctuating dipole moments of ground state atoms as adopted in London's work.

Now, using the harmonic oscillator model in Eq. (6) requires atomic values of α_0 and ω_0. Derived from the continuum theory, the Clausius-Mossotti relation connects the macroscopic matter property to the "effectively screened" microscopic property of the composing atoms [27]

$$\frac{\varepsilon(i\omega)-1}{\varepsilon(i\omega)+1}=\frac{4\pi}{3}\rho\alpha(i\omega) \tag{8}$$

Thus, by knowing the permittivity function for a single resonant peak, atomistic values for $\rho\alpha_0$ and ω_0 can be calculated, where ρ is the bulk density of the material. More details on how to use this equation to obtain atomic values from experimental data and some typical atomic values can be found in Appendix D of Ref. [51]. The validity of using the Clausius-Mossotti relation, which after all is derived from the continuum model of bulk material, in estimating the atomic properties in nanoscale materials is still an open question. As the size of clusters of atoms becomes smaller, the finite-size effects such as discrete boundary conditions become evident. Therefore, the bulk properties such as the dielectric constant or bulk density are not well defined. Although it is very important and interesting, this subject is out of scope of this chapter. Since the CDM formulation is independent from how one estimates the atomic polarizability and the characteristic frequency, we proceed without digressing. Following formulations and calculations of CDM and TCDM adopt the Clausius-Mossotti relation.

The CDM is a self-consistent method that is intrinsically atomistic and yet includes all many-body interaction terms, both of which are the characteristics of the very method, as described in Sect. 1, that can best estimate the VDW interactions among nanoscale materials. The detailed formulation is as follows [28, 49–52].

2.1 Formulation

Let us consider a system of two clusters of atoms of nano-size and of arbitrary shape as an example. By definition, the VDW interaction between two clusters A and B (with number of atoms N and N′, respectively) is the change in total energy of the system with two clusters at finite separation relative to that of the system with two clusters at infinite separation. Therefore, using CDM, one may first calculate the total energy ($V_{N+N'}$) of a system of N + N′ atoms, the total energy (V_N) of a system of N atoms, and the total energy ($V_{N'}$) of the system of (N′) atoms separately. And then, the VDW interaction energy of the system is simply obtained from subtracting the self-energy of individual clusters from the total energy of the composite system:

$$V_{vdw} = V_{N+N'} - V_N - V_{N'} \tag{9}$$

To demonstrate the CDM formulation in detail, let us start with a system containing only one cluster A of N atoms. An arbitrary atom i in the cluster will have an instantaneous dipole moment (characterized by the dynamic atomic polarizability) induced by the net local instantaneous electric field, $\vec{E}(\vec{x}_i)$:

$$\vec{p}_i = \overleftrightarrow{\alpha}_i \cdot \vec{E}(\vec{x}_i) \tag{10}$$

where the atomic polarizability is written in its general form of a tensor, which is assumed isotropic in Eqs. (5) and (6). $\vec{x}_i$ is the position vector of the atom i. The local instantaneous electric field at $\vec{x}_i$, in the absence of any externally applied field, results from the vector sum of instantaneous electric fields due to instantaneously induced dipole moments of all other atoms in the system:

$$\vec{E}(\vec{x}_i) = \sum_{\substack{j=1 \\ j\neq i}}^{N} \overleftrightarrow{T}_{ij} \cdot \vec{p}_j \tag{11}$$

where, $\overleftrightarrow{T}_{ij}$ is the usual 3-dimensional static-dipole-interaction tensor (or, a 3×3 matrix):

$$\overleftrightarrow{T}_{ij} \equiv \begin{cases} \left(\dfrac{3\hat{n}_{ij}\hat{n}_{ij} - \overleftrightarrow{I}}{r_{ij}^3} \right) & \text{for } i \neq j \\ 0 & \text{for } i = j \end{cases}. \tag{12}$$

Here, $\hat{n}_{ij} = \vec{r}_{ij}/\left|\vec{r}_{ij}\right|$ is the unit normal vector between atoms i and j, and $\overleftrightarrow{I}$ is the 3-dimensional identity tensor. Substituting Eq. (11) into Eq. (10), one obtains a set of self-consistent equations for the dipoles, one for each atom in the system:

$$\vec{p}_i - \sum_{\substack{j=1 \\ j\neq i}}^{N} \overleftrightarrow{\alpha}_i \cdot \overleftrightarrow{T}_{ij} \cdot \vec{p}_j = 0 \tag{13}$$

Note here that the indices i and j indicate the atoms, not the vector components (e.g. p_x, p_y, p_z,, etc.). The scalar product notation assumes the usual dot-product operation between vectors and tensors. The synchronizations of the dynamic group talk between ever-fluctuating atomic polarization (or, instantaneously responding dipole moment) of N atoms in the system lead to 3N normal modes, where the factor 3 originates from the three-dimensionality of atomic structure.[1]

[1] Despite the same notation adopted, the static-dipole-interaction tensor shown in Eq. (12) of this chapter has a different definition from the T-matrix used in Eq. (2) in Ref. [49], where the

Assuming now the isotropic atomic polarizability of Lorentz oscillator model (Eq. 6), one obtains,

$$\vec{p}_i - \frac{\alpha_{0i}}{\left(1 - \frac{\omega^2}{\omega_{0i}^2}\right)} \sum_{\substack{j=1 \\ j \neq i}}^{N} \overleftrightarrow{T}_{ij} \cdot \vec{p}_j = 0. \tag{14}$$

Rearranging terms, one obtains,

$$\vec{p}_i \left(1 - \frac{\omega^2}{\omega_{0i}^2}\right) - \alpha_{0i} \sum_{\substack{j=1 \\ j \neq i}}^{N} \overleftrightarrow{T}_{ij} \cdot \vec{p}_j = 0 \tag{15}$$

One then arrives at the formation of an eigenvalue problem,

$$\vec{p}_i - \alpha_{0i} \sum_{\substack{j=1 \\ j \neq i}}^{N} \overleftrightarrow{T}_{ij} \cdot \vec{p}_j = \left(\frac{\omega}{\omega_{0i}}\right)^2 \vec{p}_i \tag{16}$$

For a system composed of identical atoms (just to simplify the formulation), ω_{0i} and α_{0i} is simply replaced by ω_0 and α_0, respectively, and one may rewrite Eq. (16) as a matrix form to obtain

$$\left(\overleftrightarrow{I} + \overleftrightarrow{Q}\right) \cdot \vec{\mathrm{P}} = \left(\frac{\omega}{\omega_0}\right)^2 \vec{\mathrm{P}}. \tag{17}$$

Here, $\vec{\mathrm{P}}$ is now a 3N column matrix (x,y,z-components of each N dipoles), $\overleftrightarrow{I}$ is a $3\mathrm{N} \times 3\mathrm{N}$ identity matrix, and $\overleftrightarrow{Q} \equiv -\alpha_0 \overleftrightarrow{T}$ is a $3\mathrm{N} \times 3\mathrm{N}$ square matrix for a system composed of identical atoms. Standard linear algebra algorithms, such as the free software LAPACK (Linear Algebra PACKage) [58] or ScaLAPACK (Scalable LAPACK) [59], can be used to solve this eigenvalue problem by diagonalization of the matrix $\left(\overleftrightarrow{I} + \overleftrightarrow{Q}\right)$. There are 3N eigenvalues ($(\omega_k/\omega_0)^2$) with the integer index k runs from 1 to 3N obtained in this eigenvalue problem, which can then provide the 3N normal mode frequencies (ω_k) in units of ω_0. Since each normal mode frequency contributes a ground state energy of $\hbar\omega_k/2$ (or at finite temperature,

(Footnote 1 continued)
dimensionless T-matrix is defined as $\overleftrightarrow{T}_{ij} = -\alpha_i(\omega)\left(3\hat{n}_{ij}\hat{n}_{ij} - \overleftrightarrow{I}\right)/r_{ij}^3$ for $i \neq j$. Note that this dimensionless T-matrix adopted in Ref. [49] corresponds to the Q-matrix defined in Eq. (17) of this chapter. Accordingly, Eq. (1) in Ref. [49] should be corrected as $\vec{p}_i + \sum_{\substack{j=1 \\ j \neq i}}^{N} \overleftrightarrow{T}_{ij} \cdot \vec{p}_j = 0$, which then becomes equivalent to Eq. (13) of this chapter as it should be.

$(\hbar\omega_k/2)\coth(\hbar\omega_k/2kT)$ [60][2]), one finds the total energy of a system composed of N-Lorentz oscillators from

$$V_N = \frac{\hbar}{2}\sum_{k=1}^{3N}\omega_k \tag{18}$$

Therefore, by solving 3N linear equations self-consistently, CDM automatically provides the contributions of all many-body interactions. Note that one can also calculate the 3N eigenvectors corresponding to these eigenvalues, which corresponds to the dynamic polarizability of the cluster of atoms, i.e., the dipole moments of the system of atoms as a whole. It should be noted as well that the assumption of isotropic atomic polarizability made here is purely for computational convenience and the general form of anisotropic atomic polarizability can be adopted in CDM.

Now, let us return to our original problem of a system of two clusters, A and B, composed of N and N′ identical atoms, respectively. One obvious way to obtain the VDW interaction energy between two clusters is to strictly follow the method described above and solve for 3N eigenvalues of the cluster A. And then use the same procedure to solve for 3N′ eigenvalues of the cluster B, and then again use the same procedure to solve for 3(N + N′) eigenvalues for the composite system of (N + N′)-atom system of two clusters A and B. That is, one would diagonalize the matrix in the eigenvalue problem (Eq. (17)) to solve for all eigenvalues, and sum up the zero-point energy contributions of each and every eigenmode as shown in Eq. (18) to obtain the total energy of each cluster (V_N and $V_{N'}$, separately) and that of the total system ($V_{N+N'}$) [49–52]. The final step is to use Eq. (9) to obtain V_{vdw}. If one were to estimate the VDW energy of a system composed of many (N + N′, for example) atoms, instead of the VDW energy between two clusters of atoms, one simply needs to replace V_N and $V_{N'}$ in Eq. (9) by a sum of self-energies of individual atoms $3N\hbar\omega_0/2$ and $3N'\hbar\omega_0/2$, respectively. In Eq. (9) the subtraction of self-energies of individual atoms were not explicitly shown since the atomic self-energy terms from single-cluster systems A and B cancel those from the whole system (A + B). In other words, while the total energies in Eq. (9), V_N, $V_{N'}$ and $V_{N+N'}$, include the self-energies of composing atoms, the final VDW interaction energy V_{vdw} is free from the self-energies of composing atoms or clusters. The VDW interaction energy obtained this way includes all the higher-order many-body interaction terms, and thus is exact within dipole approximation. This method of solving CDM has been successfully implemented in various quantum chemistry calculations based on density-functional theory (DFT) [13, 61–63] and computer simulations [64]. In principle, one may extend this method to include the higher-order multipole interactions by including

[2] Lifshitz EM (1956) The theory of molecular attractive forces between solids. Sov. Phys. JETP 2:73-83; The dispersion force can be treated for finite temperature by substituting $\frac{\hbar\omega}{2}$ by $\frac{\hbar\omega}{2} + \frac{\hbar\omega}{e^{\hbar\omega/kT}-1} = \frac{\hbar\omega}{2}\coth\left(\frac{\hbar\omega}{2kT}\right)$ to account for the thermal excitation of the modes.

more characteristic frequencies than one per each harmonic oscillator [2] as mentioned in the introduction.

In a quantum chemistry study based on DFT with CDM implemented [63], it is observed that the computational cost of calculating the many-body VDW interactions using CDM is negligible compared to the underlying electronic structure calculation. It should be, however, noted here that these quantum chemistry calculations are done for systems composed of very small number of atoms (less than 200). On the other hand, computational material scientists [64] are making an effort to implement CDM into the Massively Parallel Monte Carlo (MPMC). MPMC is an open-source computer code [65] developed to study the interactions of gas phase sorbates and various metal-organic materials (MOF). As the number of particle N increases, the increasing size of matrices renders the matrix diagonalization in the eigenvalue method of CDM impractical. The bottle-neck of this challenge of implementing the CDM for large size systems comes from both memory size and run time since these grow as N^2 and N^3, respectively [28]. To circumvent the challenge of memory size, many studies involving large number of particles such as computer simulation studies in Ref. [64] calculates the many-body VDW interaction by solving the eigenvalue problem iteratively. Another way to meet this challenge of memory size is to parallelize the computer code so that diagonalization of large-size matrix can be performed. Successfully parallelized codes of CDM have been demonstrated to study systems of large (unlimited, in principle) number of particles [28, 64]. The challenge in the computational run time of using the eigenvalue method of CDM for a large system remains unresolved thus far.

There is, however, a way to circumvent both of these challenges. This procedure, which will be called as Trace-CDM (TCDM) from here on, does not require matrix diagonalization and allows one to obtain the VDW interaction energy V_{vdw} with much better efficiency and still with great accuracy. TCDM simply uses the fact that the trace of a square matrix is equal to the sum of its eigenvalues [37, 66, 67]. According to this, once the eigenvalue problem is written in its matrix form, the sum of all the eigenvalues equals the trace of the square matrix of the eigenvalue problem. This would be a great alternative way for those who are not interested in obtaining the individual eigenvalues, but just the sum of those as shown in Eq. (18). Unfortunately for those, however, the eigenvalue problem shown in Eq. (17) has eigenvalues as $\left(\omega_k^2/\omega_0^2\right)$, not (ω_k/ω_0). Therefore, the trace of the square matrix $\left(\overleftrightarrow{I}+\overleftrightarrow{Q}\right)$ in Eq. (17) is equal to $\sum_{k=1}^{3N}\left(\omega_k^2/\omega_0^2\right)$, not to $\sum_{k=1}^{3N}(\omega_k/\omega_0)$. Only if one could set up the eigenvalue problem so that the eigenvalues are not $\left(\omega_k^2/\omega_0^2\right)$, but (ω_k/ω_0), the VDW interaction energy of the system can be obtained straightforwardly according to Eq. (18). For that, the eigenvalue problem in matrix form in Eq. (17) may be rewritten as:

$$\left(\overleftrightarrow{I}+\overleftrightarrow{Q}\right)^{1/2}\cdot\vec{P}=\left(\frac{\omega}{\omega_0}\right)\vec{P} \tag{19}$$

and from this, one obtains the total VDW energy of the system of N particles:

$$Tr\left(\overleftrightarrow{I}+\overleftrightarrow{Q}\right)^{1/2}=\sum_{k=1}^{3N}\left(\frac{\omega_k}{\omega_0}\right)=V_N/(\hbar\omega_0/2), \tag{20}$$

where Tr denotes the invariant trace, the sum of eigenvalues. Finding the square root of a matrix, $\left(\overleftrightarrow{I}+\overleftrightarrow{Q}\right)^{1/2}$, however, is not an easy task and requires expensive linear algebra computations which would not offer any advantage over solving for all the eigenvalues of the original matrix, $\left(\overleftrightarrow{I}+\overleftrightarrow{Q}\right)$, as would have been done in the eigenvalue method of CDM.

To meet our original purpose of finding the memory-efficient and cost-effective way to solve CDM, let us take a detour route similar to the one taken in the perturbation theory. In short, noticing the form of the matrix in Eq. (20), let us expand it in the power series in terms of $\overleftrightarrow{Q}^n$ using the usual binomial expansion as follows [37, 67]:

$$\left(\overleftrightarrow{I}+\overleftrightarrow{Q}\right)^{1/2}=\sum_{n=0}^{\infty}c_n\overleftrightarrow{Q}^n. \tag{21}$$

Here c_n are the coefficients found in the following binomial expansion

$$(1+y)^{1/2}=\sum_{n=0}^{\infty}c_n y^n \tag{22}$$

$$c_0 = 1$$

$$c_n = c_{n-1}\left(\frac{3}{2n}-1\right) \qquad \text{for n} \neq 0$$

Then, using the properties of trace in linear algebra, the trace of the square-root of the matrix in Eq. (20) may be represented as a sum of traces of the multiples of the matrix ($\overleftrightarrow{Q}^n$) as follows:

$$Tr\left[\left(\overleftrightarrow{I}+\overleftrightarrow{Q}\right)^{1/2}\right]=Tr\left[\sum_{n=0}^{\infty}c_n\overleftrightarrow{Q}^n\right]=\sum_{n=0}^{\infty}c_n Tr\left[\overleftrightarrow{Q}^n\right] \tag{23}$$

Now, the total VDW interaction energy of the system of a cluster of N atoms is,

$$V_N=\frac{\hbar\omega_0}{2}Tr\left[\left(\overleftrightarrow{I}+\overleftrightarrow{Q}\right)^{1/2}\right]=\frac{\hbar\omega_0}{2}\sum_{n=0}^{\infty}c_n Tr\left[\overleftrightarrow{Q}^n\right]\equiv\frac{\hbar\omega_0}{2}\sum_{n=0}^{\infty}W_N^{(n)} \tag{24}$$

Obtaining the traces of $\overleftrightarrow{Q}^{n}$ is not much of a computational burden. As discussed later in this section, each n-th order term in this power series expansion indeed corresponds to the n-th order term in the perturbation theory. Also, according to the exact CDM calculations done in Ref. [49], including up to 10-th order terms in this power series expansion reproduces the exact CDM result for varying shapes and orientations of nanomaterials within much less than 1 % error (see Table 1). Figure 2 of Ref. [49] is reproduced in Fig. 1. The fractional percent energy is defined as

$$\text{Fractional \% Energy of order M} = \frac{\frac{\hbar\omega_0}{2}\sum_{n=0}^{M} W_N^{(n)}}{V_N}\times 100 = \frac{\sum_{n=0}^{M} W_N^{(n)}}{\sum_{n=0}^{\infty} W_N^{(n)}}\times 100 \quad (25)$$

Also, the numeric values of the fractional percent energy are listed in Table 1 up to the interaction order M = 15. Same symbols as adopted in Fig. 1 are used in Table 1 to indicate the corresponding configuration (orientations and types) of clusters to the ones shown in the inset of Fig. 1.

Therefore, TCDM provides a great alternative way for those who are interested only in obtaining the final VDW interaction energies (V_{vdw}), and not in obtaining the individual eigenvalues (ω_k's) and/or the corresponding eigenvectors. Returning back to our original problem of a system of two clusters A and B, composed of N and N′ identical atoms, respectively, to solve it with TCDM one starts with the self-consistent equations for '*all*' atoms in the system (N + N′) that can be expressed in matrix form

Table 1 Numerical values of the fractional percent energy of order M to the total VDW interaction energy as a function of the highest-order (M) of many-body terms included. Each symbol indicates the cluster configuration shown in the *inset* of Fig. 1

M	Decamer △	Decamer ●	Cubes ◇
2	48.66	93.93	99.33
3	65.96	77.22	95.29
4	81.93	92.87	101.03
5	89.64	94.37	99.41
6	94.42	97.37	100.23
7	96.94	98.44	99.87
8	98.36	99.19	100.07
9	99.12	99.56	99.96
10	99.53	99.77	100.03
11	99.75	99.87	99.98
12	99.87	99.93	100.02
13	99.93	99.96	99.99
14	99.96	99.98	100.01
15	99.98	99.99	99.99

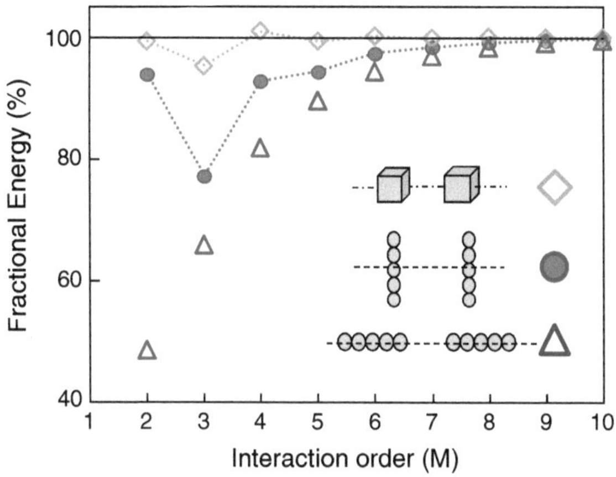

Fig. 1 Fraction of total VDW interaction energy as a function of the highest-order (M) of many-body terms included. Triangles (*circles*) correspond to the VDW interaction between two decamers at separation r/a = 50, where a is the separation between nearest atoms in the linear chain. Diamonds corresponds to the VDW interaction between two 27-atom cubic clusters at r/a = 10. Atoms in each cubic cluster are arranged in a simple cubic lattice structure with lattice constant a. The *dashed lines* are drawn to guide the eye. Refer to Ref. [49] for more details

$$\overleftrightarrow{\Omega} \cdot \vec{P} = \lambda^2 \vec{P}, \tag{26}$$

where the eigenvalues λ^2 of the square matrix

$$\overleftrightarrow{\Omega} = \begin{pmatrix} \overleftrightarrow{I} + \overleftrightarrow{Q} & \overleftrightarrow{M} \\ \overleftrightarrow{M}' & \overleftrightarrow{I}' + \overleftrightarrow{Q}' \end{pmatrix} \tag{27}$$

are the squared eigen frequencies (i.e.,$\lambda^2 = (\omega/\omega_0)^2$) of the composite system A + B. Here, $\overleftrightarrow{Q}$ denotes a 3N × 3N symmetric, traceless matrix, made of dimensionless dipole tensors connecting two atoms within cluster A and $\overleftrightarrow{I}$ is the 3N × 3N identity matrix, as already have been used in Eq. (19) for the system of N particles. $\overleftrightarrow{Q}'$ and $\overleftrightarrow{I}'$ are the corresponding matrices for the N′ atoms within cluster B. $\overleftrightarrow{M}$ is a 3N × 3N′ matrix representing the dipole interaction tensors connecting one atom in A to another atom in B. $\overleftrightarrow{M}'$ is the transpose matrix (a 3N′ × 3N matrix) of matrix $\overleftrightarrow{M}$ representing the dipole tensor connecting one atom in B to another atom in A. The eigenvalue problems of $\overleftrightarrow{I} + \overleftrightarrow{Q}$ and $\overleftrightarrow{I}' + \overleftrightarrow{Q}'$ correspond to that of each cluster A and B (see Eq. 19), respectively. As mentioned before, in TCDM, the exact non-retarded VDW dispersion energy from each eigenvalue problem is obtained by taking the trace of the square matrices:

$$W_{vdw} = Tr\left[\overset{\leftrightarrow}{\Omega}^{1/2}\right] - Tr\left[\left(\overset{\leftrightarrow}{I}+\overset{\leftrightarrow}{Q}\right)^{1/2}\right] - Tr\left[\left(\overset{\leftrightarrow}{I}'+\overset{\leftrightarrow}{Q}'\right)^{1/2}\right]. \quad (28)$$

This equation is essentially identical with Eq. (9), except that W_{vdw} is now a dimensionless VDW energy and is equal to $V_{vdw}/(\hbar\omega_0/2)$ when all the atoms in the system are identical.

The second term in the right hand side of Eq. (28) is the matrix for a system of a cluster (A) of N particles, and has already been expanded earlier in Eq. (23). Similarly, the third term in the right hand side of Eq. (28) may be expanded as:

$$Tr\left[\left(\overset{\leftrightarrow}{I}'+\overset{\leftrightarrow}{Q}'\right)^{1/2}\right] = \sum_{n=0}^{\infty} c_n Tr\left[\overset{\leftrightarrow}{Q}'^n\right]. \quad (29)$$

The power law expansion of the first term in the right hand side of Eq. (28) can also be done as follows:

$$Tr\left[\overset{\leftrightarrow}{\Omega}^{1/2}\right] = Tr\left[\left\{\begin{pmatrix}\overset{\leftrightarrow}{I} & 0\\ 0 & \overset{\leftrightarrow}{I}'\end{pmatrix} + \begin{pmatrix}\overset{\leftrightarrow}{Q} & \overset{\leftrightarrow}{M}\\ \overset{\leftrightarrow}{M}' & \overset{\leftrightarrow}{Q}'\end{pmatrix}\right\}^{1/2}\right] = \sum_{n=0}^{\infty} c_n Tr\left[\begin{pmatrix}\overset{\leftrightarrow}{Q} & \overset{\leftrightarrow}{M}\\ \overset{\leftrightarrow}{M}' & \overset{\leftrightarrow}{Q}'\end{pmatrix}^n\right] \quad (30)$$

Consequently, the dimensionless total interaction energy in Eq. (28) can now be written as a sum of all $\{W^{(n)}\}$ contributions, which again is equivalent to Eq. (9) in units of $\hbar\omega_0/2$:

$$W_{vdw} = \sum_{n=0}^{\infty} W^{(n)} \quad (31)$$

where, $W^{(n)}$ is the n-th order contribution to the dispersion interaction:

$$W^{(n)} = c_n\left\{Tr\left[\begin{pmatrix}\overset{\leftrightarrow}{Q} & \overset{\leftrightarrow}{M}\\ \overset{\leftrightarrow}{M}' & \overset{\leftrightarrow}{Q}'\end{pmatrix}^n\right] - Tr\left[\overset{\leftrightarrow}{Q}^n\right] - Tr\left[\left(\overset{\leftrightarrow}{Q}'\right)^n\right]\right\} \quad (32)$$

Incidentally, since the eigenvalues of $\overset{\leftrightarrow}{\Omega}$ and $\overset{\leftrightarrow}{I}+\overset{\leftrightarrow}{Q}$ (or $\overset{\leftrightarrow}{I}'+\overset{\leftrightarrow}{Q}'$) are necessarily of the form $\lambda_i^2 = 1 + x_i$ and $\lambda_{0i}^2 = 1 + x_{0i}$, respectively, these traces are readily expressed in terms of x_i and x_{0i}. Therefore we obtain yet another form of the general n-th order contribution to the dispersion interaction:

$$W^{(n)} = c_n\left[\sum_{i=1}^{3(N+N')}(x_i)^n - \sum_{\substack{j=1\\ j\in A}}^{3N}(x_{0j})^n - \sum_{\substack{k=1\\ k\in B}}^{3N'}(x_{0k})^n\right] \quad (33)$$

This expansion is the equivalent power-series (or perturbation) expansion in terms of $\rho\alpha_0$, as demonstrated in Ref. [49] using the eigenvalue method of the exact CDM, to which all $\overleftrightarrow{M}$ and $\overleftrightarrow{Q}$ matrices are proportional.

In Eq. (32), since the power series expansion is in matrix form, it is easier to identify where each n-th order many-body term originates from. For example, $W^{(0)} = 0$ due to the cancellation of atomic zero-point energies; $W^{(1)} = 0$ due to the tracelessness of Q matrices for a neutral non-polar system; $W^{(2)} = 2c_2\,\mathrm{Tr}\,(MM')$ due to the pairwise interaction terms with one atom in A and another in B; $W^{(3)} = 3c_3\,\mathrm{Tr}\,(QMM' + Q'M'M)$ due to the triplets with one atom in A and two atoms in B or vice versa; $W^{(4)} = 4c_4\,\mathrm{Tr}(QQMM' + Q'Q'MM' + QMQ'M' + M'MM'M/2)$, etc. A quick review of linear algebra reveals that the matrix in a trace of a product can be switched allowing tr(AB) = tr(BA). Equivalently, the trace is invariant under cyclic permutations, i.e., tr(ABCD) = tr(BCDA) = tr(CDAB) = tr(DABC). This gives the factors of 2 and 3 in the above expressions of $W^{(2)}$ and $W^{(3)}$, respectively. In Fig. 2, schematic diagrams are drawn to aid the intuitive understanding of the n-th order many-body interaction contributions.

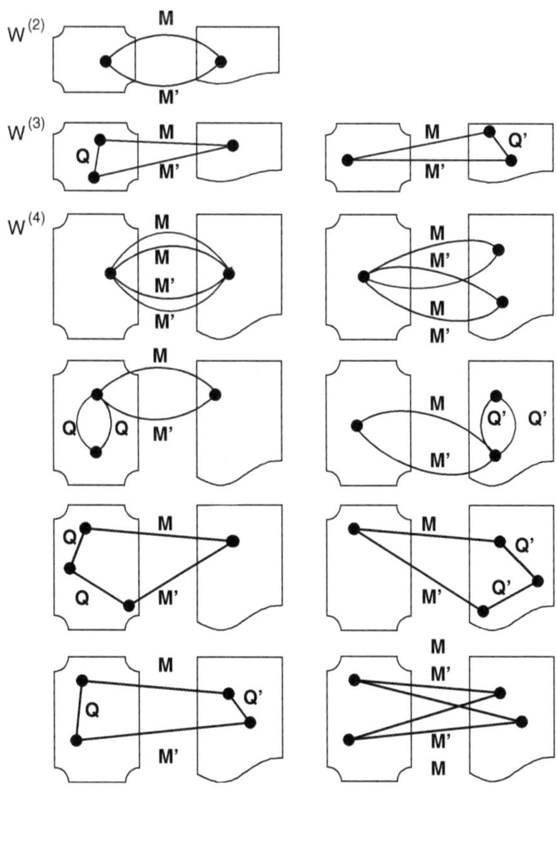

Fig. 2 Schematic diagrams of the n-th order many-body interaction contributions

To gain a solid understanding of how the TCDM works and what makes it so efficient, we will lay out four examples of calculating the VDW interaction energies for a couple of simple systems using both ways: (1) The eigenvalue method of exact CDM which adds all the eigenvalues obtained from diagonalizing the square matrix in the eigenvalue problem, and (2) the alternative way of TCDM which calculates the trace of the square-root of the matrix by expanding it in binomial power series. We will show that these two methods give same results as they should.

2.2 Example 1: Calculation of the VDW Interaction Energy of a Pair of Identical Non-polar Atoms Using TCDM

Let us consider a pair of identical non-polar atoms at a distance of r (Fig. 3). This problem is the simplest case and the eigenvalue problem becomes:

$$\begin{pmatrix} 1 & 0 & 0 & \gamma & 0 & 0 \\ 0 & 1 & 0 & 0 & \gamma & 0 \\ 0 & 0 & 1 & 0 & 0 & -2\gamma \\ \gamma & 0 & 0 & 1 & 0 & 0 \\ 0 & \gamma & 0 & 0 & 1 & 0 \\ 0 & 0 & -2\gamma & 0 & 0 & 1 \end{pmatrix} \begin{pmatrix} p_{1x} \\ p_{1y} \\ p_{1z} \\ p_{2x} \\ p_{2y} \\ p_{2z} \end{pmatrix} = \left(\frac{\omega}{\omega_0}\right)^2 \begin{pmatrix} p_{1x} \\ p_{1y} \\ p_{1z} \\ p_{2x} \\ p_{2y} \\ p_{2z} \end{pmatrix} \tag{34}$$

where, $\gamma = \alpha_0/r^3$.

To remain closely related to the general formulations discussed previously for the VDW interactions between two clusters of atoms, purely for the purpose of demonstration, let us assume that these atoms belong to two different clusters A and B. Then, comparing Eq. (34) to Eqs. (26) and (27), one obtains $\overleftrightarrow{Q} = \overleftrightarrow{Q}' = 0$ and

$$\overleftrightarrow{M} = \overleftrightarrow{M}' = \begin{pmatrix} \gamma & 0 & 0 \\ 0 & \gamma & 0 \\ 0 & 0 & -2\gamma \end{pmatrix}. \tag{35}$$

From Eq. (32), then, one obtains the n-th order contribution to the dispersion interaction:

$$W^{(n)} = c_n Tr\left\{ \begin{pmatrix} 0 & \overleftrightarrow{M} \\ \overleftrightarrow{M}' & 0 \end{pmatrix}^n \right\}. \tag{36}$$

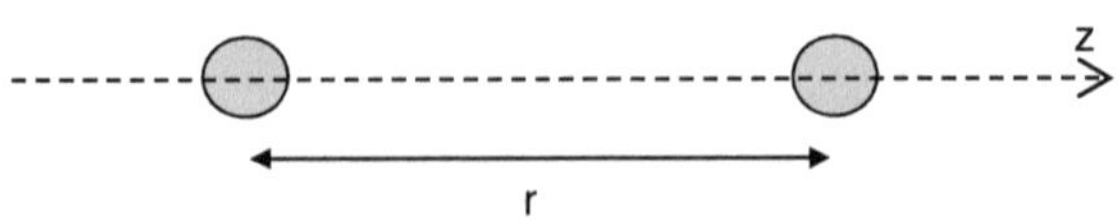

Fig. 3 A pair of identical non-polar atoms at separation r

Note that, alternatively, one could have considered this problem as a cluster of two atoms, instead of two clusters composed of one atom each. In that case, the eigenvalue problem of Eq. (34) corresponds to Eq. (19) with

$$\overleftrightarrow{Q} = \begin{pmatrix} 0 & \overleftrightarrow{M} \\ \overleftrightarrow{M}' & 0 \end{pmatrix} \tag{37}$$

and thus regaining Eq. (36) from Eq. (23). Both perspectives of the same physical system are of course equivalent.

As stated earlier, both $W^{(0)}$ and $W^{(1)}$ are zero due to the cancellation of self-energies of each atom and the tracelessness of Q matrix, respectively. When n = 2, $c_2 = -1/8$, and

$$Tr\left\{\begin{pmatrix} 0 & \overleftrightarrow{M} \\ \overleftrightarrow{M}' & 0 \end{pmatrix}\begin{pmatrix} 0 & \overleftrightarrow{M} \\ \overleftrightarrow{M}' & 0 \end{pmatrix}\right\} = Tr\begin{pmatrix} \overleftrightarrow{M}\overleftrightarrow{M}' & 0 \\ 0 & \overleftrightarrow{M}'\overleftrightarrow{M} \end{pmatrix} = Tr\left[\overleftrightarrow{M}\overleftrightarrow{M}' + \overleftrightarrow{M}'\overleftrightarrow{M}\right]$$
$$= 2\,Tr\left[\overleftrightarrow{M}\overleftrightarrow{M}\right] \tag{38}$$

The last step is taken using M = M′ since the two atoms are identical. Also, from Eq. (35), $Tr(MM) = 6\gamma^2 = 6(\alpha_0/r^3)^2$. Consequently, one obtains the first surviving term (n = 2) in the many-body power series expansion,

$$V_{vdw}^{(2)} = \left(\frac{1}{2}\hbar\omega_0\right)W^{(2)} = \left(\frac{1}{2}\hbar\omega_0\right)\left(-\frac{1}{8}\right)(2)\left(6\frac{\alpha_0^2}{r^6}\right) = -\frac{3}{4}\hbar\omega_0\frac{\alpha_0^2}{r^6} \tag{39}$$

Including only till the second-order term, we now recover the London's formula, Eq. (7), for two identical non-polar atoms at separation r. Let us further explore the higher-order terms. Because all the diagonal elements are zero in the matrix in Eq. (36), the trace of the n-th power of the matrix is zero for all odd n. Therefore, the next non-zero term is when n = 4, $c_4 = -5/128$ which gives

$$W^{(4)} = c_4 Tr\left\{\begin{pmatrix} 0 & \overleftrightarrow{M} \\ \overleftrightarrow{M}' & 0 \end{pmatrix}^4\right\} = c_4 Tr\begin{bmatrix} \overleftrightarrow{M}\overleftrightarrow{M}'\overleftrightarrow{M}\overleftrightarrow{M}' & 0 \\ 0 & \overleftrightarrow{M}'\overleftrightarrow{M}\overleftrightarrow{M}'\overleftrightarrow{M} \end{bmatrix}$$
$$= 2c_4 Tr\left[\overleftrightarrow{M}\overleftrightarrow{M}\overleftrightarrow{M}\overleftrightarrow{M}\right] = 36c_4\left(\frac{\alpha_0}{r^3}\right)^4 \tag{40}$$

Or, the 4-th order VDW interaction term is obtained as

$$V_{vdw}^{(4)} = \left(\frac{1}{2}\hbar\omega_0\right)W^{(4)} = -\frac{45}{64}\hbar\omega_0\frac{\alpha_0^4}{r^{12}}. \tag{41}$$

Here, note that the power expansion coefficient c_n is always a negative number for even n due to the $(-1)^{n-1}$ factor in it. On the other hand, $Tr[M^n] = \gamma^n\,[1 + 1 + (-2)^n]$ is always positive for even n. Therefore, one can conclude that all the non-zero contribution of $W^{(n)}$ is negative. Therefore, including higher-order terms in the VDW interaction between two atoms increases the attraction. Note that this next higher order term of n = 4 in the VDW interaction of two atoms also shows up as one of the 4-body interaction terms in the VDW interaction between two clusters of many atoms (see $W^{(4)}$ in Fig. 2).

It should be noted that the n-th order contribution of Eqs. (32) and (33) in the "many-body" power series expansion of Eq. (31) is not exactly the same as the "n-body" interaction. This example of VDW interaction between "two" atoms (therefore, two-body) demonstrates that the power expansion always is an infinite order expansion [68]. The reasoning behind the conventional use of London's formula for the VDW interaction energy for the two-atom system is based on the usual assumption of perturbation theory that the contributions of the higher-than-the-second-order terms will be smaller than the first nonvanishing second-order term. This assumption is valid for certain systems such as the one considered in this example, however is not always the case in general as demonstrated in our previous work of VDW interaction between two odd-shaped clusters [49, 50].

2.3 Example 2: Calculation of the VDW Interaction Energy of a Pair of Identical Non-polar Atoms Using the Eigenvalue Method of Exact CDM

The eigenvalues of Eq. (34) in the general form of

$$\overleftrightarrow{\Omega} \cdot \vec{P} = u\vec{P} \tag{42}$$

can be numerically found from solving the characteristic equation of the eigenvalue problem:

$$\det\left[\overleftrightarrow{\Omega} - u\overleftrightarrow{I}\right] = 0 \tag{43}$$

After some lines of derivation, one obtains:

$$(1-u)^6 - 6\gamma^4(1-u)^4 + 9\gamma^4(1-u)^2 - 4\gamma^6 = 0 \tag{44}$$

which gives six solutions,

$$\omega_z^{\pm} = \omega_0(1 \pm 2\gamma)^{1/2}; \quad \omega_x^{\pm} = \omega_y^{\pm} = \omega_0(1 \pm \gamma)^{1/2} \tag{45}$$

Due to symmetry, the x- and y-direction normal mode frequencies are degenerate. The corresponding normal modes (eigenvectors) are listed in Table 2. Here, the dot and cross symbols indicate out-of-plane and into-the-plane direction, respectively.

Table 2 Eigenvalues and corresponding eigenmodes in a system of two atoms at separation r

	$\omega_z^+ = \omega_0(1+2\gamma)^{1/2}$
	$\omega_x^+ = \omega_0(1+\gamma)^{1/2}$
	$\omega_y^+ = \omega_0(1+\gamma)^{1/2}$
	$\omega_x^- = \omega_0(1-\gamma)^{1/2}$
	$\omega_y^- = \omega_0(1-\gamma)^{1/2}$
	$\omega_z^- = \omega_0(1-2\gamma)^{1/2}$

Now, the VDW interaction between two atoms can be obtained following Eq. (18) after subtracting the zero-point self-energy of two atoms when they are at infinite separation:

$$V_{vdw} = \frac{\hbar\omega_0}{2}\left[(1+2\gamma)^{1/2} + (1-2\gamma)^{1/2} + 2(1+\gamma)^{1/2} + 2(1-\gamma)^{1/2} - 3 - 3\right] \quad (46)$$

Here, the 3-dimensional harmonic oscillator of each atom has zero-point self-energy of $3\hbar\omega_0/2$. For $\gamma \equiv \alpha_0/r^3 < 1$, one can expand the square-root terms as done in Eq. (22) to obtain,

$$V_{vdw} = \frac{\hbar\omega_0}{2}\left[\left(-\frac{1}{8}\right)\left(12\gamma^2 + \cdots\right)\right] = -\frac{3}{4}\hbar\omega_0\frac{\alpha_0^2}{r^6} - \frac{45}{64}\hbar\omega_0\frac{\alpha_0^4}{r^{12}} - \cdots \quad (47)$$

The first non-zero term gives again the London's VDW representation. Each of the higher order terms with γ^n with $n > 2$ equals to the n-th order contribution to the dispersion interaction $W^{(n)}$. This power expansion of VDW interaction indeed corresponds to Eq. (33) for the simplest case of a two-atom system and demonstrates that the x_i shown in Eq. (33) is proportional to γ in the Drude atomic model.

2.4 Example 3: Calculation of VDW Interaction Energy of a Linear Chain of Four Identical Atoms Using TCDM

The eigenvalue problem is formulated following Eqs. (26) and (27):

$$\overleftrightarrow{\Omega} \cdot \vec{P} = \lambda^2 \vec{P}$$

with

$$\overleftrightarrow{\Omega} = \begin{pmatrix} \overleftrightarrow{I} + \overleftrightarrow{Q} & \overleftrightarrow{M} \\ \overleftrightarrow{M}' & \overleftrightarrow{I} + \overleftrightarrow{Q}' \end{pmatrix}$$

Here, $\overleftrightarrow{Q} = \begin{pmatrix} 0 & \overleftrightarrow{M}_{12} \\ \overleftrightarrow{M}'_{12} & 0 \end{pmatrix}$, $\overleftrightarrow{Q}' = \begin{pmatrix} 0 & \overleftrightarrow{M}_{34} \\ \overleftrightarrow{M}'_{34} & 0 \end{pmatrix}$, $\overleftrightarrow{M} = \begin{pmatrix} \overleftrightarrow{M}_{13} & \overleftrightarrow{M}_{14} \\ \overleftrightarrow{M}_{23} & \overleftrightarrow{M}_{24} \end{pmatrix}$, and $\overleftrightarrow{M}' = \begin{pmatrix} \overleftrightarrow{M}'_{13} & \overleftrightarrow{M}'_{23} \\ \overleftrightarrow{M}'_{14} & \overleftrightarrow{M}'_{24} \end{pmatrix}$

with $\overleftrightarrow{M}_{ij} = \overleftrightarrow{M}'_{ij} = \begin{pmatrix} \gamma_{ij} & 0 & 0 \\ 0 & \gamma_{ij} & 0 \\ 0 & 0 & -2\gamma_{ij} \end{pmatrix}$ and $\gamma_{ij} \equiv \frac{\alpha_0}{r_{ij}^3} = \frac{\alpha_0}{z_{ij}^3}$.

The distance between two atoms i and j is noted as r_{ij}, and is equal to z_{ij} when the linear chain lies along z-axis as shown in Fig. 4. Now, the n-th order contribution to the dispersion interaction in the chain system becomes:

$$W^{(n)} = c_n Tr\left\{ \begin{pmatrix} \overleftrightarrow{Q} & \overleftrightarrow{M} \\ \overleftrightarrow{M}' & \overleftrightarrow{Q}' \end{pmatrix}^n \right\} \tag{48}$$

Since both $\overleftrightarrow{Q}$ and $\overleftrightarrow{Q}'$ are traceless 6 × 6 matrices,

$$Tr\, \overleftrightarrow{Q}^n = 2\,(2 + 2^n)\gamma_{12}^n \quad \text{for even n } (n \geq 2)$$
$$Tr\, \overleftrightarrow{Q}^n = 0 \quad \text{for odd n } (n \geq 1)$$

And similarly,

$$Tr\left\{\overleftrightarrow{Q}'^n\right\} = 2\,(2 + 2^n)\gamma_{34}^n \quad \text{for even n } (n \geq 2)$$
$$Tr\left\{\overleftrightarrow{Q}'^n\right\} = 0 \quad \text{for odd n } (n \geq 1)$$

On the other hand, since $\overleftrightarrow{M}_{ij}$ is a 3 × 3 diagonal matrix,

$$Tr\left\{\overleftrightarrow{M}_{ij}^n\right\} = (2 + (-2)^n)\,\gamma_{ij}^n \quad \text{for all n},$$

Fig. 4 A linear chain of four identical atoms

which can be delineated as:

$$Tr\left\{\overleftrightarrow{M}_{ij}\right\} = 0 \quad \text{for n} = 1$$
$$Tr\left\{\overleftrightarrow{M}_{ij}^{n}\right\} > 0 \quad \text{for even n}$$
$$Tr\left\{\overleftrightarrow{M}_{ij}^{n}\right\} < 0 \quad \text{for odd n } (>1)$$

Again, as stated earlier, both $W^{(0)}$ and $W^{(1)}$ are zero due to the cancellation of self-energies of each atom and the tracelessness of the Q and Q′ matrices, respectively.

When n = 2, $c_2 = -1/8$, and

$$Tr\left[\begin{pmatrix} \overleftrightarrow{Q} & \overleftrightarrow{M} \\ \overleftrightarrow{M}' & \overleftrightarrow{Q}' \end{pmatrix}^2\right] = Tr\begin{pmatrix} \overleftrightarrow{Q}\overleftrightarrow{Q} + \overleftrightarrow{M}\overleftrightarrow{M}' & \overleftrightarrow{Q}\overleftrightarrow{M} + \overleftrightarrow{M}\overleftrightarrow{Q}' \\ \overleftrightarrow{M}'\overleftrightarrow{Q} + \overleftrightarrow{Q}'\overleftrightarrow{M}' & \overleftrightarrow{M}'\overleftrightarrow{M} + \overleftrightarrow{Q}'\overleftrightarrow{Q}' \end{pmatrix}$$
$$= Tr\left[\overleftrightarrow{Q}\overleftrightarrow{Q} + \overleftrightarrow{M}\overleftrightarrow{M}' + \overleftrightarrow{M}'\overleftrightarrow{M} + \overleftrightarrow{Q}'\overleftrightarrow{Q}'\right] \tag{49}$$

Here, $Tr\,\overleftrightarrow{Q}^2 = 12\gamma_{12}^2$, $Tr\,\overleftrightarrow{Q}'^2 = 12\gamma_{34}^2$, and $Tr\left\{\overleftrightarrow{M}\overleftrightarrow{M}'\right\} = Tr\left\{\overleftrightarrow{M}'\overleftrightarrow{M}\right\} = 6\left(\gamma_{13}^2 + \gamma_{23}^2 + \gamma_{14}^2 + \gamma_{24}^2\right)$.

Therefore, one obtains the VDW energy of a chain of four identical atoms at arbitrary separations:

$$W_{chain}^{(2)} = c_2 Tr\left[\begin{pmatrix} \overleftrightarrow{Q} & \overleftrightarrow{M} \\ \overleftrightarrow{M}' & \overleftrightarrow{Q}' \end{pmatrix}^2\right] = -\frac{3}{2}\left(\gamma_{12}^2 + \gamma_{34}^2 + \gamma_{13}^2 + \gamma_{23}^2 + \gamma_{14}^2 + \gamma_{24}^2\right) \tag{50}$$

Note that this second-order contribution is a simple sum of London's pair interaction terms over all possible distinct pairs of atoms in the system. This VDW result corresponds to the VDW interaction calculated under the 2-body pair-sum approximation described in the introduction of the present chapter. Six distinct pairs listed in Eq. (50) are schematically drawn in the Fig. 5.

In the next higher-order term with n = 3, $c_3 = 1/16$ and

$$Tr\left[\begin{pmatrix} \overleftrightarrow{Q} & \overleftrightarrow{M} \\ \overleftrightarrow{M}' & \overleftrightarrow{Q}' \end{pmatrix}^3\right] = Tr\left[\overleftrightarrow{Q}^3 + \overleftrightarrow{Q}'^3 + \overleftrightarrow{M}\overleftrightarrow{M}'\overleftrightarrow{Q} + \overleftrightarrow{Q}\overleftrightarrow{M}\overleftrightarrow{M}' + \overleftrightarrow{M}'\overleftrightarrow{Q}\overleftrightarrow{M} + \overleftrightarrow{Q}'\overleftrightarrow{M}'\overleftrightarrow{M} + \overleftrightarrow{M}'\overleftrightarrow{M}\overleftrightarrow{Q}' + \overleftrightarrow{M}\overleftrightarrow{Q}'\overleftrightarrow{M}'\right]. \tag{51}$$

Here, $Tr\,\overleftrightarrow{Q}^3 = Tr\,\overleftrightarrow{Q}'^3 = 0$. Also, $Tr\left\{\overleftrightarrow{M}\overleftrightarrow{M}'\overleftrightarrow{Q}\right\} = Tr\left\{\overleftrightarrow{M}'\overleftrightarrow{Q}\overleftrightarrow{M}\right\} = Tr\left\{\overleftrightarrow{Q}\overleftrightarrow{M}\overleftrightarrow{M}'\right\}$ and $Tr\left\{\overleftrightarrow{Q}'\overleftrightarrow{M}'\overleftrightarrow{M}\right\} = Tr\left\{\overleftrightarrow{M}'\overleftrightarrow{M}\overleftrightarrow{Q}'\right\} = Tr\left\{\overleftrightarrow{M}\overleftrightarrow{Q}'\overleftrightarrow{M}'\right\}$ due to the cyclic invariance of the trace of the multiples of matrices [66]. After some calculations, one obtains

Fig. 5 Schematic diagrams of the 2nd-order many-body interaction contributions listed in Eq. (50), W$^{(2)}$

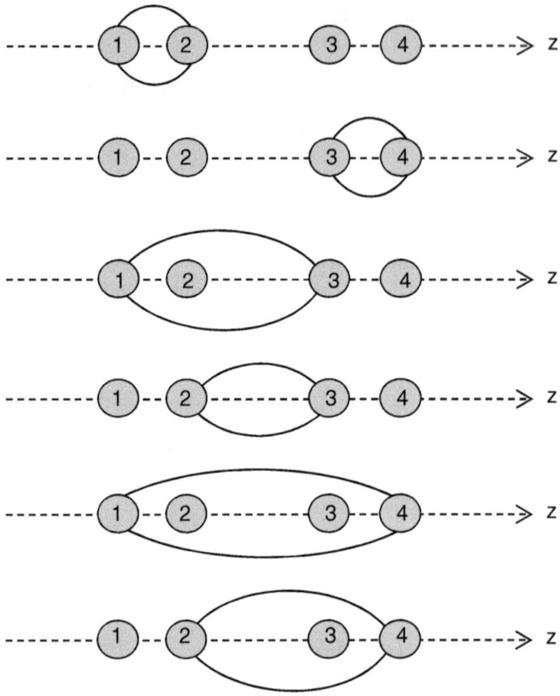

$$Tr\left\{\overleftrightarrow{M}\overleftrightarrow{M}'\overleftrightarrow{Q}\right\} = -12(\gamma_{13}\gamma_{32}\gamma_{21} + \gamma_{14}\gamma_{42}\gamma_{21})$$

and

$$Tr\left\{\overleftrightarrow{Q}'\overleftrightarrow{M}'\overleftrightarrow{M}\right\} = -12(\gamma_{34}\gamma_{41}\gamma_{13} + \gamma_{34}\gamma_{42}\gamma_{23}).$$

Therefore, the third-order contribution becomes

$$\begin{aligned} W^{(3)}_{chain} &= c_3 Tr\left[\begin{pmatrix} \overleftrightarrow{Q} & \overleftrightarrow{M} \\ \overleftrightarrow{M}' & \overleftrightarrow{Q}' \end{pmatrix}^3\right] \\ &= -\frac{9}{4}(\gamma_{13}\gamma_{32}\gamma_{21} + \gamma_{14}\gamma_{42}\gamma_{21} + \gamma_{34}\gamma_{41}\gamma_{13} + \gamma_{34}\gamma_{42}\gamma_{23}) \end{aligned} \quad (52)$$

Note here that this third-order contribution of the VDW interaction of this system corresponds to the sum of Axilrod-Teller-Muto (ATM) 3-body VDW interaction terms [18–20] for all possible distinct set of three atoms. The ATM 3-body term of one set of trio (A, B, C) then gives:

$$\Delta V_{vdw}^{ATM} = \frac{9}{8}\alpha_0^3 \frac{(1 + 3\cos\theta_A \cos\theta_B \cos\theta_C)}{r_{AB}^3 r_{BC}^3 r_{CA}^3}\left(\frac{1}{2}\hbar\omega_0\right) = -\frac{9}{4}\frac{\alpha_0^3}{r_{AB}^3 r_{BC}^3 r_{CA}^3}\left(\frac{1}{2}\hbar\omega_0\right) \tag{53}$$

Here, the angles θ_A, θ_B and θ_C are the inner angles in the triangle formed by three atoms A, B, and C, as shown in Fig. 6. In the linear configuration of the present example, since all the atoms lie in a line, two of these angles will be zero and one angle will be 180°, for example $\theta_A = \theta_B = 0$ and $\theta_C = 180°$, making the term $(1 + 3\cos\theta_A \cos\theta_B \cos\theta_C) = -2$ for any combination of three atoms A, B and C. That allows us to take the last step in the above equation.

In the system of a linear chain of four atoms, there are four sets of trio made of distinct three atoms ({1, 2, 3}, {1, 2, 4}, {3, 4, 1}, and {3, 4, 2}) as listed in Eq. (52) of $W_{chain}^{(3)}$, which are schematically drawn in Fig. 7.

When n = 4, $c_4 = -5/128$, and

$$Tr\left[\begin{pmatrix} \overleftrightarrow{Q} & \overleftrightarrow{M} \\ \overleftrightarrow{M}' & \overleftrightarrow{Q}' \end{pmatrix}^4\right] = Tr\left[\begin{array}{r} \overleftrightarrow{Q}^4 + \overleftrightarrow{Q}'^4 + \overleftrightarrow{Q}\overleftrightarrow{Q}\overleftrightarrow{M}\overleftrightarrow{M}' + \overleftrightarrow{Q}\overleftrightarrow{M}\overleftrightarrow{M}'\overleftrightarrow{Q} + \overleftrightarrow{M}\overleftrightarrow{M}'\overleftrightarrow{Q}\overleftrightarrow{Q} + \overleftrightarrow{M}'\overleftrightarrow{Q}\overleftrightarrow{Q}\overleftrightarrow{M} \\ + \overleftrightarrow{Q}'\overleftrightarrow{Q}'\overleftrightarrow{M}'\overleftrightarrow{M} + \overleftrightarrow{Q}'\overleftrightarrow{M}'\overleftrightarrow{M}\overleftrightarrow{Q}' + \overleftrightarrow{M}'\overleftrightarrow{M}\overleftrightarrow{Q}'\overleftrightarrow{Q}' + \overleftrightarrow{M}\overleftrightarrow{Q}'\overleftrightarrow{Q}'\overleftrightarrow{M}' \\ + \overleftrightarrow{Q}\overleftrightarrow{M}\overleftrightarrow{Q}'\overleftrightarrow{M}' + \overleftrightarrow{M}\overleftrightarrow{Q}'\overleftrightarrow{M}'\overleftrightarrow{Q} + \overleftrightarrow{Q}'\overleftrightarrow{M}'\overleftrightarrow{Q}\overleftrightarrow{M} + \overleftrightarrow{M}'\overleftrightarrow{Q}\overleftrightarrow{M}\overleftrightarrow{Q}' \\ + \overleftrightarrow{M}\overleftrightarrow{M}'\overleftrightarrow{M}\overleftrightarrow{M}' + \overleftrightarrow{M}'\overleftrightarrow{M}\overleftrightarrow{M}'\overleftrightarrow{M} \end{array}\right]$$

Here, $Tr\,\overleftrightarrow{Q}^4 = 36\gamma_{12}^4$, $Tr\,\overleftrightarrow{Q}'^4 = 36\gamma_{34}^4$, and $Tr\left(\overleftrightarrow{M}_{ij}\right)^4 = 18\gamma_{ij}^4$. Also,

$Tr\left[\overleftrightarrow{Q}\overleftrightarrow{Q}\overleftrightarrow{M}\overleftrightarrow{M}'\right] = Tr\left[\overleftrightarrow{Q}\overleftrightarrow{M}\overleftrightarrow{M}'\overleftrightarrow{Q}\right] = Tr\left[\overleftrightarrow{M}\overleftrightarrow{M}'\overleftrightarrow{Q}\overleftrightarrow{Q}\right] = Tr\left[\overleftrightarrow{M}'\overleftrightarrow{Q}\overleftrightarrow{Q}\overleftrightarrow{M}\right]$,
$Tr\left[\overleftrightarrow{Q}'\overleftrightarrow{Q}'\overleftrightarrow{M}'\overleftrightarrow{M}\right] = Tr\left[\overleftrightarrow{Q}'\overleftrightarrow{M}'\overleftrightarrow{M}\overleftrightarrow{Q}'\right] = Tr\left[\overleftrightarrow{M}'\overleftrightarrow{M}\overleftrightarrow{Q}'\overleftrightarrow{Q}'\right] = Tr\left[\overleftrightarrow{M}\overleftrightarrow{Q}'\overleftrightarrow{Q}'\overleftrightarrow{M}'\right]$,
$Tr\left[\overleftrightarrow{Q}\overleftrightarrow{M}\overleftrightarrow{Q}'\overleftrightarrow{M}'\right] = Tr\left[\overleftrightarrow{M}\overleftrightarrow{Q}'\overleftrightarrow{M}'\overleftrightarrow{Q}\right] = Tr\left[\overleftrightarrow{Q}'\overleftrightarrow{M}'\overleftrightarrow{Q}\overleftrightarrow{M}\right] = Tr\left[\overleftrightarrow{M}'\overleftrightarrow{Q}\overleftrightarrow{M}\overleftrightarrow{Q}'\right]$, and
$Tr\left[\overleftrightarrow{M}\overleftrightarrow{M}'\overleftrightarrow{M}\overleftrightarrow{M}'\right] = Tr\left[\overleftrightarrow{M}'\overleftrightarrow{M}\overleftrightarrow{M}'\overleftrightarrow{M}\right]$ due to cyclic invariance of the trace of multiples

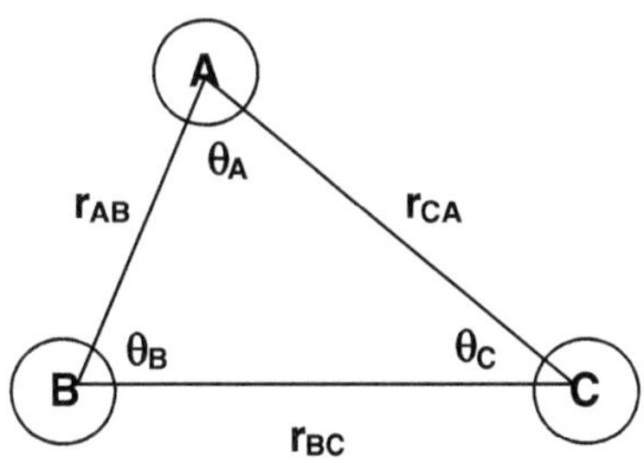

Fig. 6 The triangle in the Axilrod-Teller-Muto triple-dipole interaction

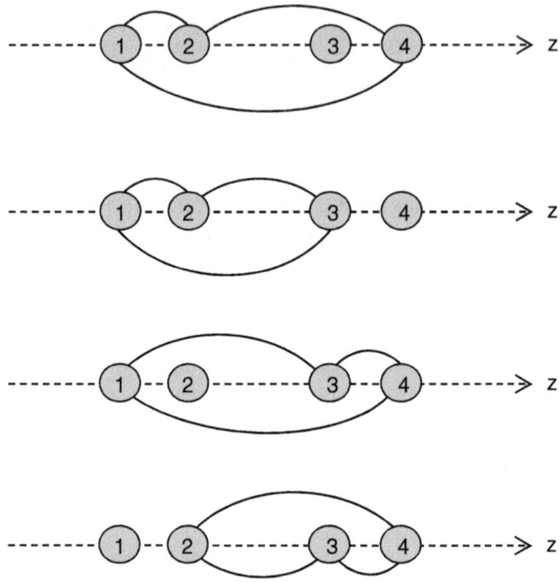

Fig. 7 Schematic diagrams of the 3rd-order many-body interaction contributions listed in Eq. (52), $W^{(3)}$

of matrices. These four distinct traces are obtained after some calculations as follows.

$$Tr\left[\overleftrightarrow{Q}\overleftrightarrow{Q}\overleftrightarrow{M}\overleftrightarrow{M}'\right] = Tr\left[M_{12}M'_{21}M_{13}M'_{31} + M_{12}M'_{21}M_{14}M'_{41} + M_{12}M'_{21}M_{23}M'_{32} + M_{12}M'_{21}M_{24}M'_{42}\right]$$
$$= 18\left(\gamma_{21}^2\gamma_{13}^2 + \gamma_{21}^2\gamma_{14}^2 + \gamma_{12}^2\gamma_{23}^2 + \gamma_{12}^2\gamma_{24}^2\right)$$

$$Tr\left[\overleftrightarrow{Q}'\overleftrightarrow{Q}'\overleftrightarrow{M}'\overleftrightarrow{M}\right] = Tr\left[M_{34}M'_{43}M'_{13}M_{31} + M'_{34}M_{43}M'_{14}M_{41} + M_{34}M'_{43}M'_{23}M_{32} + M'_{34}M_{43}M'_{24}M_{42}\right]$$
$$= 18\left(\gamma_{43}^2\gamma_{31}^2 + \gamma_{34}^2\gamma_{41}^2 + \gamma_{43}^2\gamma_{32}^2 + \gamma_{34}^2\gamma_{42}^2\right)$$

$$Tr\left[\overleftrightarrow{Q}\overleftrightarrow{M}\overleftrightarrow{Q}'\overleftrightarrow{M}'\right] = 2\,Tr\left[M_{12}M_{23}M_{34}M'_{41} + M_{12}M_{24}M'_{43}M'_{31}\right]$$
$$= 36\left(\gamma_{12}\gamma_{23}\gamma_{34}\gamma_{41} + \gamma_{12}\gamma_{24}\gamma_{43}\gamma_{31}\right)$$

$$Tr\left[\overleftrightarrow{M}\overleftrightarrow{M}'\overleftrightarrow{M}\overleftrightarrow{M}'\right] = Tr\begin{bmatrix} M_{13}^4 + M_{14}^4 + M_{23}^4 + M_{24}^4 \\ +2M_{31}^2M_{14}^2 + 2M_{13}^2M_{32}^2 + 2M_{14}^2M_{42}^2 + 2M_{32}^2M_{24}^2 + 4M_{13}M_{32}M_{24}M_{41} \end{bmatrix}$$
$$= 18\left(\gamma_{13}^4 + \gamma_{14}^4 + \gamma_{23}^4 + \gamma_{24}^4 + 2\gamma_{31}^2\gamma_{14}^2 + 2\gamma_{13}^2\gamma_{32}^2 + 2\gamma_{14}^2\gamma_{42}^2 + 2\gamma_{32}^2\gamma_{24}^2 + 4\gamma_{13}\gamma_{32}\gamma_{24}\gamma_{41}\right)$$

Thereby, the fourth-order contribution of the many-body VDW interaction becomes:

$$W_{chain}^{(4)} = -\frac{45}{32}\left[\begin{array}{l} \{\gamma_{12}^4 + \gamma_{13}^4 + \gamma_{14}^4 + \gamma_{23}^4 + \gamma_{24}^4 + \gamma_{34}^4\} \\ +2\{\gamma_{21}^2\gamma_{13}^2 + \gamma_{21}^2\gamma_{14}^2 + \gamma_{12}^2\gamma_{23}^2 + \gamma_{12}^2\gamma_{24}^2 + \gamma_{43}^2\gamma_{31}^2 + \gamma_{34}^2\gamma_{41}^2 \\ \qquad + \gamma_{43}^2\gamma_{32}^2 + \gamma_{34}^2\gamma_{42}^2 + \gamma_{31}^2\gamma_{14}^2 + \gamma_{13}^2\gamma_{32}^2 + \gamma_{14}^2\gamma_{42}^2 + \gamma_{32}^2\gamma_{24}^2\} \\ +4\{\gamma_{12}\gamma_{23}\gamma_{34}\gamma_{41} + \gamma_{12}\gamma_{24}\gamma_{43}\gamma_{31} + \gamma_{13}\gamma_{32}\gamma_{24}\gamma_{41}\} \end{array}\right] \tag{54}$$

Note here that this fourth-order contribution of the VDW interaction agrees with its general from of Eq. (1) in Ref. [36] which is derived from the fourth-order perturbation theory. Each term in the fourth-order many-body interaction contributions corresponds to distinct set of two atoms, three-atoms, and four atoms in this four-atom system. The combinations of atoms involved in Eq. (54) are schematically drawn in Fig. 8 in the order each term appears in the equation.

2.5 *Example 4: Calculation of the VDW Interaction Energy of Two Dimers Lying Along the Connecting Line Using TCDM*

To demonstrate the application of TCDM for the calculation of VDW interaction energy between two or more clusters of atoms, we will look at the system of four identical atoms located along a line from a different perspective. Let us consider this system of four atoms as two dimers lying along the connecting line (Fig. 9). A practical example of this system would be the linear alignment of hydrogen molecules (H_2) studied in Ref. [20]. Now, the VDW interaction energy between the two dimers can be obtained by subtracting the self-energy of each dimer from the total energy of the system. Recall that the self-energy of each dimer has already been obtained in Examples 1 and 2.

The eigenvalue problem of this system will be of the same form as derived in Example 3 for the system of a liner chain of four atoms. To estimate the VDW interaction between two dimer, however, the Eq. (48) should be modified, since now the self-energy of each dimer should be subtracted from the total energy following the general expression given in Eq. (32):

Thereby, the first nonvanishing term is the second-order many-body contribution:

$$\begin{aligned} W_{\substack{\text{two}\\ \text{dim ers}}}^{(2)} &= c_2 Tr\left\{\begin{pmatrix} \overset{\leftrightarrow}{Q} & \begin{matrix}\overset{\leftrightarrow}{M} \\ \overset{\leftrightarrow}{M'}\end{matrix} & \overset{\leftrightarrow}{Q'} \end{pmatrix}^2\right\} - c_2 Tr\left\{\overset{\leftrightarrow}{Q}^2\right\} - c_2 Tr\left\{\overset{\leftrightarrow}{Q}'^2\right\} \\ &= -\frac{3}{2}\left(\gamma_{13}^2 + \gamma_{23}^2 + \gamma_{14}^2 + \gamma_{24}^2\right) \end{aligned} \tag{55}$$

Note that this second-order contribution is a simple sum of London's pair interactions over all possible pairs of atoms, with one atom from the first dimer (A)

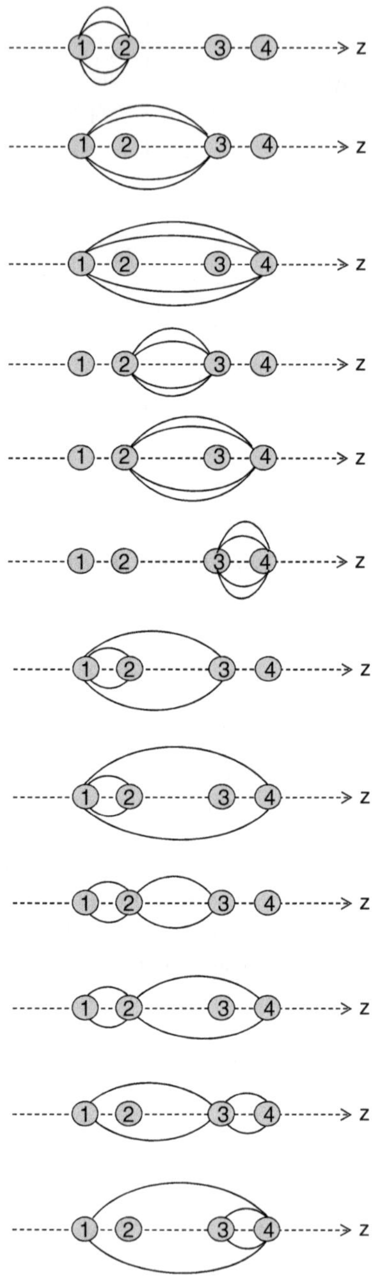

Fig. 8 Schematic diagrams of the 4th-order many-body interaction contributions listed in Eq. (54), $W^{(4)}$

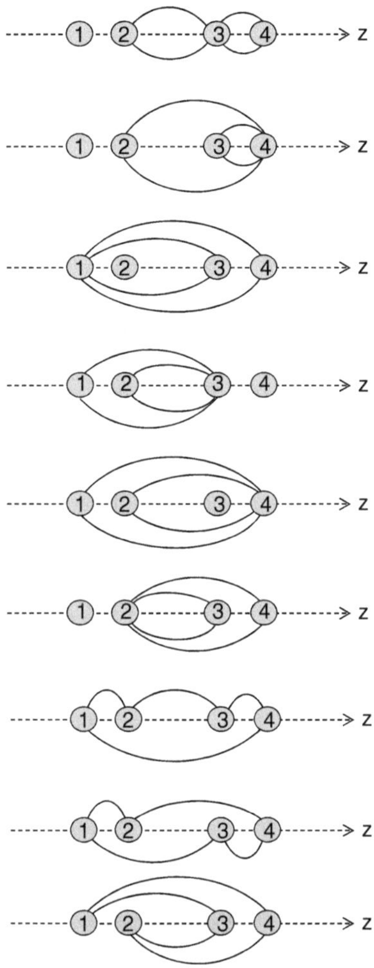

Fig. 8 continued

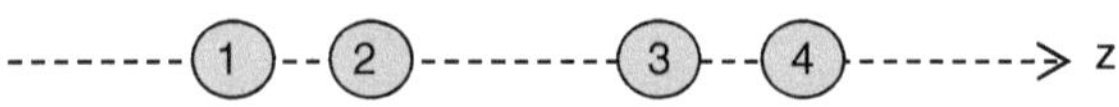

Fig. 9 Two identical dimers lying along the z-axis

and another from the second dimer (B), in the system. This again corresponds to the VDW interaction from calculations based on pair-sum approximation. Note that the two-body interaction terms involving two atoms that belong to the same cluster (γ_{12}^2 and γ_{34}^2 terms, in this example) are subtracted as self-energies of each cluster (or dimer, in this example).

It should be noted here that the third-order contribution of the VDW interaction between two dimers remains the same as that in the chain of four atoms which was considered in Example 3. This is because of the fact that $Tr\,\overleftrightarrow{Q}^3 = Tr\,\overleftrightarrow{Q}'^3 = 0$. Therefore, $W^{(3)}_{\substack{two\\dimers}} = W^{(3)}_{chain}$. In fact, $Tr\,\overleftrightarrow{Q}^n = Tr\,\overleftrightarrow{Q}'^n = 0$ for all odd n due to the tracelessness of Q and Q′ matrices. Therefore, the n-th order contribution of the VDW between two dimers or of a chain of four atoms remains the same, $W^{(n)}_{\substack{two\\dimers}} = W^{(n)}_{chain}$, for odd n.

3 Discussion and Concluding Remarks

The aim of this chapter is to present an overview of the CDM method which is intrinsically atomistic method which includes all the many-body interaction terms self-consistently. This rare combination of characteristics makes CDM an ideal method to calculate the VDW interaction energy involved in nanoscale systems. However, to obtain the VDW interaction energy using the eigenvalue method of exact CDM, one first needs to obtain all the eigenvalues in the eigenvalue problem of matrix size (3N × 3N) for a system composed of N particles. Usually the eigenvalue problems are solved by direct diagonalization or iterative methods. As a result, there was a concern for solving the eigenvalue problems of exact CDM for systems composed of many atoms and molecules such as those found in computer simulation studies. The computational burden is from both in memory size and in run time. As N increases, the memory size and the run time increases as N^2 and N^3, respectively. The problem with memory size can be resolved by parallelization. The burden from the run time still remains. Thereby, a memory-efficient and cost-effective computational method, TCDM, is presented.

TCDM, an alternative way to execute CDM, is to obtain VDW interaction energy by calculating the trace of the square-root of the 3N × 3N square matrix, rather than its eigenvalules. It is demonstrated in this chapter that the power series expansion in TCDM is indeed equivalent to that of the perturbation theory. Therefore, one will have to terminate the series and determine how many higher-order terms to include, unlike the eigenvalue method of exact CDM discussed above which intrinsically includes all the many-body terms. It is however shown, in our previous studies [49], that including terms less than 10 ($n < 10$) gives an excellent convergence (much less than 1 %) to the exact value of VDW Interaction energy for various configurations as listed also in Fig. 1 and Table 1 in this chapter.

A question might be raised. If TCDM is based on the power series expansion that is equivalent to that of the traditional perturbation theory, why do we use TCDM? From the perturbation theory, in principle, one can derive all the higher-order many-body interaction terms. However, most researchers have studied and/or included the contribution of three-body interactions in addition to the pair-wise sum approximation as the only many-body correction term. There has been no quantitative estimation of many-body interaction terms beyond the three-body term done using the expression derived from the perturbation theory, to the best knowledge of the author. I believe that this is because the analytic formula of the n-th order many-body term that can be derived from the perturbation theory becomes impractically cumbersome as the order n increases beyond three. Only the two-body (London formula) and three-body (ATM formula) terms have been adopted by researchers. On the other hand, in TCDM, one just needs to set up the initial 3N × 3N matrix of the eigenvalue problem of the system. The matrix multiplications are executed in the computer code which would be much more cost-effective than diagonalization. One does not need to explicitly derive the analytic form of each n-th order term. This feature of TCDM will make it practical for any interested future users to include increasingly many higher-order n-body terms as desired.

It should be noted that, although, in this chapter, two simple systems (composed of two and four identical atoms, respectively, aligned in a line) are considered as examples to explicitly demonstrate the actual execution of CDM and TCDM, the use of both methods can be extended for accurate estimates of the VDW energy in any nanoscale systems composed of dielectric materials. The formulation of exact CDM and efficient TCDM can readily be generalized for systems composed of non-identical atoms.

Finally, I would like to conclude this chapter with the following list of a few unresolved issues that can further improve our understanding of the VDW interaction:

(1) Retardation should be included as a function of inter-atomic and/or inter-cluster separation. The CDM presented in this chapter calculates the non-retarded VDW interaction. As the separation increases, larger than about 10 nm, the retardation effect gradually increases due to the finite speed of light. Even though this retardation effect is receiving increasing attention due to rapid advances in the nanotechnology in both fundamental/applied sciences and engineering, calculations have been limited to either non-retarded VDW or fully-retarded VDW interactions. One may refer chapter 6 of ref. [2], for example, for a detailed discussion of both relativistic and semi-classical methods to calculate the retarded dispersion forces at intermediate separations, but the extension of these methods to many-body systems is not a trivial problem.
(2) The higher-order multipole dispersion interaction terms beyond the dipole approximation should be investigated. Just like the higher-order many-body interaction terms, the higher-order multipole dispersion interaction terms have been commonly assumed as negligible compared to the dipole interaction terms. As the common assumption on the higher-order many-body interaction terms being small is shown incorrect for some odd-shaped nanoparticle systems [49], it is critical to test this assumption of the small higher-order multipole interaction terms in nanoscale systems.

(3) The validity of using the Clausius-Mossotti (CM) relation (see Eq. 8) in nanoscale materials should be investigated as mentioned in Sect. 2. The "effective" atomic polarizability is required as an input in calculating the VDW interaction. There has been a considerable effort made in developing the effective non-local atomic polarizability tensor using a self-consistent quantum mechanical calculation using DFT [61], which includes both bonded and non-bonded many-body interaction terms. It would be interesting to see the extension of such calculation for the larger nanoscale systems of varying shape and size to test the validity of the CM relation.

Achievement of these important and challenging tasks will allow us to accurately predict the VDW forces in nanoscale systems at varying separations as well as in those of varying shape, size and composition.

Acknowledgments I am grateful to all my collaborators in research projects involving CDM—Milton Cole, Darrell Velegol, Amand Lucas, Jorge Sofo, and Paul Kent. I most gratefully acknowledge the significant contribution of Amand Lucas in guiding me through the initial formulations of CDM and TCDM. I thank the Louisiana Board of Regents-RCS grant (LEQSF(2012-15)-RD-A-19) for support.

References

1. Rowlinson JS (2005) Cohesion: a scientific history of intermolecular forces. Cambridge University Press, Cambridge
2. Margenau H, Kestner NR (1972) Theory of intermolecular forces. Pergamon, Oxford
3. Langbein D (1974) Springer tracks in modern physics, vol 72. Springer, Berlin
4. Parsegian VA (2006) van der Waals forces: a handbook for biologists, chemists, engineers, and physicists. Cambridge University, New York
5. Israelachvili J (1991) Intermolecular and surface forces, 2nd edn. Academic, New York
6. Eisenschitz R, London F (1930) Über das Verhältnis der van der Waalsschen Kräfte zu den Homöopolaren Bindungskräften. Z Physik 60:491–527 (in German)
7. London F (1930) Zur Theorie und Systematik der Molekularkräfte. Z Physik 63:245–2791930 (in German)
8. London F (1930) Über einige Eigenschaften und Anwendungen der Molekularkräfte. Z Phys Chem B11:222–251 (in German)
9. Hamaker HC (1937) The London-Van der waals attraction between spherical particles. Physica 4:1058–1072
10. London F (1937) The general theory of molecular forces. Trans Faraday Soc 33:8–26
11. Bishop KJM, Wilmer CE, Soh S, Grzybowski BA (2009) Nanoscale force and their use in self-assembly. Small X:1–31
12. Sernelius BE, Román-Velázquez CE (2008) Beyond the simple proximity force approximation: geometrical effects on the nonretarded Casimir interaction. Phys Rev A 78:032111
13. Marom N, DiStasio Jr. RA, Atalla V, Levchenko S, Reilly AM, Chelikowsky JR, Leiserowitz L, Tkatchenko A (2013) Many-body dispersion interactions in molecular crystal polymorphism. Angew Chem Int Ed 52:1–5
14. Bergström L (1997) Hamaker constants of inorganic materials. Adv Colloid Interface Sci 70:125–169

15. Girard C, Bouju X (1991) Coupled electromagnetic modes between a corrugated surface and a thin probe tip. J. Chem. Phys. 95:2056–2064
16. Nel AE, Mädler L, Velegol D, Xia T, Hoek EMV (2009) Understanding biophysicochemical interactions at the nano-bio interface. Nature Mat. 8:543–557
17. Bruch LW, Diehl RD, Venables JA (2007) Progress in the measurement and modeling of physisorbed layers. Rev Mod Phys 79:1381–1454
18. Sushkov AO, Kim WJ, Dalvit DAR, Lamoreaux SK (2011) Observation of the thermal Casimir force. Nature Phys. 7:230–233
19. Berland K, Borck Ø, Hyldgaard P (2011) Van der Waals density functional calculations of binding in molecular crystals. Comput Phys Commun 182:1800–1804
20. Liu R-F, Ángyán JG, Dobson J (2011) Dispersion interaction in hydrogen-chain models. J Chem Phys 134:114106
21. Mo Y, Turner KT, Szlufarska I (2009) Friction laws at the nanoscale. Nature 457:1116–1119
22. Velegol D, Holtzer GL, Radović-Moreno A, Cuppett JD (2007) Force Measurements between Sub-100 nm colloidal particles. Langmuir 12:1275–1280
23. Pio JM, Taylor MA, van der Veer WE, Bieler CR, Cabrera JA, Janda KC (2010) Real-time dissociation dynamics of the Ne_2Br_2 van der Waals complex. J. Chem. Phys. 133:014305
24. Wang ZH, Wei J, Morse P, Dash JG, Vilches OE, Cobden DH (2010) Phase transitions of adsorbed atoms on the surface of a carbon nanotube. Science 327:552–555
25. Krim J, Solina DH, Chiarello R (1991) Nanotribology of a Kr monolayer: a quartz-crystal microbalance study of atomic-scale friction. Phys Rev Lett 66:181–184
26. Stifter T, Marti O, Bhushan B (2000) Theoretical investigation of the distance dependence of capillary and van der Waals forces in scanning force microscopy. Phys Rev B62:667–673
27. Böttcher CJF (1973) Theory of electric polarization, 2nd edn, vol 1. Dielectrics in static fields, Chap. V. Elsevier Scientific Publishing Co. Amsterdam
28. Kim H-Y, Kent PRC (2009) van der Waals forces: accurate calculation and assessment of approximate methods in dielectric nanocolloids up to 16 nm. J Chem Phys 131:144705
29. Margenau H (1938) Quadrupole contributions to London's dispersion forces. J Chem Phys 6:896
30. Hirschfelder JO, Meath WJ (2007) The nature of intermolecular forces (1967). Adv Chem. Phys.: Intermol Forces 12 (Published on line)
31. Axilrod BM, Teller E (1943) Interaction of the van der Waals type between three atoms. J Chem Phys 11:299–300
32. Axilrod BM (1951) Triple-dipole interaction. J Chem Phys 19:719–724
33. Muto Y (1943) Force between non-polar molecules. Proc Phys Math Soc Jpn 17:629–631 in Japanese
34. Bade WL (1957) Drude-model calculation of dispersion forces. I General theory. J Chem Phys 27:1280–1284
35. Bade WL, Kirkwood JG (1957) Drude-model calculation of dispersion forces. II The linear lattice. J Chem Phys 27:1284–1288
36. Bade WL (1958) Drude-model calculation of dispersion forces. III The fourth-order contribution. J Chem Phys 28:282–284
37. Lucas AA (1967) Collective contributions to the long-range dipolar interaction in rare-gas crystals. Physica 35:353–368
38. McLachlan AD (1964) Van der Waals forces between an atom and a surface. Mol Phys 7:381–388
39. Barker JA (1976) Interatomic potentials for inert gases from experimental data. In: Klein ML, Venables JA (eds) Rare gas solids, vol 1, Chap. 4. Academic Press, New York, pp 212–264
40. Klein ML, Koehler TR (1976) Lattice dynamics of rare gas solids. In: Klein ML, Venables JA (eds) Rare gas solids. Academic Press, New York

41. Meath WJ, Aziz RA (1984) On the importance and problems in the construction of many-body potentials. Mol Phys 52:225–243
42. Dzyaloshinskii IE, Lifshitz EM, Pitaevskii LP (1961) The general theory of van der Waals forces. Adv Phys 10:165–209
43. Schmeits M, Lucas AA (1983) Physical adsorption and surface plasmons. Prog Surf Sci 14:1–52
44. Hamaker HC (1937) The London-van der Waals attraction between spherical particles. Physica 4:1058–1072
45. Hunter J (1986) Foundations of colloids science. In: White LR (ed) The theory of van der Waals forces, vol 1, Chap. 4. Clarendon/Oxford University, New York
46. Munday JN, Capasso F, Parsegian VA (2009) Measured long-range repulsive Casimir-Lifshitz force. Nature 457:170–173
47. Kim H-Y, Sofo JO, Velegol D, Cole MW, Mukhopadhyay G (2005) Static Polarizabilities of dielectric nanoclusters. Phys Rev A 72:053201
48. Kim H-Y, Sofo JO, Velegol D, Cole MW (2006) Fully retarded van der Waals interaction between dielectric nanoclusters. J Chem Phys 125:174303
49. Kim H-Y, Sofo JO, Velegol D, Cole MW, Lucas AA (2006) van der Waals forces between nanoclusters: importance of many-body effects. J Chem Phys 124:074504
50. Kim H-Y, Sofo JO, Velegol D, Cole MW (2007) Van der Waals dispersion forces between dielectric nanoclusters. Langmuir 23:1735–1740
51. Cole MW, Velegol D, Kim H-Y, Lucas AA (2009) Nanoscale van der Waals interactions. Mol Simul 35:849–866
52. Cole MW, Gergidis LN, McNutt JP, Velegol D, Kim H-Y, Bond ZK (2010) Many body VDW forces involving chains. SPIE J Nanophotonics 4:041560
53. Hohm U, Kerl K (1987) Temperature dependence of measured mean molecular polarizability on account of translational motion. Mol Phys 61:1295–1298
54. Bohren CF, Huffman DR (1983) Absorption and scattering of light by small particles. Wiley-VCH Verlag GmbH & Co, Weinheim
55. Born M, Wolf E (1975) Principles of optics: electromagnetic theory of propagation, interference and diffraction of light, 5th edn. Pergamon Press, Oxford
56. McLachlan AD (1963) Retarded dispersion forces between molecules. Proc Roy Soc A271:387–401
57. McLachlan AD (1963) Retarded dispersion forces in dielectrics at finite temperatures. Proc Roy Soc A274:80–90
58. http://www.netlib.org/lapack
59. http://www.netlib.org/scalapack
60. Lamoreaux SK (2011) Casimir force, scholarpedia 6:9746. doi:10.4249/scholarpedia.9746
61. Gobre VV, Tkatchenko A (2013) Scaling laws for van der Waals interactions in nanostructured materials. Nat Commun 4:2341. doi:10.1038/ncomms3341
62. Tkatchenko A, Scheffler M (2009) Accurate Molecular Van Der waals interactions from ground-state electron density and free-atom reference data. Phys Rev Lett 102:073005. doi:10.1103/PhysRevLett.102.073005
63. Tkatchenko A, DiStasio Jr. RA, Car R, Scheffler M (2012) Accurate and efficient method for many-body van der Waals interactions. Phys Rev Lett 108:236402. doi:10.1103/PhysRevLett.108.235402
64. Tudor B, Space B (2013) Solving the many-body polarization problem in GPUs: application to MOF's. J Comp Sci Educ 4:30–34
65. Massively Parallel Monte Carlo (MPMC) http://code.google.com/p/mpmc/
66. Saad Y (2011) Numerical methods for large eigenvalue problems. Society for Industrial and Applied Mathematics, http://www.siam.org/books/c166)

67. Lucas AA (1966) Contributions electrostatiques Collectives et corrections Radatives à l'Energie de Van der Waals des Cristaux Moléculaires. Thèse de doctorat. University of Liege (unpublished: Private communication)
68. Stone A (2013) The theory of intermolecular forces, 2nd edn. Oxford University Press, Oxford

Adsorption of Gases in Nanomaterials: Theory and Simulations

M.T. Mbaye, S.M. Maiga and S.M. Gatica

Abstract Physical adsorption (physisorption) is the study of atoms or molecules weakly bound to material surfaces. Physisorption-related investigations raise critical questions concerning phase transitions, fractals, wetting transitions, two-dimensional superfluidity, and Van der Waals interactions. This chapter focuses on adsorption of gases (e.g. Ar, Kr, H_2, CO_2, and CH_4) in nanomaterials, and in particular the authors describe equilibrium properties of the gases adsorbed in carbon nanotubes, graphene and Metal Organic Frameworks (MOFs). The adsorption potential used for developing the theoretical model for studying physisorption involves the summing of two-body interactions, and several important properties of adsorbates can be obtained via simulations, namely equilibrium properties, thermal characteristics, selectivity, wetting features, and structure and phase of the adsorbed monolayer. Applications of physisorption include the separation of cryogenic gases, their storage and their use as a surface characterization tool.

Keywords Adsorption · Physisorption · Nanomaterials · Molecular dynamics simulation · Van der Waals interactions · Carbon nanotubes · Graphene · Metal organic frameworks (MOFs) · Wetting transitions

1 Introduction to Physical Adsorption

"Physical adsorption" (physisorption) is a term applied to atoms or molecules that are weakly bound to surfaces. Physisorption has been extensively explored for more than half of a century because of interest in both potential applications and basic science. These applications include the separation of cryogenic gases, their storage and their use as surface characterization tools, such as the measurement of surface

M.T. Mbaye · S.M. Maiga · S.M. Gatica (✉)
Howard University, Washington, DC, USA
e-mail: sgatica@howard.edu

P. Misra (ed.), *Applied Spectroscopy and the Science of Nanomaterials*,
Progress in Optical Science and Photonics 2, DOI 10.1007/978-981-287-242-5_6

area of porous media by nitrogen adsorption. The science of physisorption encompasses a wide variety of fundamental questions, including many related to phase transitions. For example, adsorption provided the first definitive confirmation of landmark calculations of Onsager, Lee and Yang concerning the Ising model's prediction of critical exponents for the two-dimensional (2D) liquid-vapor transition (measured in CH_4/graphite experiments of Kim and Chan [1]). Similarly, the measured exponents of the order-disorder transition of He isotopes on graphite confirmed the predictions of the three-state Potts model [2–4]. Other basic principles that are tested by physisorbed films include ideas concerning fractal surfaces (e.g. in porous media), wetting transitions (first seen for He on Cs) and 2D superfluidity (on numerous surfaces), as well as the theories of van der Waals interactions. The many successes in this field and the numerous remaining questions concerning both the forces and resulting behavior of monolayer films are reviewed in a monograph by Bruch et al. [5] and a recent article by Bruch et al. [6].

The remarkable capability of physical adsorption to explore and analyze these diverse phenomena *quantitatively* derives from the weak binding (binding energy < 0.3 eV). This has several significant consequences. One is that the adsorbed film represents just a small perturbation of the underlying surface, simplifying theoretical analysis enormously. Often the substrate is assumed to be perfectly rigid, unaffected by the film (although we, and other groups, have studied the limitations of that assumption) [7]. Another consequence is that the adsorbed gases are not significantly altered from their 3D gas phase electronic state, meaning that their mutual interactions are relatively well known. A particularly valuable consequence of the small binding energy is that a vapor coexists with the film (except at very low T), so that one can determine experimentally the chemical potential μ of both phases; they are equal in the case of equilibrium, which occurs relatively quickly for physisorbed gases. Letting $\beta^{-1} = k_B T$, the relationship is

$$\mu_{film}(\theta, T) = \mu_{vapor}(P, T) = k_B T \ln(P\beta\lambda^3) \quad (1)$$

(The right-most expression assumes that the vapor is ideal and monatomic). Here, the 2D film density $\theta \equiv N/A$, where A is the surface area and N is the coverage; the quantity λ is the de Broglie thermal wavelength of the gas atoms, each of mass m:

$$\lambda = (2\pi\beta\hbar^2/m)^{1/2} \quad (2)$$

μ and T together provide two of the three intensive thermodynamic variables (the third being θ or the spreading pressure) entering the film's equation of state.

This chapter concerns adsorption in nanomaterials. In particular, we describe equilibrium properties of gases adsorbed in carbon nanotubes (NTs), graphene and Metal Organic Frameworks (MOFs). The behavior of adsorbed matter in the nanoscale is dramatically different from the corresponding bulk, due in part to the reduced dimensionality. In the following sections, we describe the main results of our simulations; we also predict novel phases not observed yet, presenting intriguing challenges and the opportunity to make significant breakthroughs in understanding the distinct phenomena.

The most important ingredient of simulations and theory of adsorption is the model adopted for the potential. The simplest way to obtain the adsorption potential for studying physisorption phenomena is by summing two-body interactions.

Most theoretical studies on fluids with porous materials have assumed that the substrate is rigid. This is a good approximation at low temperature, and if the film-substrate attraction is weak, as is the case for most systems. In many cases, the masses of the adsorbate is much smaller than the substrate atoms, or the substrate's atoms are grasped by strong cohesive forces. Due to the rigid substrate assumption, there are instances that result in qualitative errors. This occurs when the energy difference between competing phases of the adsorbate is very small, which make the system's behavior particularly sensitive to the rigidity approximation. We recently found [8, 9] that a very small expansion or contraction of the substrate may have striking consequences on the thermodynamic properties of the adsorbate. For example, the quasi-1D ^{3}He system in a very small, rigid carbon nanotube (NT) is found in the gaseous state, while in a non-rigid NT forms a liquid state. Moreover, several authors have addressed the non-rigidity of the substrate in simulations of adsorption, mostly in slit pores, carbons, and zeolites [10–14]. This flexibility is a characteristic that distinguishes MOFs from other materials. Fletcher et al. [15] studied the impact of MOFs flexibility on adsorption. They concluded that structural changes might occur, including gating and trapping of adsorbates. Greathouse et al. [16] found only minor effects on the adsorption of noble gases. However, they suggest that gases that interact stronger with the host, such as CO_2 or water, lead to significant and non-negligible effects.

Fluid–fluid interactions also have to be carefully modeled. Spherically symmetric molecules (monoatomic, CF_4, CH_4) are well described by the Lennard Jones (LJ) potential given by,

$$U(r) = 4\varepsilon\left(\left(\frac{\sigma}{r}\right)^{12} - \left(\frac{\sigma}{r}\right)^{6}\right) \tag{3}$$

where ε is the depth of the potential well, σ is the finite distance at which the inter-particle potential is zero and r is the distance between the particles.

Diatomic molecules (H_2, CO, NO etc.) may be characterized as a rigid 2-point LJ model. The H_2–H_2 interaction is well represented by the Silvera-Goldman model, [17] although quantum corrections need to be addressed at low temperature.

The CO_2 molecule is commonly modeled as a three-site linear rigid molecule with partial charges in each site, $q_O = -0.332e$ and $q_C = +0.664e$ set to mimic the quadrupole moment, $Q = -4.3$ B. The bond length of the molecule is 0.116 nm. Hence its molecular interaction is a superposition of LJ and Coulomb potentials. We have tested this model in our study of adsorption of CO_2 in carbon nanotubes and MOFs [18].

For water–water interactions, the TIP4P potential [19] is frequently adopted. This is a traditional model where water is considered a rigid molecule with one LJ site located at the oxygen atom and three point charges, two of them located on the OH bonds (+q) and one negative (−q) along the bisector of the two OH bonds. This potential has been tested to give the saturated liquid density in good agreement with experiments.

The LJ parameters (ε and σ) are obtained by fitting physical properties of the gases [20, 21]. The mixed parameters of different species are estimated using the semi-empirical mixing rules given by [5]

$$\begin{aligned}\sigma_{ia} &= \frac{(\sigma_i + \sigma_a)}{2} \\ \varepsilon_{ia} &= \sqrt{\varepsilon_i \varepsilon_a}\end{aligned} \tag{4}$$

Computational methods have been used to study adsorption for many decades. In the numerical study of adsorption, the following experiment is simulated: the substrate is exposed to a vapor at constant P and T. After some time passed, long enough to reach equilibrium, the uptake is measured under those P, T conditions. This can be simulated numerically by the method Grand Canonical Monte Carlo (GCMC), in which the calculation of the uptake is based on statistical mechanics theory. In the GCMC method, the uptake is obtained at any given temperature and chemical potential (not pressure) of the adsorbate. The chemical potential of the adsorbate is identical to that of the vapor since both are in thermal equilibrium, and the last one is related to the pressure of the vapor through the equation of state. This technique has been extensively used in studies of adsorption in carbon nanotubes and other porous media [22–26].

The method of Molecular Dynamics (MD) is normally used as an alternative approach to study kinetic and transport properties. This technique gives insights on the mechanisms of adsorption and diffusion. In the simulation, the time dependence of physical properties of the adsorbate is calculated, from where diffusion and transfer rates can be obtained [27].

Mixtures can be simulated by the commonly used "Ideal Adsorbed Solution Theory" (IAST) [28] IAST has been found to be accurate for most cases, but it is known to fail for mixtures that differ strongly in size or polarity. Some authors have confirmed that the theory is reliable in the case of CH_4/CO_2 mixtures in IRMOFs [29]. Regardless, the accuracy of IAST has to be tested.

Several properties of the adsorbate can be obtained from the simulations:

1.1 Equilibrium Properties

The main direct result from a GCMC simulation is the "adsorption isotherm", i.e. the average number of molecules adsorbed at a given temperature and vapor pressure. Typically, we run simulations to obtain the isotherms at several temperatures and a wide range of pressures, from a minimum value P_0 to saturated vapor pressure. The value of P_0 is such that for each adsorbate/temperature, the transition from vapor to finite coverage is obtained. The average energy per particle, with details of the gas-surface and gas-gas contributions is also a direct output of the GCMC. In summary, $\langle n \rangle$ (P, T), $\langle E_{gs} \rangle$ (P, T) and $\langle E_{gg} \rangle$ (P, T) are computed for all adsorbates.

1.2 Thermal Properties

The isosteric heat of adsorption is defined as

$$q_{st}(n,T) = k_b T^2 \left(\frac{\partial lnP}{\partial T} \right)_n \tag{5}$$

where P is the vapor pressure and the derivative is made at constant coverage n. The q_{st} can be calculated from the adsorption isotherms, $\langle n \rangle$ (P, T), obtained from GCMC simulations at different temperatures (typically 5 or 10 K apart). The calculation of q_{st} allows various tests: it is a quantity usually measured, so it permits comparison with experiments; from its low coverage limit we can evaluate the binding energy (BE), given by BE = $q_{st} - bK_B T$ where b is a factor that depends on the dimensionality (b = 2, 3/2 or 1 for 1, 2 and 3 dimensions respectively); at high coverage q_{st} approaches the latent heat of vaporization. It has also been found that on carbon nanotubes and nanohorns, the q_{st} of CO_2 increases at high coverage, opposite to the behavior of spherical adsorbates [30].

1.3 Selectivity

Adsorption selectivity in a binary mixture of components i and j is defined as

$$S_{i/j} = \frac{(x_i/x_j)}{(y_i/y_j)} \tag{6}$$

where x_i and y_i are the concentration of species i in the adsorbed phase and vapor phase respectively. From results of pure gas adsorption, we can evaluate the selectivity using the IAST theory. At low coverage, the relation between the molar concentrations and pressure is given by Raoult's law, $P\,y_i = P_i^0 x_i$, where P_i^0 is the equilibrium vapor pressure of the i pure component at the same spreading pressure. From these equations, we obtain the selectivity to be $S_{1/2} = P_2^0/P_1^0$. For low coverage, we can approximate P_i^0 by the pressure of uptake. Hence, under this approximation the selectivity is given by the ratio of the uptake pressures, resulting in high selectivity for those gases that are adsorbed at the lower vapor pressure. In the case of mixtures containing water where IAST may be inaccurate, the method of choice would be two-components GCMC.

1.4 Structure and Phase of the Adsorbed Monolayer

In the GCMC simulations, millions of configurations of the adsorbate are obtained for each P, T. With these configurations that contain the position of all the adsorbed molecules, one can examine the structure and phase of the adsorbed layer. In particular, it is possible to find commensurate lattices of adsorbates. We have done research of commensurate solid phases of Ar and Kr on nanotubes, [26] and there are experimental data for nanotubes [31] and graphite [5]. To search for commensuration we first inspect the E_{gs} versus P at fixed T: a sudden decrease in the gas-surface energy as P increases may indicate a transition to a commensurate phase. If that is the case, we analyze sample configurations near the given pressure. The analysis consists on visual inspection, and the calculation of the structure factor to test the commensuration. If the phase results incommensurate, we evaluate the radial distribution function, to distinguish between a liquid or solid (incommensurate) layer.

1.5 Wetting Transitions

A wetting transition (first proposed by Cahn [32]) and, independently, by Ebner and Saam [33]) is a first order transition, along a line in the P–T plane, ending with a critical point. It is manifested as a discontinuous jump in film coverage, as a function of pressure P (below saturated vapor pressure P_{svp} and above the *wetting temperature,* T_w, [34] or else at P_{svp} by observing the contact angle formed by a liquid droplet in contact with the surface.

As discussed recently in Ref. [35], this transition has been observed previously only for two very different kinds of adsorption systems: (a) cryogenic fluids- He, Ne and H_2 films on alkali metals, at low T, and (b) Hg films at very high T (>1,300 °C) and P (>120 MPa), close to its critical point. In 2004, Gatica et al. argued that a similar transition ought to occur for water on graphite and a number of other surfaces [36]. Quite recently, Taborek, of UC Irvine, has found preliminary evidence of such a wetting transition-the first such transition for water [37]. His preliminary result for water on graphite shows the same phenomenon at somewhat higher T_w (unspecified). The so-called "simple model" predicted a transition temperature $T_w \sim 230$ C on graphite, but the estimation carries the large uncertainty in both the model and the H_2O-graphite adsorption potential used. This exciting, but uncertain situation, is a strong motivation for more careful study of this transition. The wetting behavior can be evaluated from the adsorption isotherms obtained in GCMC simulations.

In the next sections, we show examples of simulations on carbon nanotubes, graphene and MOFs. In Sect. 2, we describe a system that has recently exhibited quite novel behavior experimentally; it is concerned with adsorption of gases on a *single* carbon nanotube. In Sect. 3, we discuss the adsorption of gases on graphene. Section 4 is dedicated to adsorption in MOFs, in particular its application to separation of gases. We omit from this chapter the extensive case of adsorption in bundles of carbon NTs and refer the interested reader to previous publications, for example Refs. [22, 38].

2 Adsorption on a Suspended Carbon Nanotube

In 2010, Wang, Cobden and Vilches at the University of Washington measured the adsorption of Kr on a suspended carbon NTs [39]. They deduced the coverage from the adsorption-induced frequency shift of a high Q resonance of the vibrating nanotube. Their data revealed the presence of a commensurate phase with 1/6 fractional coverage, which coincides with the value observed for Kr on the surface of *planar graphite*. That coverage arises in the $\sqrt{3} \times \sqrt{3}$ R30° phase, in which an adatom is localized on just one out of every three hexagons of graphite. The very same ordered phase dominates the monolayer phase diagrams of both He and hydrogen isotopes on graphite [5].

The appearance of these experimental results stimulated us to undertake quantitative calculations aimed at understanding the phenomenon.

We theoretically studied the phases of Ar and Kr adsorbed on zigzag and armchair NTs of different radii from 4.74 to 28.18 Å. We run GCMC simulations of the adsorption on the exterior of the NT only, as we assume that the NTs are closed like in most experiments. In our simulation, we used the anisotropic LJ potential, adapted from the graphite-He interaction proposed by Cole et al. [40, 41]

$$U_2^{aniso}(x) = 4\varepsilon_{aC}\left[\left(\frac{\sigma_{aC}}{x}\right)^{12}\left\{1+\gamma_R\left[1-\frac{6}{5}\cos^2\phi\right]\right\} - \left(\frac{\sigma_{aC}}{x}\right)^{6}\left\{1+\gamma_A\left[1-\frac{3}{2}\cos^2\phi\right]\right\}\right] \tag{7}$$

where φ is the angle between the interatomic separation vector and the normal to the plane, and γ_A, γ_R are parameters of the model. This potential is adequate for spherical or monatomic molecules; it includes the anisotropy of the C atoms on graphene, which originates in the anisotropic π-bonds.

Our results show that when the vapor pressure exceeds the "uptake pressure" (P_u), a monolayer condenses around the NT. The monolayer forms at a distance of approximately 3.2–3.6 Å from the NT surface (Fig. 1); its phase may be a fluid, a commensurate solid (CS) or incommensurate solid (IS), depending on the size and chirality of the NT. The value of P_u depends on the temperature, the adsorbate and the radius of the NT, but not on the chirality.

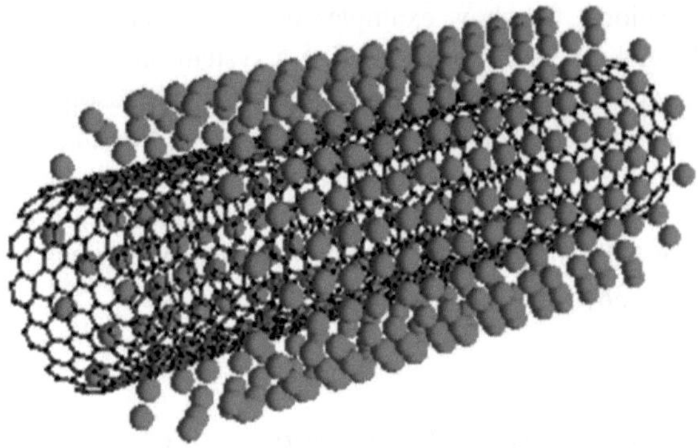

Fig. 1 Schematic view of the Kr monolayer covering a carbon NT

Our results indicate that Ar forms incommensurate phases on all the NTs considered [42]. For Kr, on the other hand, we found three commensurate solid phases of fractional coverage 1/3, 1/4 and 1/6. The 1/3 phase is the only commensuration found in an armchair NT. It forms on a 4.74 Å radius (7, 7) NT, the smallest NT considered, and it does not appear on the same-size zigzag NT (12, 0). The ¼ phase appears in the zigzag NTs of intermediate width (6.7–17.6 Å radii), while the phase is incommensurate in the same-radii armchair NTs. This result agrees with predictions by Kim et al. [43] showing that the corrugation of the gas surface interactions depends on the chiral angle; the chiral angle of zigzag and armchair NTs are 0 and 30° respectively. In wider zigzag NTs (25.83–28.18 Å), Kr forms a low density 1/6 phase. This can be explained observing that, for zigzag NTs (n, 0) with n > 40, the energy of the 1/6 lattice is lower than the energy of the ¼ lattice [26]. The occurrence of the 1/3 and 1/4 phases in narrow NTs is a consequence of the curvature of the surface. There are four distinctive "sites of adsorption" on the surface of the NT, labeled A, S, SP1 and SP2 in Fig. 2. Sites S, located above

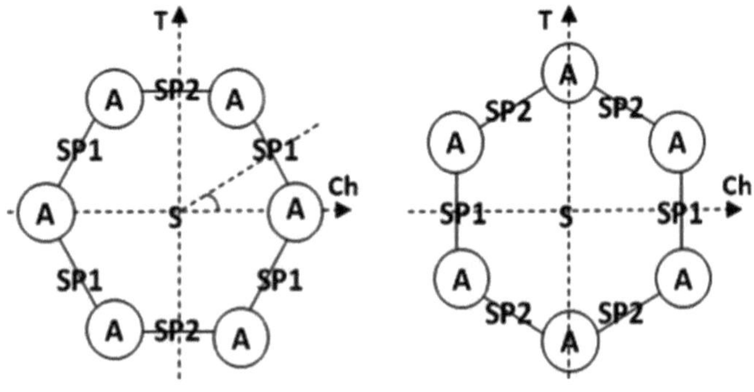

Fig. 2 Adsorption sites on an armchair (*left*) and zigzag (*right*) NTs. Sites A and S are located on top of the carbon atoms and at the center of the hexagons, respectively. The T axis is parallel to the axis of the NT (Adapted from Ref. [43])

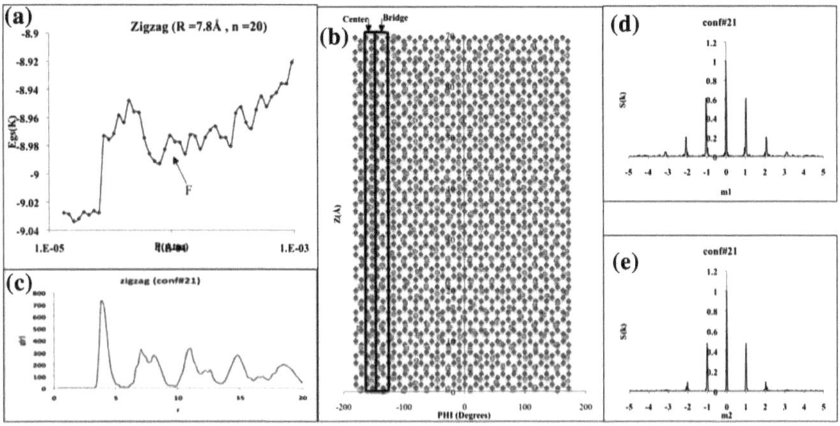

Fig. 3 **a** (*upper left*) Egs versus P(atm); **b** (*center*) Unrolled NT (*brown*) overlapped with the Kr adatoms (*green*); center (S) and bridge (SP1) sites are boxed; **c** (*lower left*) radial distribution function; **d** (*upper right*) Structure function-helical order ($m_2 = 0$) and **e** (*lower right*) longitudinal order ($m_1 = 0$)

the center of the hexagons are the most attractive ones, for all radii and chirality, including graphene or flat graphite. For NTs, the next favorable are SP2 followed by A sites, which are located on top of the carbon atoms, and SP1 are the least attractive. Notice that on graphene or graphite, the SP1 and SP2 sites are identical. The energy difference between the S and SP1 sites, usually referred as the "corrugation" of the potential, is small (approximately 50 K) compared to the Kr–Kr LJ energy parameter (171 K). Hence, the formation of commensurate phases responds to a delicate balance between the Kr-surface and Kr–Kr energies, modulated by the temperature. The 1/3 and ¼ phases are arrangements of S and SP1 sites. The 1/6 phase, on the other hand, is formed by S sites only.

We exemplify with the results of the simulation of Kr at 77.4 K on a (20, 0) zigzag NT, of radius 7.8 Å. In Fig. 3a, we show the gas-surface energy as a function of the pressure. We observe that the E_{gs} is a non-monotonic function of the pressure. It steps up as s result of a vapor to liquid transition that occurs at the uptake pressure P_u, and steps down at a higher pressure where the transition to a CS takes place (marked as F in the figure). In Fig. 3b, we examine the unrolled adatoms' lattice overlapped with the unrolled NT. The Kr atoms take positions at sites S (center) and SP1 (bridge). The commensuration was quantitatively confirmed by calculating the radial distribution function (Fig. 3c) and the structure function (Fig. 3d, e).

The *tour de force* experiments of the Cobden-Vilches group at the University of Washington (UW) provided an unprecedented opportunity to test our understanding of adsorption of gases on a *single* carbon nanotube [44]. Prior to their study, many theory/simulation investigations of adsorption on nanotubes, including our own, could be compared only with data taken from either nanotube bundles or nanotube mats, with attendant uncertainties about the geometry, the distribution of nanotube sizes and the effects of heterogeneity [22].

The UW data were remarkable, in particular, insofar as they exhibited vertical steps (within experimental error) in the adsorption isotherms, consistent with an essentially *perfect* single nanotube. Thus, the challenge was to see whether our simulations would yield results consistent with their data, which consisted of careful measurements of the number of adsorbed atoms (Ar or Kr, in their initial study) as a function of P and T.

3 Adsorption on Graphene

Recent advances in the synthesis of graphene have encouraged research of physical adsorption on graphene-based materials. Many groups have advanced on studies related to possible applications, including gas storage and separation. For example, Kutcha et al. [45] studied carbon-based nanospaces for the storage of H_2. They found that graphene-like porous carbons could meet the DOE goals even at room temperature if the substrate–adsorbate interaction is strengthened by a modification of the surface properties.

Palmer et al. [46] investigated numerically the separation of CO_2/CH_4 mixtures in Nanoporous Carbons at ambient temperature, finding optimal conditions for a slit pore of 0.61 nm and a (8, 8) single walled carbon nanotube bundle.

Although many theoretical studies had predicted that H_2 uptake would be enhanced by separating the carbon layers in graphite, it was not until recently that the group of T. Yildirim at NIST was able to design such a material; it was accomplished by pillaring graphene oxide (GO) planes [47, 48]. Furthermore, GO layers are linked together to form a novel layered structure by using the well-known reactions between boronic acids and hydroxyl groups. Such structures can have tunable pore widths, volumes, and binding sites depending on the linkers chosen. They can also exhibit interesting gas sorption properties. All these proposed materials are made of one or many layers of graphene with tunable separation.

To perform GCMC simulations we carefully select the model of the interactions:

3.1 Fluid-Graphene Interaction

Interactions with graphene layers may be described by continuous, atomistic or hybrid models. These models assume that the graphene lattice is rigid.

3.1.1 Continuous Model

The analogous of Steel's 10-4-3 potential for graphite [49] is the result of the integration of the LJ fluid-carbon interaction over a smooth sheet of carbon of areal density θ_c. This simple-model potential depends only on the distance to the graphene plane (z),

$$U^{cont}(z) = 2\pi\theta_c\varepsilon_{aC}\sigma_{aC}^2[(\frac{2}{5})(\frac{\sigma_{aC}}{z})^{10} - (\frac{\sigma_{aC}}{z})^4] \quad (8)$$

This simple version of the potential ignores corrugation and anisotropy; hence it is adequate in cases where those effects are not relevant. For example, when studying a multilayer substrate, the lower layers may be described by it.

3.1.2 Atomistic Model

The atomistic potential is computed as the sum of atom-carbon interactions:

$$U^a(r) = \sum_i U_{aC}(|r - R_i|) \quad (9)$$

where the vectors r and R_i are the position of the adatom and ith carbon atom of graphene, respectively. The function U_{aC} has to be chosen for each adsorbate.

3.1.3 Hybrid Model

In the hybrid model, the interaction with the nearest carbon atoms (within a distance r_{nn} to be determined) in the first layer is computed with an atomistic potential. The interaction with next carbon atoms in the first layer (ΔU^H) is computed as the integration over the remaining graphene sheet,

$$\Delta U^H = 2\pi\theta_c \int_{r_{nn}}^{\infty} U_{aC}(z^2 + \rho^2)\rho d\rho \quad (10)$$

Assuming that U_{aC} is the LJ potential, Eq. (10) becomes a function on the distance to the layer z,

$$\Delta U^H(z) = 2\pi\theta_c\varepsilon_{aC}\sigma_{aC}^6\left[\frac{2\sigma_{aC}^6}{5(z^2 + r_{nn}^2)^5} - \frac{1}{(z^2 + r_{nn}^2)^2}\right] \quad (11)$$

The second and following layers interact with the continuous potential given in Eq. (8) (Fig. 4). The parameter r_{nn} is determined by testing the hybrid potential against the completely atomistic one ($r_{nn} >> \sigma$). For $r_{nn} = 0$, the hybrid potential becomes identical to U^{cont}.

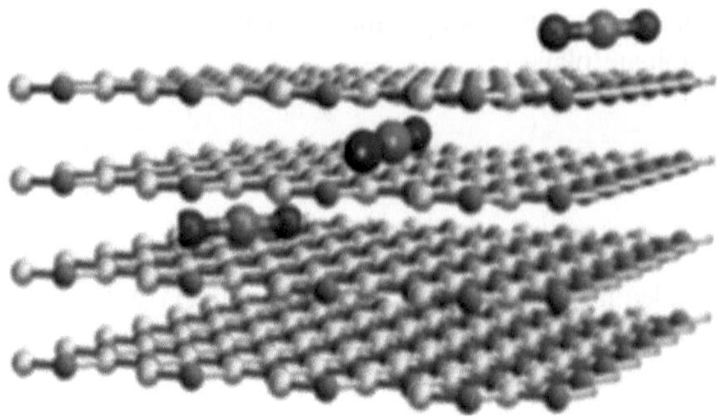

Fig. 4 Schematic picture of multilayer graphene and CO_2 adsorbed

3.2 Potential U_{aC} for Each Kind of Adsorbate

As in the case of carbon nanotubes, the anisotropic LJ potential given in Eq. (7) is adequate for spherical or monatomic molecules.

For CO_2, the electrostatic interaction with the quadrupole moment of the carbon atoms is added. Vernov and Steele [50] estimated the quadrupole moment of graphite from values of the quadrupole moments of benzene, naphthalene, and anthracene. Later, Whitehouse and Buckingham measured a value 30 % smaller than Vernov's (-3.03×10^{-40} Cm^2 per carbon atom) [51]. We note that the LJ dispersion energy dominates the CO_2-graphene interaction, therefore assumptions and approximations made for the value of the quadrupole moment, as well as the anisotropy parameters, are not critical.

For the case of water that is a polar molecule, one extra term is included, representing the interaction with the induced dipole of the carbon atoms. Zhao et al. [52] demonstrated that the polar terms are important for strongly polar fluids. The angle-averaged interaction is given by

$$\mu_{\mu i}(r) = -\frac{\mu_f^2 \alpha_c}{(4\pi\varepsilon_0)^2 r^6} \tag{12}$$

where ε_0 is the vacuum permittivity, μ_f is the dipole moment of the water molecule (1.85 D) and α_c is the angle-averaged polarizability of carbon atoms in graphene ($\alpha_c = 0.878$ Å^3 [53]). Karapetian et al. [54] studied the properties of a water cluster on graphite and concluded that using an anisotropic polarizability on the carbon atoms gave similar structures and binding energies to those obtained using an isotropic polarizability

As an example, we show the results of simulations of Kr on a rigid graphene sheet. The graphene is located in the bottom of the simulation cell; the boundary conditions are reflective in the top wall and periodic in the transverse walls, to simulate a 2D infinite substrate. The Kr-graphene energy is computed by the atomistic potential, and the Kr–Kr interaction is LJ. The vapor is assumed an ideal gas. In Fig. 5, we display the adsorption isotherms in the form of fractional

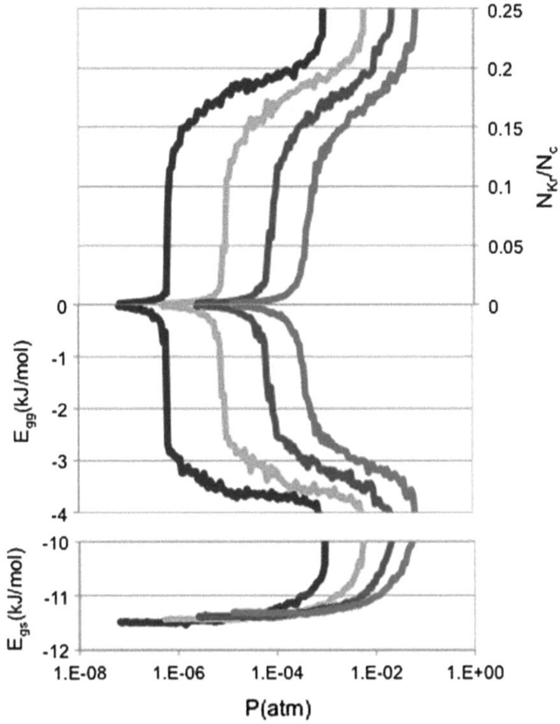

Fig. 5 Adsorption isotherms (*right axis*), gas–gas energy and gas–surface energy (*left axis*) for Kr on graphene at T = 77 K (*blue*), 87 K (*green*), 97 K (*purple*) and 107 K (*red*)

coverage, defined as N_{kr}/N_c, gas-gas energy and gas-surface energy as a function of the vapor pressure, for four temperatures from 77 to 107 K. At the lowest temperature, we observe a discontinuous raise in the coverage at the uptake pressure P_u. Simultaneously, the E_{gg} decreases (i.e. becomes more negative), due to the growing population of the Kr monolayer. The E_{gs}, on the other hand, takes the minimum value at zero coverage. The vertical step in coverage observed at T = 77 K indicates that the low-density vapor (V) and condensed liquid (L) phases coexist at P_u. The density gap between both phases becomes smaller as T increases until it disappears at T = 107 K, which corresponds to the critical temperature of the V → L transition. The uptake pressure P_u increases with the temperature, implying that the atoms need a stronger "push" to get adsorbed from the vapor. The minimum E_{gs} also increases with the temperature; at 0 K, it is equal to the minimum of the Kr-graphene potential energy. The density of the monolayer continuously increases from approximately 0.1 to 0.22. In Fig. 6, we show a snapshot of the Kr atoms overlapped with the carbon atoms. The data shows a triangular lattice of density $N_{kr}/N_c = 1/6$, we believe the phase is commensurate.

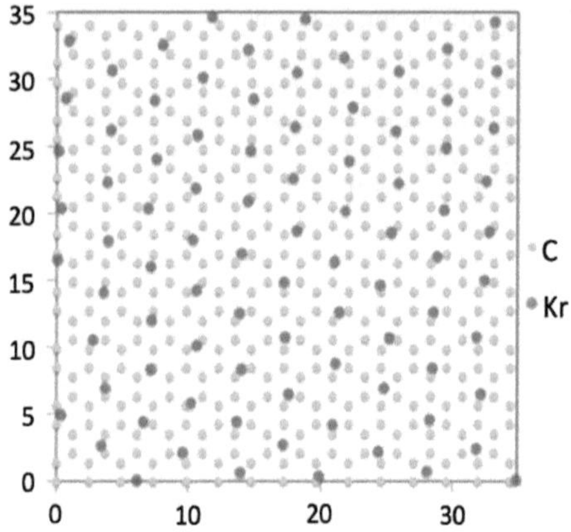

Fig. 6 Snapshot of the Kr atoms overlapped with the carbon atoms at T = 77 K, P = 5.0 × 10^{-6} atm

4 Gas Separation by Physical Adsorption in Metal Organic Frameworks

Gas separation by adsorption can be accomplished by three basic physical mechanisms: equilibrium, steric and kinetic [55]. Equilibrium mechanisms are based on the strength of the attraction of the gases to the substrate; the kinetic mechanisms are those based on the differences in the rates of adsorption and/or transport of the gas on and through the substrate; the steric mechanisms are based on the incompatibility between the size or shape of the pores in the substrate and those of the adsorbate gas molecules. Many materials have been used for this purpose during the past half-century, including carbon nanotubes, activated carbons and zeolites. In recent years, attention has been drawn increasingly to a wider variety of materials, the MOFs. MOFs consist of metal centers and/or metal clusters connected by organic linkers forming 3-D porous structures with 1D or 2D channels. The structure and pore diameters of MOFs are well determined and can be controlled by design, showing many potential advantages over other nanoporous materials. MOFs have exceptional properties due to their flexibility, and can be tailor-made, opening the possibility of designing materials that target specific mixtures of gases.

Although adsorption is one of the most promising methods for energy-efficient separation of gas mixtures, the use of MOFs for that purpose has been explored to a limited extent [56–60]. For example, Duren et al. [56] studied the extraction of n-butane from mixtures at 298 K, and proposed new, not yet synthesized materials with higher selectivity. Yang et al. [55] studied separation of $CO_2/N_2/O_2$ from flue gases in CuBTC ($Cu(BTC)_2(H_2\,O)_3$, (BTC) = benzene-1,3,5-tricarboxylate), finding a selectivity of approximately 20 up to 5 MPa and at 295 K; Wang et al. [57] studied mixtures of CO_2/CO, C_2H_4/CO_2 and C_2H_4/C_2H_6 in CuBTC, finding selectivity of approximately 20, 2 and 1 for CO_2/CO, $C2H_4/C_2H_6$, $CO_2/C2H_4$, respectively. Adsorption studies also help to increase the understanding we have of substrates when they are correlated with experimental results. The properties of adsorbates are thus one of the aspects of nanomaterials that need to be assessed to enable its applications and help bring closer its use in new technologies.

MOFs consist of metal ions connected by organic molecules to form scaffolding-like structures similar to a molecular Tinker toy. Behaving as molecular-scale sponges, these MOFs have wide-ranging potential uses for filtering, capturing or detecting molecules, such as carbon dioxide, or hydrogen storage for fuel cells.

MOFs are synthesized by a self-assembly process in which metal or metal oxide vertices are connected by functionalized organic linker molecules to form extended crystalline porous materials. Different combinations of vertices and linkers lead to materials with varying topologies, pore sizes and functionalities. These materials may have extensive open-framework structures resulting in the inclusion of guest species during synthesis. These species may be removed via desolvation and the resulting empty framework may maintain structural integrity giving a porous MOF material with a large apparent surface area. Other guest molecules can then be adsorbed into such porous structure. The MOF connectivities and topologies are controlled by the coordination preferences of the metal and ligand. Many groups currently investigate MOFs, experimentally, theoretically and computationally [61–66]. A number of reviews describe the experimental synthesis, characterization and applications of MOFs [67–69]. Since too many different MOFs have been synthesized to date, atomic simulations are needed to identify those that have required properties for targeted applications.

IRMOFs (isoreticular MOFs) are a "family" of MOFs with different linkers and unchanged metallic centers. Yaghi and coworkers developed IRMOFs [70]. These materials consist of zinc–oxygen complexes connected by carboxylate-terminated linkers and form a regular cubic lattice. All IRMOFs have a topology adapted from the prototype IRMOF-1 (also called MOF5) in which an oxide-centered Zn_4O tetrahedron is edged-bridged by six carboxylates to reticulate into a three-dimensional porous network (See Fig. 7). The density of these materials varies in small increments (~ 0.1) in the range 1.00 g/cm^3 for IRMOF-5 to 0.21 g/cm^3 for IRMOF-16. The percent free volume varies from 55.8 to 91.1 %. Remarkably, the lowest percent free volume obtained in this series exceeds that found in most open zeolites, such us faujasite [71] in which it is 45–50 %. Adsorption properties of IRMOFs have also received considerable attention [56, 64, 66]. Although most of these adsorption studies concentrate on CO_2 at room temperature, other gases have also been studied. For example, Dubbeldam et al. [64] reported NVT simulations of Ar and N_2 in IRMOF-1 at temperatures in the range 30–90 K. Duren et al. [56] and Keskin et al. [67] investigated the separation of methane/n-butane and CO_2/methane mixtures in IRMOFs at room temperature, suggesting that these are promising materials for such separations.

Our numerical study of adsorption of CO_2, CH_4 and H_2 in IRMOF-1 confirms that the material is suitable for gas separation [18]. We run GCMC simulations and applied the IAST approximation to calculate the selectivity of CO_2. We modeled the MOF cell as a cube with adsorption sites at the corners and on the edges (see Fig. 8). At each adsorption site, we located a point particle, with LJ parameters chosen to fit experimental data of the IRMOF-1. We also assigned electric dipole

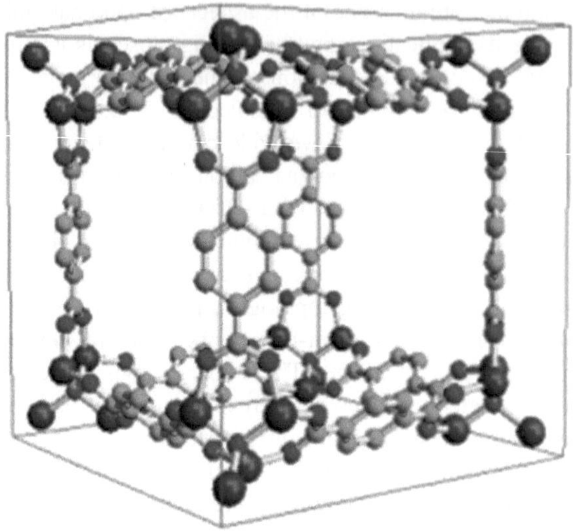

Fig. 7 The IRMOF-1 cubic cell has 8 Zn_4O corner clusters connected with 24 linker molecules; the Zn_4O tetrahedrons are linked by 1,4-benzene dicarboxylates. The size of the box is 1.3 nm. Zn in *purple*, O in *red*, C in *green*. Hydrogen is not shown for clarity

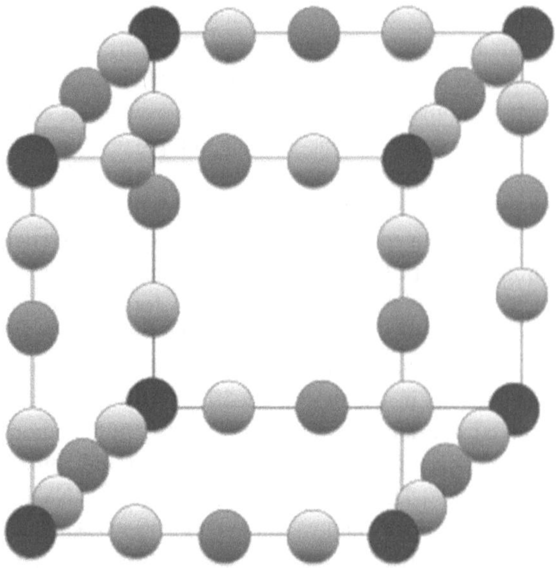

Fig. 8 Simple model of the IRMOF-1. *Red*, *blue* and *orange dots* represent three distinct sites of adsorption

and quadrupole moments to the point particles, to imitate the charge distribution of the IRMOF. We tested the effect on the selectivity of the temperature and variations of the size and multipoles. In Fig. 9 we compare the adsorption isotherms of CO_2 and CH_4 in a MOF with a cubic cell containing only LJ sites. In the left panel, we observe that in a compressed MOF the uptake pressure of CO_2 is significantly lower than the one of CH_4. The pressure gap is larger at 140 K than at 300 K.

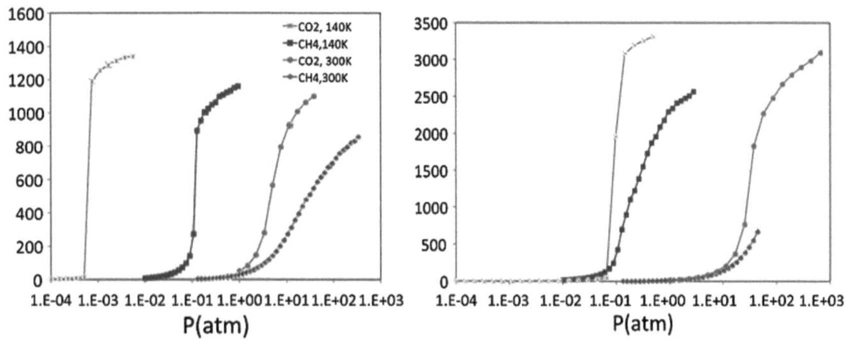

Fig. 9 Compared adsorption of CO_2 and CH_4 (in arbitrary units) in a 10 % compressed MOF (*left panel*) and 20 % expanded MOF (*right panel*), at T = 140 and 300 K as indicated in the legend (Adapted from Ref. [18])

In the expanded MOF (right panel), on the other hand, the pressure gap almost vanishes. Hence, the selectivity of CO_2 is boosted in the low temperature compressed MOF. We also found that inserting dipoles at the corner sites favors the selectivity. Adding quadrupoles at the edges, on the other hand, results insignificant. We conclude that three mechanisms can improve the MOF's capability to separate CO_2 from CH_4 and H_2: reducing the temperature, slightly compressing the MOF and inserting molecules with a finite dipole.

Acknowledgments We are grateful for the support of NSF DMR1006010, DMR1205608 and HRD1208880

References

1. Kim HK, Chan MHW (1984) Experimental determination of a two-dimensional liquid-vapor critical-point exponent. Phys Rev Lett 53:170–173
2. Alexander S (1975) Lattice gas transition of He on grafoil-continuous transition with cubic terms. Phys Lett A 54:353
3. Schick M (1980) Theory of helium monolayers. In: Dash JG, Ruvalds J (eds) Phase transitions in surface films. Plenum, New York, pp 65–113
4. Bretz M (1977) Ordered helium films on highly uniform graphite- finite size effects, critical parameters and 3-state Potts model. Phys Rev Lett 38:501–505
5. Bruch LW, Cole MW, Zaremba E (2007) Physical adsorption: forces and phenomena. Dover, Mineola
6. Bruch LW, Diehl RD, Venables JA (2007) Pogress in the measurement and modeling of physisorbed layers. Rev Mod Phys 79:1381–1454
7. Cole MW, Gatica SM, Kim H-Y, Lueking AD, Sircar S (2012) Gas adsorption in novel environments, including effects of pore relaxation. J Low Temp Phys 166:231–241
8. Kim H-Y, Gatica SM, Stan G, Cole M (2009) Effects of substrate relaxation on adsorption in Pores. J Low Temp Phys 156:1–8

9. Gatica SM, Kim H-Y (2009) Condensation of fluids confined in non-rigid nanopores: with a little help from the substrate. J Low Temp Phys 157:382–394. published on line
10. Kowalczyk P, Ciach A, Neimar A (2008) Adsorption-induced deformation of microporous carbons: pore size distribution effect. Langmuir 24:6603
11. Longhurst M, Quirke N (2007) Pressure dependence of the radial breathing mode of carbon nanotubes: the effect of fluid adsorption. Phys Rev Lett 98:145503
12. Do D, Nicholson D, Do H (2008) Effects of adsorbent deformation on the adsorption of gases in slitlike graphitic pores: a computer simulation study. J Phys Chem C 112:14075
13. Ravikovitch P, Neimark A (2006) Density functional theory model of adsorption on amorphous and microporous silica materials. Langmuir 22:11171
14. Gunther G, Schoen M (2009) Sorption strains and their consequences for capillary condensation in nanoconfinement. Molec Sim 35:138–150
15. Fletcher AJ, Thomas KM, Rosseinsky MJ (2005) Flexibility in metal-organic framework materials: impact on sorption properties. J Solid State Chem 178:2491
16. Greathouse JA, Kinnibrugh TL, Allendorf MD (2009) Adsorption and separation of noble gases by IRMOF-1: grand canonical monte carlo simulations. Ind Eng Chem Res 48:3425–3431
17. Simonyan1 VV, Diep1 P, Johnson JK (1999) Molecular simulation of hydrogen adsorption in charged single-walled carbon nanotubes. Chem Phys 111:9778. http://dx.doi.org/10.1063/1.480313
18. Maiga S, Medina M, Durodola OJ, Gatica SM (2013) Simulations of adsorption of CO_2 in MOFs and analysis of hypothetical MOFs. J Low Temp Phys 172(3–4):274–288
19. Jorgensen WL, Chandrasekhar J, Madura JD, Impey RW, Klein ML (1983) J Chem Phys 79:926
20. Watts RO, McGee IJ (1976) Liquid state chemical physics. Wiley, New York
21. Maitland GC, Rigby M, Smith EB, Wakeham WA (1987) Intermolecular forces: their origin and determination. Clarendon, Oxford, pp 565–566. Tables A 3.1 and A 3.2
22. Calbi MM, Cole MW, Gatica SM, Bojan MJ, Stan G (2001) Condensed phases of gases inside nanotube bundles. Rev Modern Phys 73:857
23. Gatica SM, Bojan MJ, Stan G, Cole MW (2001) Quasi-one and two-dimensional transitions of gases adsorbed on nanotube bundles. J Chem Phys 114:3765
24. Gatica SM, Cole MW (2005) Capillary condensation in cylindrical nanopores. Phys Rev 72:041602
25. Gatica SM, Li HI, Trasca RA, Cole MW, Diehl RD (2008) Xe adsorption on a C60 monolayer on Ag(111). Phys Rev B 77:045414
26. Kim HY, Cole MW, Mbaye M, Gatica SM (2011) Phase behavior of Ar and Kr films on carbon nanotubes. J Phys Chem A 115:7249–7257 (J. P. Toennies Festschrift)
27. Rapaport DC (2004) The art of molecular dynamics simulation. Cambridge University Press, Cambridge
28. Myers AL, Prausnitz JM (1965) Thermodynamics of mixed-gas adsorption. AIChE J 11:121
29. Babarao R, Hu Z, Jiang J, Chempath S, Sandler SI (2007) Langmuir 23:659
30. Krungleviciute V, Ziegler CA, Banjara SR, Yudasaka M, Iijima S, Migone AD (2013) Langmuir 29(30):9388–9397
31. Wang Z, Wei J, Morse P, Dash JG, Vilches OE, Cobden DH (2010) Phase transitions of adsorbed atoms on the surface of a carbon nanotube. Science 327:552–555
32. Cahn JW (1977) Critical-point wetting. J Chem Phys 66:3667–3672
33. Saam WF (2009) Wetting, capillary condensation and more. J Low Temp Phys 157:77–100
34. Shi W, Johnson JK, Cole MW (2003) Wetting transitions of hydrogen and deuterium on the surface of alkali metals. Phys Rev B 68:125401-1–125401-7
35. Gatica SM, Cole MW (2009) To wet or not to wet: that is the question. J Low Temp Phys 157:111–136
36. Gatica SM, Xiongce Z, Johnson JK, Cole MW (2004) Wetting transition of water on graphite and other surfaces. J Phys Chem B108:11704–11708
37. Friedman SR, Khalil M, Taborek P (2013) Private communication. Phys Rev Lett 111:226101

38. Calbi MM, Cole MW, Gatica SM, Bojan MJ, Johnson JK (2008) Adsorbed gases in bundles of carbon nanotubes: theory and simulation. In: Bottani EJ, Tascón JMD (eds) Chapter 9 of adsorption by carbons. Elsevier Science Publishing, Amsterdam, pp 369–402. ISBN: 0-08-044464-4
39. Wang ZH, Wei J, Morse P, Dash JG, Vilches OE, Cobden DH (2010) Phase transitions of adsorbed atoms on the surface of a carbon nanotube. Science 327:552
40. Carlos WE, Cole MW (1979) Anisotropic He–C pair interaction for a He atom near a graphite surface. Phys Rev Lett 43:697–700
41. Carlos WE, Cole MW (1980) Interaction between a He atom and a graphite surface. Surf Sci 91:339–357
42. Mbaye MT (2014) Ph.D. thesis. Howard University
43. Kim H-Y, Booth EC, Mbaye MT, Gatica SM (2014) Contribution of chirality to the adsorption of a Kr atom on a single wall carbon nanotube. J Low Temp Phys 175:590–603
44. Lee H-C, Vilches OE, Wang Z, Fredrickson E, Morse P, Roy R, Dzyubenko B, Cobden DH (2011) Kr and ^{4}He adsorption on individual suspended single-walled carbon nanotubes. J Low Temp Phys. doi:10.1007/s10909-012-0642-3
45. Kuchta B, Firlej L, Pfeifer P, Wexler C (2010) Numerical estimation of hydrogen storage limits in carbon based nanospaces. Carbon 48:223–231
46. Palmer JC, Moore JD, Roussel TJ, Brennan JK, Gubbins KE (2011) Adsorptive behavior of CO2, CH4 and their mixtures in carbon nanospace: a molecular simulation study. Phys Chem Chem Phys 13:3985–3996
47. Burress JW, Gadipelli S, Ford J, Simmons JM, Zhou W, Angew TY (2010) Graphene oxide framework materials suggested for hydrogen storage and carbon capture. Chem Int Ed 49:8902–8904
48. Srinivas G, Burress JW, Ford J, Yildirim T (2011) Porous graphene oxide frameworks: synthesis and gas sorption properties. J Mater Chem 21:11323
49. Steele WA (1973) Surf Sci 36:317
50. Vernov A, Steele WA (1992) The electrostatic field at a graphite surface and its effect on molecule: solid interactions. Langmuir 8:155–159
51. Whitehouse DB, Buckingham AD (1993) Experimental determination of the atomic quadrupole moment of graphite. J Chem Soc Faraday Trans 89:1909–1913
52. Zhao X, Johnson JK (2005) Mol Simul 31(1):1–10
53. Chang T-M, Dang LX (1996) J Chem Phys 104:6772
54. Karapetian K, Jordan KD (2003) Properties of water clusters on a graphite sheet. Springer, New York, p 139
55. Yang RT (1997) Gas separation by adsorption processes. Chem Eng 1:352
56. Düren T, Snurr RQ (2004) Assessment of isoreticular metal-organic frameworks for adsorption separations: a molecular simulation study of methane/*n*-butane mixtures. J Phys Chem B 108:15703
57. Wang S, Yang Q, Zhong C (2008) Adsorption and separation of binary mixtures in a metal-organic framework Cu-BTC: a computational study. Sep Purif Technol 60:30
58. Yang Q, Zhong C (2006) Molecular simulation of carbon dioxide/methane/hydrogen mixture adsorption in metal-organic frameworks. J Phys Chem B 110:17776
59. Babarao R, Hu Z, Jiang J, Chempath S, Sandler SI (2007) Storage and Separation of CO_2 and CH_4 in Silicalite, C168 Schwarzite, and IRMOF-1: a comparative study from monte carlo simulations. Langmuir 23:659
60. Krungleviciute V, Lask K, Migone AD (2008) Kinetics and equilibrium of gas adsorption on RPM1-Co and Cu-BTC metal-organic frameworks: potential for gas separation applications AIChE J 54:918
61. Krungleviciute V, Lask K, Heroux L, Migone AD, Lee J-Y, Li J, Skoulidas A (2007) Argon Adsorption on Cu_3(Benzene-1,3,5-tricarboxylate)$_2$(H2O)$_3$ metal-organic framework. Langmuir 23:3106
62. Garberoglio G (2007) Computer simulation of the adsorption of light gases in covalent organic frameworks. Langmuir 23:12154

63. Walton KS, Millward AR, Dubbeldam D, Frost H, Low JJ, Yaghi OM, Snurr RQ (2008) Understanding inflections and steps in carbon dioxide adsorption isotherms in metal organic frameworks. J Am Chem Soc 130:406
64. Dubbeldam D, Frost H, Walton KS, Snurr RQ (2007) Molecular simulation of adsorption sites of light gases in the metal-organic framework IRMOF-1. Fluid Phase Equilib 261:152
65. Keskin S, Sholl DS (2007) Screening metal-organic framework materials for membrane-based methane/carbon dioxide separations. J Phys Chem C 111:14055
66. Greathouse JA, Kinnibrugh TL, Allendorf MD (2009) Adsorption and separation of noble gases by IRMOF-1: grand canonical monte carlo simulations. Ind Eng Chem Res 48:3425–3431
67. Keskin S, Liu J, Rankin RB, Johnson JK, Sholl DS (2009) Progress, opportunities, and challenges for applying atomically detailed modeling to molecular adsorption and transport in metal-organic framework materials. Ind Eng Chem Res 48:2355
68. Rowsell JLC, Yaghi OM (2004) Metal-organic frameworks: a new class of porous materials. Microporous Mesoporous Mater 73:3
69. Mueller U, Schubert M, Teich F, Puetter H, Schierle-Arndt K, Pastre J (2006) Metal organic frameworks—prospective industrial applications. J Mater Chem 16:626
70. Eddaoudi M, Kim J, Rosi N, Vodak D, Wachter J, O'Keeffe M, Yaghi OM (2002) Systematic design of pore size and functionality in isoreticular MOFs and their application in methane storage. Science 295:479
71. Bennet MJ, Smith JV (1968) Mater Res Bull 3:633

Atom-Precise Metal Nanoclusters

Anu George and Sukhendu Mandal

Abstract A nanocrystal is a crystallite with size greater than about 2 nm. Nanoclusters are non-crystalline nanoparticles that are typically small and composed of a specific number of metal atoms in the core, which are protected by a shell of ligands. Optical properties of large metal nanoparticles in external electromagnetic fields are a function of their size, free-electron density and dielectric function relative that of the surrounding medium. The ultra-small size of nanoclusters allows them to exhibit distinct quantum confinement effects, which in turn results in their discrete electronic structure and molecular-like properties, such as HOMO-LUMO electronic transitions, enhanced photoluminescence, and intrinsic magnetism, to name a few of the characteristics. Metal nanoclusters play an important bridging role between nanochemistry and molecular chemistry. A basic understanding of the structure, electronic and optical properties, as the materials evolve from the atomic state to nanoclusters to fcc-structured nanocrystals, constitutes a major evolution across length scales, and leads to fundamental insights into the correlation between the structure and key characteristics of metal nanoclusters.

Keywords Nanocrystal · Metal nanoclusters · Nanostructures · Quantum confinement · Nanochemistry

1 Introduction

Nanoscale materials are generally classified based on three different length scales into three size domains: nanoparticles, nanocrystal and clusters [1, 2]. Nanoparticles are broadly defined as nanoscale particles ranging from 1 to 100 nm [3]. The term “nanocrystal” is defined as a nanoscale crystallite with size >2 nm.

A. George · S. Mandal (✉)
Indian Institute of Science Education and Research, Thiruvananthapuram, Kerala, India
e-mail: sukhendu@iisertvm.ac.in

P. Misra (ed.), *Applied Spectroscopy and the Science of Nanomaterials*,
Progress in Optical Science and Photonics 2, DOI 10.1007/978-981-287-242-5_7

Nanoclusters (NCs) are those non-crystalline nanoparticles, which are typically very small and are composed of a specific number of metal atoms in the metal core, which are protected by shell of ligands. NCs are non-crystallographic in nature due to lack of translational symmetry (presence of fivefold symmetry destroys the translational symmetry) [4]. Optical properties of large metal nanoparticles to external electromagnetic fields depend on their sizes, free-electron density and therefore their nearly bulk-like dielectric function relative to that of the surrounding medium. This behaviour can be quantitatively described by Mie theory. When particle size decreases and approaches to the electron beam free path ($\sim$50 nm for Au and Ag), the dielectric function and refractive indices become strongly size dependent. In that case Mie theory provides an adequate explanation with proper modification. When particle size approaches to sub nanometer size regime (the Fermi wavelength of an electron, or $\sim$0.5 nm for Au and Ag)—the optical, electronic, and chemical properties of metal NCs differ dramatically from the other two size regimes [1, 2, 5–7]. The ultra-small size of these NCs induces distinctive quantum confinement effects, which result in discrete electronic structure and molecular-like properties [8–10], such as HOMO-LUMO electronic transition, enhanced photoluminescence, intrinsic magnetism, and many others [11–13].

These NCs bridge small organometallic complexes and larger crystalline metal nanoparticles, which involves lot of transitions in terms of electronic structure and material properties. Understanding the evolution as a function of cluster size is one of the grand challenges in chemistry and physics in recent time.

In this book chapter, we discuss the bridging roles of metal NCs between molecular chemistry and nanochemistry. Fundamental understanding of the evolution of the structure, electronic, and optical properties as the materials evolve from the atomic state to NCs to fcc-structured nanocrystals, is of paramount importance and constitutes a major task in nano and materials science research. The evolution across length scales will lead to the fundamental insights into structure–property correlation.

Providing this missing link between atomic and nanoparticle behavior in noble metals, highly fluorescent gold NCs smoothly link the optical- and electronic-structure transitions from atoms to nanoparticles with observable free-electron behavior, and likely serve as a guide in understanding the even brighter emission from other noble NCs.

In this book chapter we will start with different syntheses methods, then structural aspect of these materials followed by fluorescence behavior.

2 Synthetic Methods

As we go to the sub-nanometer size, well known for various surface properties and applications, the controlled synthesis of the metal clusters become an important criteria because of the sensitivity of the cluster's properties to the number of atoms. Generally, the synthesis of metal NCs is approached by two different routes.

One is the bottom-up approach where the metal precursor ions are reduced by suitable reducing agent to form zero valent atoms and the nucleation of these atoms forming atom-precise clusters. The other way is the top-down approach which includes the ligand induced etching of larger nanoparticles giving clusters. Experimental parameters and reaction conditions such as metal to ligand ratio, nature of reducing agents, structure of the protecting ligands, reaction temperature, pH and time are adjusted to control the core size and surface properties of the metal NCs.

These cluster materials were synthesized by both gas phase and solution methods. Gas phase clusters which are formed when a bulk metal is heated to near its melting point, the emitted atomic vapor contains dimers (M_2), trimers (M_3), tetramers (M_4) and larger ones M_n, with n up to even a few hundred, but the yield drastically drops with increasing size. In contrast to the gas phase clusters which are bare (without ligand protection), solution phase clusters are protected with various ligands (e.g., phosphines, amines, thiols, etc.) which form a monolayer on the cluster surface and prevent the clusters from aggregation in solution or in the solid phase.

Now we will elaborate few synthetic methods in solution phase. These are below.

2.1 Modified Brust–Schiffirin Methods

Brust et al. in 1994 provided an easy and effective method for the synthesis of monolayer protected metal cluster which was later widely applied to various areas of nanoscience and nanotechnology [14]. In the past decade different types of modified Brust–Schiffirin methods have been developed to prepare metal and semiconductor NCs. In this method, first metal precursor is dissolved in an aqueous solution and then transferred to an organic solvent by a phase transferring agent such as tetraoctylammonium bromide and the subsequent addition of organic ligands and the reducing agent generate metal NCs. The above method is two phases method but Brust et al. also developed one phase method that is carried out in a polar solvent. By using Brust–Schiffirin methods (one phase and two phases), Au NCs [15] with different core sizes and then other metal clusters such as Pt [16], Ag [17], Cu [18] have been successfully synthesized with different protecting ligands and reducing agents.

2.2 Template-based Syntheses

Template-based synthesis exhibits various advantages for the preparation of fluorescent sub-nanometer clusters such as precise cluster size control, excellent stability of the clusters against aggregation, biocompatibility etc. This synthesis method makes use of polymers, polyelectrolytes, proteins, dendrimers and DNA as the templates. These templates provide predetermined environment favorable to produce NCs with well controlled size and shape. Moreover, these clusters are widely used for the catalysis due to their available active surface area. Polyamidoamine (PAMAM) as template had made considerable attention in the synthesis of highly fluorescent Cu, Au, Ag, Pt and other transition metal clusters. The core shell structure of PAMAM dendrimers provides an ideal template to produce and stabilize metal NCs [19]. In this particular method, the metal ion precursor and the dendrimer is co-dissolved in a solvent in order to trap the metal ions into the interior of the dendrimer and is chemically reduced using suitable reducing agents with suitable pH of the medium.

2.3 Precursor- or Ligand-Induced Etching Method

The ligands used for stabilizing the NCs such as thiols, amines, phosphines, polymers etc. play multiple roles in determining the cluster size, optical and electronic properties. The metal core can be etched by excessive ligands due to the strong interaction between the metal atoms and the protecting ligands. Size focusing methods on the basis of ligand induced etching have been used to synthesize atomically monodisperse metal NCs [20]. With the ligand etching process, the smallest metal NCs can also be synthesized. In addition to ligand etching based size focusing, a precursor-induced etching method have also been developed for metal NC syntheses [21]. In this method, the metal nanoparticles were synthesized in an organic medium like toluene and were etched into small NCs by adding metal precursor solution drop by drop to the particle solution. The most common one is the thiol etching method which produced atomically monodispersed metal NCs.

2.4 Microemulsion Methods

Microemulsions are thermodynamically stable, isotopic liquid mixtures of oil, water, surfactant, frequently in combination with co-surfactant. The three basic types of microemulsions are direct (oil dispersed in water), reversed (water dispersed in oil) and bi-continuous which provide a liquid core shell structure. The reversed microemulsion has been demonstrated as an effective and reproducible system to synthesize photoluminescent NCs with a narrow size distribution [22]. The homogeneous metal NCs could be formed and restricted inside the liquid core, and the cluster size can be easily tuned by adjusting the liquid core dimensions of microemulsions.

2.5 Electrochemical Syntheses

Electrochemical method like other methods has also been widely used to prepare metal NCs. The process of this method was introduced by Reetz et al. [23]. The metal source used here is a sacrificial anode and the metal ions produced are reduced at the cathode giving rise to metal NCs stabilized by surfactants in the electrolytic solution. The advantage of this method over the chemical reduction is low reaction temperature, large scale yield and easy manipulation of the cluster size by tuning current, voltage, electrolyte concentration of stabilizers etc. NCs synthesized by this method are very stable over several years.

2.6 Solid State Syntheses

Apart from the synthesis in aqueous or organic solutions, solid state synthesis has been recently employed in the formation of metal NCs. This method has various advantages over solution phase syntheses. The solid state route overcomes the difficulty in controlling the particle size due to the fast reduction process in solvents. This process of synthesizing cluster is simple and involved three steps. Recently Ag_9 quantum cluster [24] was synthesized through the solid state route by grounding a mixture of silver nitrate and mercaptosuccinic acid in the solid state until a change of color was observed. Sodium borohydride in solid form was added, and the mixture was ground, resulting in the formation of brownish black powder that showed strong affinity to water. The cluster solution was formed with a strong effervescence and was precipitated by the addition of excess ethanol. The highly stable nanoclusters under inert atmosphere can be obtained in gram scale by the solid state route.

2.7 Synthetic Methods of NCs

See Table 1.

3 Single Crystal Structure

Bulk metal and metal nanocrystals have a face-centered cubic (fcc) structure. NCs do not exhibit translational symmetry in terms of atomic arrangement, indeed, many of them possess fivefold symmetry (e.g., icosahedron-based), which destroys the translational symmetry and hence such clusters are non-crystallographic and adopt icosahedron motif. The icosahedral structure of M_{13} clusters can indeed be transformed from the fcc structure through the formation of cuboctahedral structure.

Table 1 Different synthetic methods of metal nanoclusters

Compounds	Method of synthesis	Reagents used	Reaction condition	References
1. Au_8, Au_{13}	Surfactant free synthesis	$HAuCl_4.4H_2O$, DMF	140 °C	Langmuir (2010) 26(8):5926
			6 h	
2. $Au_{16–31}$		$HAuCl_4.3H_2O$, EtOH, CH_2Cl_2, phenylethanethiol, $NaBH_4$	0 °C	J Phys Chem C (2010) 114 (2):741
			24 h	
1. Au_{25}	Phase transfer synthesis	$HAuCl_4.3H_2O$, TOAB, toluene, $PhCH_2CH_2SH$, $NaBH_4$	0 °C	J Am Chem Soc (2008) 130:1138
2. Au_{39}, Au_{40}			~12 h	
		$HAuCl_4.3H_2O$, TOAB, toluene, $PhCH_2CH_2SH$, $NaBH_4$	0 °C	Nanoscale Res Lett (2012) 7:277
			~12 h	
1. Au_{25}	One step synthesis	$HAuCl_4.3H_2O$, THF, $PhCH_2CH_2SH$, $NaBH_4$	0 °C for 3 h	J Mater Chem (2009) 19:622
			RT for 60 h	
2. $Ag_7(DMSA)_7$		$AgNO_3$, EtOH, DMSA, $NaBH_4$	0 °C for 4 h	J Am Chem Soc (2009) 131:16672
			RT for 12 h	
3. $Au_{25}(BSA)$		$HAuCl_4.3H_2O$, BSA, NaOH	37 °C for 12 h	J Am Chem Soc (2009) 131:888–889
4. $Au_{19}SG_{13}$		$HAuCl_4.3H_2O$, EtOH, $PhCH_2CH_2SH$, toluene, borane tert-butylamine complex	0 °C for 30 min	J Am Chem Soc (2011) 133:9670
			RT for 60 h	
5. $Au_{25}SG_{18}$		$HAuCl_4.3H_2O$, EtOH, $PhCH_2CH_2SH$, toluene, $NaBH_4$	0 °C for 30 min	J Am Chem Soc (2011) 133:9670
			RT for 60 h	

(continued)

Table 1 (continued)

Compounds	Method of synthesis	Reagents used	Reaction condition	References
1. $Ag_{44}(SeR)_{30}$	Solid state synthesis	$AgNO_3$, Benzeneselenol, $NaBH_4$	RT 15 min	J Phys Chem Lett (2013) 4:3351
2. $Ag_{32}(SG)_{19}$		$AgNO_3$, GSH, $NaBH_4$	RT 15 min	Nanoscale (2013) 5:9404
1. Au_8	From nanoparticles	$HAuCl_4$, CH_3OH, MSA, $NaBH_4$, GSH	0 °C for 30 min	Nano Res (2008) 1:333
			55 °C for 12 h	
2. Au_{15}	In cavities	$HAuCl_4$, CH_3OH, GSH, $NaBH_4$, Cyclodextrin	0 °C for 30 min	Chem Mater (2011) 23:4
			70 °C for 48 h	
1. Au(BSA)	Microwave irradiation	$HAuCl_4$, BSA, NaOH	300 W 6 min	J Mater Chem (2012) 22:1000

Transformation from cuboctaheron to icosehadron structure leads to an appreciable energy gain, which further stabilizes the icosahedral structure. Presence of both the radial and peripheral bonds is of importance in stabilizing the icosahedral structure. The icosahedron possesses twofold, threefold, and fivefold rotation axes. An interesting issue is at what threshold size icosahedral structure will disappear. The structural transformation from icosahedral to fcc structure leads to reduction of the number of peripheral M–M bonds. Thus, when the surface energy of NCs becomes optimum at a certain size, this transformation would spontaneously occur, but the threshold size is still unknown.

It is challenging but exciting to determine the crystal structures of NCs. It seems that every size of NCs comes as a surprise in terms of their structure. To determine the crystal structure of NCs, the best and most reliable approach is still X-ray crystallography. Unfortunately, to grow high-quality single crystals of NCs is, in general, very difficult, albeit breakthroughs have been made recently in the cases of Au_{102}, Au_{25}, Au $_{23}$ etc. clusters with thiols or phosphine based ligands. The single crystal structure can reveal information about the structural relation with NC and larger counterpart (e.g., bulk or nanocrystals). The crystal structure of Au_{102}(p-MBA)$_{44}$, where p-MBA = SPhCOOH, shows that it contains Au_{49} decahedral kernel and the rest of the 53 Au atoms can be divided into three groups: (1) 30 Au atoms form two symmetry-equivalent 15-atom rhombicosidodecahedral fragments and face-cap the Au_{49} kernel under pseudo D_{5h} symmetry, forming the Au_{79} kernel; (2) 10 Au atoms cap the 10 square-shaped Au facets of the Au_{79} kernel under pseudo D_{5h} symmetry, forming a protecting layer; (3) 13 Au atoms are multiply connected as part of a steric/electronic coating to the Au atoms of the Au_{79} kernel. The 44 p-MBA ligands interact not only with gold but also with one another, forming a rigid surface layer.

X-Ray crystallographic analysis of Au_{25} cluster structure features a centered icosahedral Au_{13} core (kernel), which is further capped by a second shell comprised of the remaining twelve Au atoms. The entire Au_{25} cluster is protected by 18-SR ligands. As Au nanocrystals (>2 nm) typically exhibit fcc structure (proved by powder X-ray diffraction pattern), the icosahedral structure (observed in Au_{25}) must disappear at some threshold size with increasing cluster size and transform to close-packed layer structures. It also remains to be seen whether other types of Au–S bonding modes appear in other sized $Au_n(SR)_m$ clusters. On the other hand, whether gold thiolate clusters can adopt hollow cages or helical structures is an interesting question. Some attempts have made through theoretical calculation but far from the experimental results.

4 Optical Properties of Noble Metal Clusters

Drude has explained the electrical and thermal conductivities of metals by modifying the kinetic theory of gases to account for the highly increased electron densities and interactions of metals. The Drude model considers the valence electrons of metals to be free electrons, due to strong electron-screening effects. These free electrons are delocalized in bulk metals and do not reside on any specific metal atoms. As a result, free electrons move in the constant potential field provided by the positively charged cores.

In case of metals the highest-electron energy of the uppermost filled level is called the Fermi energy (E_f), which is independent of the metal size. The displacement of free electrons in an electric field gives rise to polarization changes at the nanoparticle surface with a linear restoring force between electrons and positive charges, which is applicable for all size scales. Differences in size-dependent optical response mainly arise from the change in the number of free electrons. When metal nanoparticle sizes become much smaller than the electron mean free path, both the absorption frequency and line width have simple size dependences and are quantitatively related to the inverse particle radius, R^{-1}. Consequently, when the metal nanoparticle size approaches the Fermi wavelength, the continuous-band structures of metals break up into discrete energy levels. Kubo predicted quantitatively the electronic structure of very small metal clusters based on the recognition that the quasi-continuous electron energy states of bulk metals become discrete on the few-atom scale. Owing to the strong confinement of free electrons in this size regime, NCs possess multiband step-wise optical absorption behavior instead of collective plasmon excitation, and exhibit interesting molecule-like properties such as quantized charging, magnetism, and strong luminescence. The luminescence properties of the noble metal clusters are discussed below.

5 Luminescent Clusters

Nobel metal NCs have emerged as novel luminescent nanomaterials because of their better performance in many aspects like biocompatibility, photostability, and non-toxicity relative to organic dyes and semiconductor quantum dots. With the rapid development of synthesis strategies, the luminescence properties of metal NCs have fascinated the scientific community. There are a number of novel synthetic methods developed to fabricate highly luminescent gold/silver (Au/Ag) NCs. Using suitable agents for stabilizing clusters from aggregating and enhancing their fluorescence is also of key importance for obtaining highly fluorescent NCs.

The one pot synthesis of protein and thiol protected luminescent Au clusters was widely studied. A green chemical synthesis of highly fluorescent BSA@Au (Bovine Serum Albumin—BSA) clusters with red emission was reported by Xie et al. [25]. Choudhari et al. [26] synthesized the red emitting NLf@Au (Native Lactoferrin—NLf) clusters in an alkaline pH condition and studied the concentration dependent time evolution of these clusters in the protein templates. Hence conditions during the synthesis (e.g., pH) can also have significant effects on the size, structure, and other properties of the resultant protein-protected Au NCs. Accurate structural determination of the protein-protected NCs, which are crucial for the study of their luminescence properties is lacking because of difficulties in their structural characterization due to relatively wide distribution of sizes and the difficulty in obtaining single crystals of the protein-protected Au NCs for X-ray crystallography. Lie et al. [27] successfully synthesized highly red emitting BSA–Au NCs and HSA–Au NCs (Human Serum Albumin—HSA) within a very short time by microwave irradiation. Xi Yang et al. [28] reported the synthesis of water-soluble, monodispersed, and bluish green emitting Histidine@Au NC by an ultra-facile one-step reaction. These clusters are possibly used to fabricate fluorescent biosensors.

Fluorescent Au NCs stabilized with various other thiols such as tiopronin [29], phenylethylthiolate [30], thiolate α-cyclodextrin [31] and 3-mercaptopropionic acid [32] have also been reported. Nienhaus group [33] recently reported a facile synthesis of water-soluble fluorescent Au NC using a mild reductant, tetrakis (hydroxymethyl)phosphonium chloride (THPC) and D-penicillamine (DPA) as the stabilizer emitted bright orange-red fluorescence at 610 nm as shown in Fig. 1. The obtained DPA-capped Au NC was used for cellular imaging applications as in Fig. 2.

Apart from the above mentioned one-step strategy luminescent metal NCs can also be produced by etching large gold nanoparticles by thiols. Pradeep et al. [34] synthesized two fluorescent magic number clusters, Au_{25} and Au_8 by adjusting the pH in mercaptosuccinic acid protected gold nanoparticles via ligand etching with glutathione. These clusters can be viable option for in vivo studies since they are highly biocompatible due to their low metallic content. There are reports where the NCs exhibit pronounced red photoluminescence only after phase transfer to aqueous solution. Lin et al. [21] reported the synthesis of Au NCs based on precursor induced Au NP@DDAB (didodecyldimethylammonium bromide—DDAB) etching in organic phase and ligand exchange with reduced lipoic acid (dihydrolipoic acid—DHLA) to transfer the particles to aqueous solution as shown in Fig. 3.

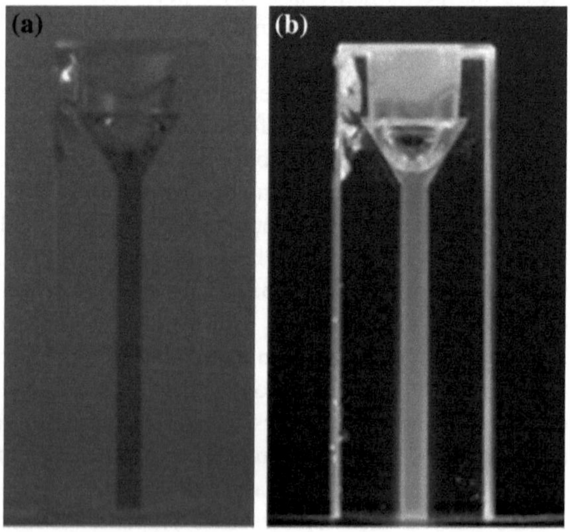

Fig. 1 *Photographs* of an as-prepared aqueous solution of DPA–AuNCs, **a** in visible light and **b** under a UV light source with wave-length 365 nm. Reprinted with permission from Ref. [33] © 2010 the Royal Society of Chemistry

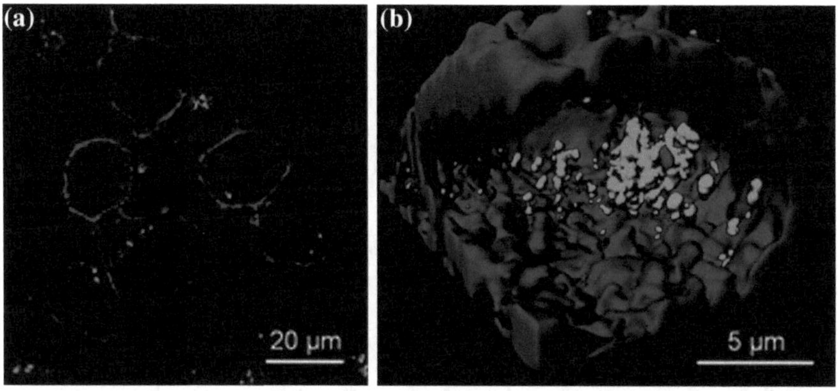

Fig. 2 **a** Confocal *image* of HeLa cells after incubation with DPA–AuNCs for 2 h. **b** Cross-section of a 3D image reconstruction, showing internalized DPA–AuNCs. Membranes were stained with the *red* dye DiD. *Images* were taken by 2-photon excitation at 810 nm. Reprinted with permission from Ref. [33] © 2010 the Royal Society of Chemistry

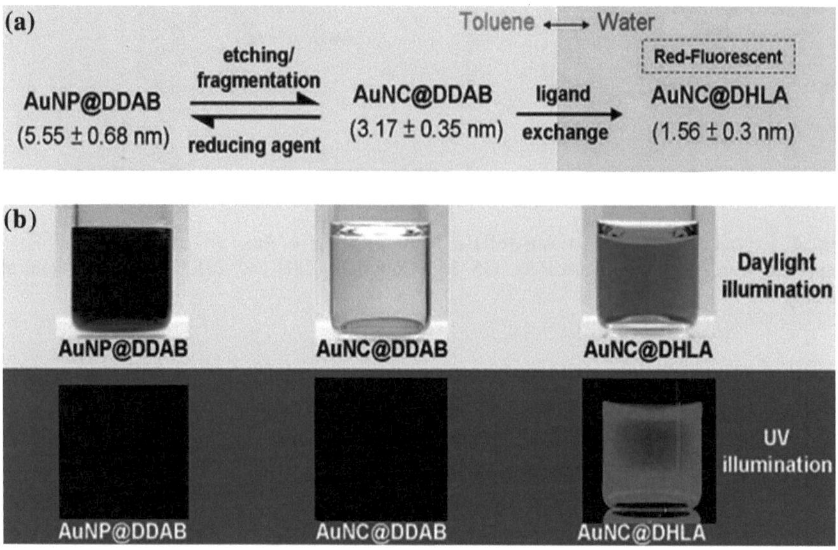

Fig. 3 **a** General strategy to fabricate water soluble fluorescent Au nanoclusters. **b** *Pictures* of particle solutions under daylight and UV light. Reprinted with permission from Ref. [21]. Copyright 2009 American Chemical Society

DHLA@Au NC as synthesized exhibited less photobleaching than organic fluorophores. However it was more prone to photobleaching than the semiconductor quantum dots, which show an excellent photostability.

Single crystal structure of several thiolate–Au NCs [e.g., $Au_{25}(SR)_{18}$ [35], $Au_{36}(SR)_{24}$, $Au_{38}(SR)_{24}$, [36] and $Au_{102}(SR)_{44}$ [37] are shown in Fig. 4 underscore

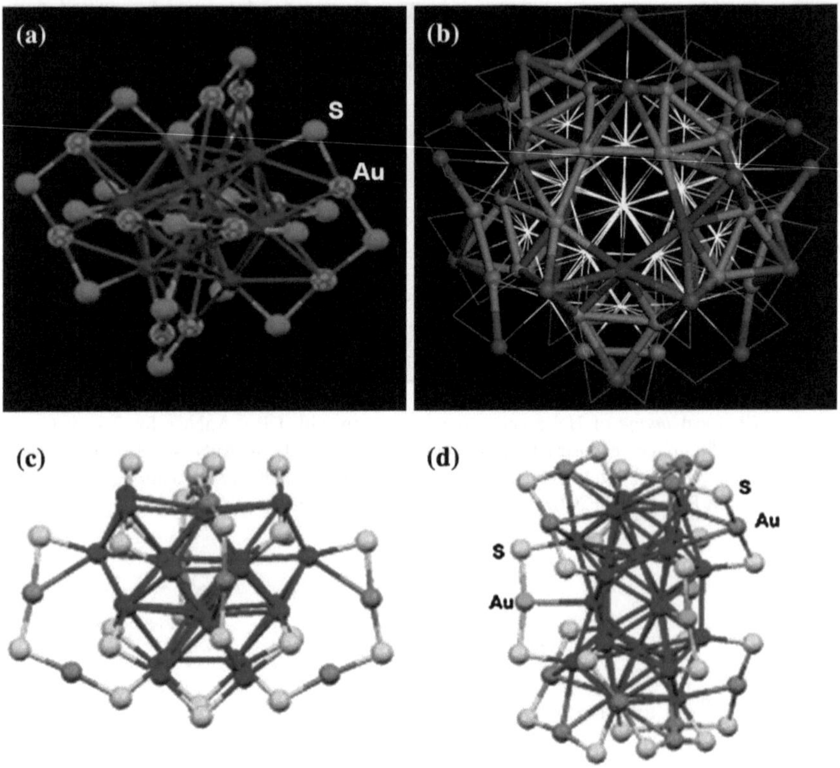

Fig. 4 Crystal structures of **a** $Au_{25}(SR)_{18}$, **b** $Au_{102}(SR)_{44}$ **c** $Au_{36}(SR)_{24}$ and **d** $Au_{38}(SR)_{24}$. Reprinted with permission from Refs. [35–37]. Copyright 2008 and 2013 American Chemical Society

the significant effects of the thiolate ligand or thiolate–Au^{1} complexes on the luminescence of the NCs. Chang et al. [38] suggested that the fluorescence of Au NCs, prepared via thiol-induced etching, mainly originated from Au NCs/polynuclear Au(1)-thiol (core/shell) complexes, based on the fact that these clusters displayed a strongly Stokes-shifted luminescence with long lifetimes, which is characteristic of thiol–Au(I) complexes that display ligand-metal charge transfer and metal–metal interactions.

There are also several recent reports on the syntheses of polymer-stabilized fluorescent Au NCs. For instance, Shuming et al. [39] reported a ligand-induced etching process for preparing highly fluorescent and water-soluble metal NCs by the ligand exchange reactions using multivalent coordinating polymers such as polyethylenimine (PEI) which are able to etch performed colloidal gold nanocrystals, leading to atomic Au clusters that are highly fluorescent upon UV light excitation. Tan et al. [40] explained the effect of polymer ligand structures on fluorescence of Au NCs. Three types of non-fluorescent tridentate thioether-terminated polymer ligands, poly(methyl methacrylate) (PTMP-PMMA), poly(*n*-butyl

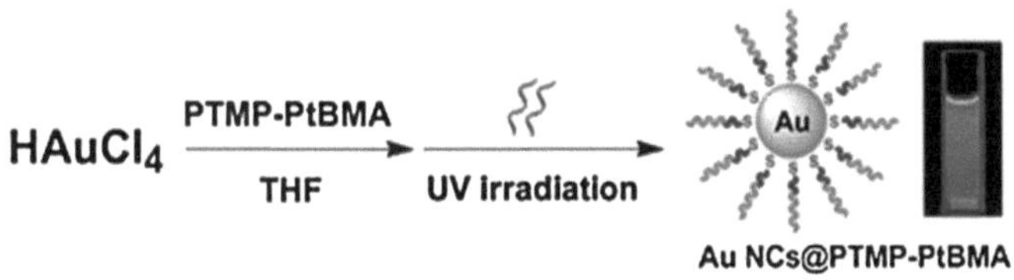

Fig. 5 Preparation of fluorescent Au NCs stabilized by tridentates polymer ligands in a homogeneous organic system. Reprinted with permission from Ref. [40] © 2010 the Royal Society of Chemistry

methacrylate) (PTMP-PBMA) and poly(*tert*-butyl methacrylate) (PTMP-P*t*BMA), PTMP—Pentaerythritol tetrakis-3-mercaptopropionate, were designed to synthesize blue emitting Au NCs by a simple facile photoreduction method. The photoreduction approach is feasible for avoiding the use of additional reducing agents, which is good for studying the photoluminescence mechanism of Au NCs. The capability of polymer ligand to donate electrons plays an important role in improving fluorescence. Among these the highest fluorescence and quantum yield by NC@PTMP-P*t*BMA is shown in Fig. 5 could be attributed to the strong electron donating compared to other polymers owing to their carbon chains.

Highly luminescent small metal NCs are also prepared by utilizing dendrimers as templates. The general procedure consists of mixing dendrimer (G4–OH or G2–OH) and gold ions ($HAuCl_4.nH_2O$) in distilled water. After adding strong reducing agent of $NaBH_4$, small fluorescent Au NCs and large nanoparticles are created simultaneously. The confined intra-space of dendrimer restricts the growth of Au NCs. Reduced gold atoms aggregate within the dendrimers to form small NCs. Dickson et al. [41] synthesized monodisperse, blue emitting Au_8 NC (Fig. 6). These are encapsulated in and stabilized by biocompatible poly(amidoamine) (PAMAM) dendrimer and exhibit a fluorescence quantum yield of ~41 % in

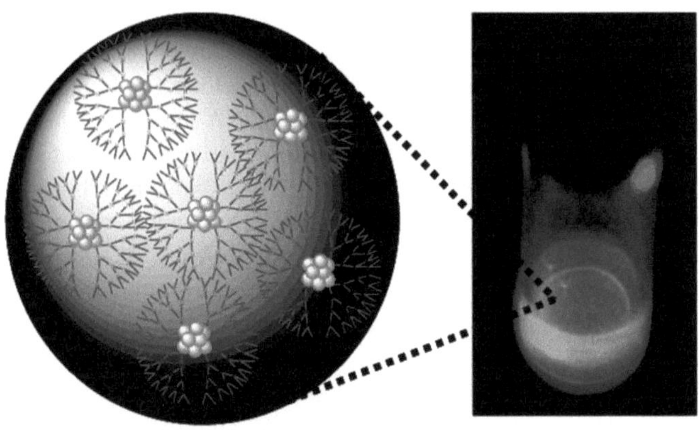

Fig. 6 Emission from Au NC under long-wavelength UV lamp irradiation. Reprinted with permission from Ref. [41]. Copyright 2003 American Chemical Society

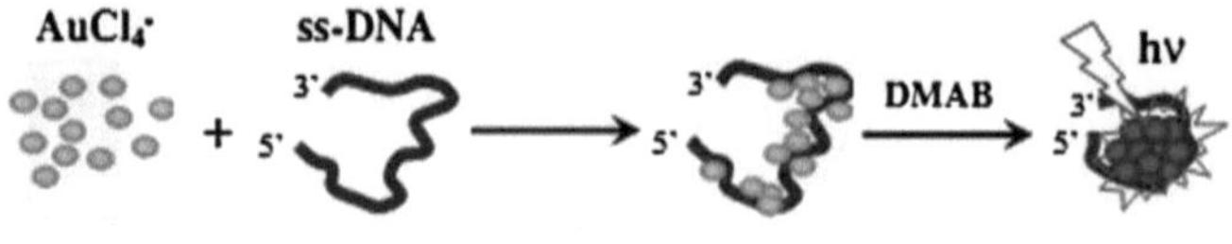

Fig. 7 Schematic illustration for the formation of Au NCs templated by DNA

aqueous solution. Practical applications of gold NC as a novel fluorophore become possible due to the 100-fold enhancement in quantum yield which probably results from the lower density of states present in very small Au_8 NCs which minimizes the internal nonradiative relaxation pathways. Additionally, the larger dendrimer cage better protects these NCs/nanodots from quenchers in solution.

Lin et al. [42] designed a simple and facile strategy for preparing Au NCs by a polarity-specific ion pair association. This strategy can enhance quantum yield of gold NC from 20 to 60 % after microwave irradiation. Martinez et al. [43] demonstrated the use of physiological temperature (37 °C) and ascorbic acid (vitamin C), a mild biologically derived reductant, for the well-defined synthesis of Au NCs using PAMAM dendrimer templates, without the formation of gold nanoparticles.

DNA sequences could be used to tune the Au NCs emission. Apart from its well-known role as the cellular store-house of information, DNA is now being used to construct rigid scaffolds in one, two, and three dimension on the nanoscale. Recently Xu et al. [44] presented a new approach for the synthesis of water-soluble, red fluorescent Au NCs, templated by single stranded DNA and dimethylamine borane as a mild reductant (Fig. 7). Twenty-three-mer single stranded DNAs with the sequences of 5′–GAGGCGCTGCCYCCACCATGAGC–3′ (named 23-Ys, Y = C, A, G, and T) were employed in this work. The as-prepared Au NCs display high stability at physiological pH condition, and thus, wide applications are anticipated for the biocompatible fluorescent Au NCs serving as nanoprobes in bioimaging and related fields.

Liu et al. [45] synthesized fluorescent Au NCs using various DNAs as hosts. By comparison among hairpin DNAs (HP–DNAs) with a pristine stem segment and varied loop sequences, they found that the emission behavior of the HP–DNA-hosted Au NCs depends on the loop sequences. The cytosine loop was found to be the most efficient host to produce fluorescent Au NCs. The DNA-hosted Au NCs can be used as probes to in situ analyze the DNA hybridization and cytosine related mutation.

DNA-templated few-atom Ag NCs with excellent biocompatibility and unique optical properties have attracted much attention in biological sensing [46], information processing [47] and cellular imaging [48]. DNA is well suited for synthesis of silver NCs since the heterocyclic DNA bases offer various binding sites for Ag(I) ions and diverse DNA sequences can be used for programmed growth of Ag NCs. Different DNA sequences and lengths can be used to modulate the emissions of DNA templated silver NCs, [49–51] (Fig. 8a). Fygenson et al. [52] found that few-atom Ag clusters attached to single-stranded DNA exhibit visible fluorescence with spectral properties that are sensitive to the sequence and secondary structure of the bases that comprise the strand.

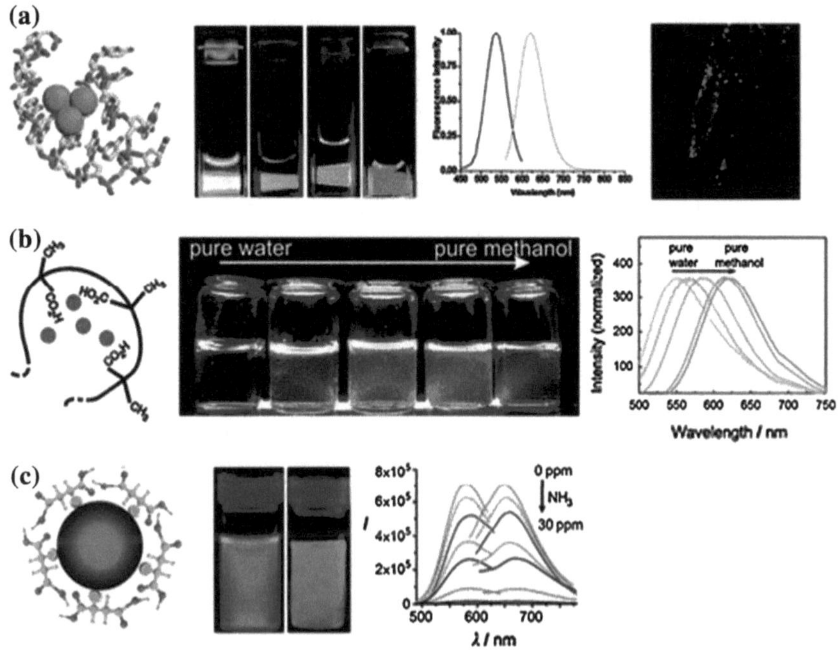

Fig. 8 **a** Representation of silver nanoclusters encapsulated in DNA oligonucleotides. *Photographs* under UV-light of samples with different oligonucleotides and hence different emitters. Emission spectra of the last sample, showing red emitters. Confocal fluorescence microscopic *image* of live cells incubated with (anti-heparin sulfate)-(DNA Oligonucleotides)-(silver nanoclusters). Reprinted with permission from Ref. [49, 50] Copyright 2008, 2004 American Chemical Society and from Ref. [51] with permission of John Wiley and Sons. **b** Schematic *drawing* of silver nanoclusters protected by carboxyl groups of poly(methacrylic acid). *Photograph* under UV-light of samples in water/methanol mixtures, from pure water on the *left* to pure methanol on the *right*. Emission spectra of the samples *imaged*. Reproduced from Refs. [53, 56] with permission of Wiley-VCH Verlag GmbH and Co. KgaA. **c** Silver nanoclusters prepared by interfacial etching from silver nanoparticles and stabilized with small molecules (i.e. mercaptosuccinic acid). Samples under UV light and fluorescence quenching by addition of NH_3. Reprinted from Ref. [59] with permission of Royal Society of Chemistry and from Ref. [58] with permission of Wiley-VCH Verlag GmbH and Co. KgaA

Linear polyacrylates such as poly(methacrylic acid) (PMAA) can act as an excellent scaffold for the preparation of silver NCs in water solution by photoreduction with visible light [53], UV light [54] or sonochemically [55]. PMAA stabilized NCs have an excellent stability and can be transferred to other scaffolds and solvents [56] and can sense the local environment. Ras et al. [50] observed that the fluorescence emission properties of PMAA-stabilized Ag NCs could be tuned to a great extent by selecting appropriate solvents (Fig. 8b). For instance, red shift in the emission peak was observed upon transfer from water to methanol.

In addition to polymers, small molecules containing either carboxylic groups or thiols have been used to stabilize silver NCs in solution. Zhou et al. [57] synthesized blue emitting silver NC using GSH as the capping agent in aqueous solution by a facile one pot sonochemical method which shows excellent sensitivity and high selectivity towards sulfide ion (S^{2-}), with a detection limit of 2 nM. Silver NCs are also prepared by the interfacial etching from silver nanoparticles and stabilized with small molecules like mercaptosuccinic acid by Pradeep et al. [58, 59]. The luminescence and the absorption are instantaneously quenched on addition of NH_3 due to the etching of the cluster core by NH_3 (Fig. 8c). Recently Pradeep et al. introduced a new solid state route to prepare luminescent Ag_9 NC protected with mercaptosuccinic acid [24]. The obtained Ag_9 cluster showed luminescence under UV light at room temperature with excitation and emission maxima at 625 and 720 nm respectively.

Dickson et al. [60] successfully synthesized water soluble Ag NCs in PAMAM dendrimers through direct photoreduction in ambient conditions. Fluorescence grows with increasing irradiation time as silver ions are photoreduced in the dendrimer host. With synthetic control of the dendrimer attachment, these NCs are likely to find use as biological labels.

6 Applications of Fluorescent Clusters

6.1 As Optical Sensors

Optical sensors are a type of analytical devices utilizing a component that can generate optical signals to detect target analysis. For the construction of an optical sensor, the optical material or probe is an indispensable component for the best performance of the sensor. Earlier various optical probes, such as organic dyes [61], conjugated polymers, semiconductor quantum dots [62] and noble metal nanoparticles [63] have been used for optical sensor development. Recent advances in the synthesis and characterizations of highly luminescent noble metal NCs provide another class of promising probes for the construction of novel optical sensors with improved features in terms of simplicity, selectivity, sensitivity, and miniaturizability.

To construct an efficient fluorescent sensor, two key components must be present in the luminescent Au/Ag NCs, i.e., a recognition component to provide specific interaction with an analyte and a transducer component for signaling the occurrence of the interaction. The luminescence properties of the NCs are highly sensitive to the local environment as well as the size and the structure of the NCs, which can provide an excellent response for signaling their interaction with analytes. “Turn-off” and “turn-on” luminescence detection are two common features in the NC-based optical sensor [64]. In a typical “turn-off” scheme, the luminescence of the NCs is quenched by the analyte due to the specific interaction between the analyte and the NC via the metal core or the ligand shell. In a typical “turn-on” scheme, the luminescence is initially held back by an inhibitory agent and the addition of the analyte can selectively remove the inhibitor from the NCs and thereby restoring the luminescence of the NCs. In a typical Au/Ag NCs both the metal core and the ligand shell can specifically interact with the analytes, and can therefore function as the recognition components.

6.1.1 Detection of Metal Ions

Hg^{2+} is a highly toxic and widespread pollutant ion which causes severe damaging effects to the brain, nervous system and the kidney even at very low concentrations [65]. Recently, metal NCs have been utilized as Hg^{2+} sensors by the quenching of their fluorescence based on the unique metallophillic interactions between Au^+ or Ag^+ and Hg^{2+} where the metal core of the NC act as the recognition component. It is well known that a strong metallophilic bond could be formed between d^{10} centers of Hg^{2+} and Au^+ owing to the large dispersion forces between the closed-shell metal atoms, which are further magnified by relativistic effects [66] which can efficiently quench the luminescence of Au/Ag NCs, which can be used for Hg^{2+} detection. Xie et al. demonstrated this recognition chemistry for Hg^{2+} detection using luminescent Au@BSA NCs [67]. Upon the addition of Hg^{2+} to the BSA–Au NCs, the red emission of the NCs was quenched within seconds owing to the specific and strong metallophilic Hg^{2+} and Au^+ interactions. These clusters were also successfully used to construct an intracellular sensor for Hg^{2+} detection. By virtue of the high quantum yield of BSA–Au NCs, the limit of detection for Hg^{2+} was estimated to be 0.5 nM, which is much lower than the maximum level of mercury in drinking water (10 nM) permitted by the U. S. Environmental Protection Agency (EPA).

High quantum yield of the NCs and high fraction of active Au^+/Ag^+ species on the NC surface are the two key factors that leads to the selectivity and sensitivity of the sensor based on $Hg^{2+}\cdots Au^+$ or $Hg^{2+}\cdots Ag^+$ metallophilic interactions (Fig. 9).

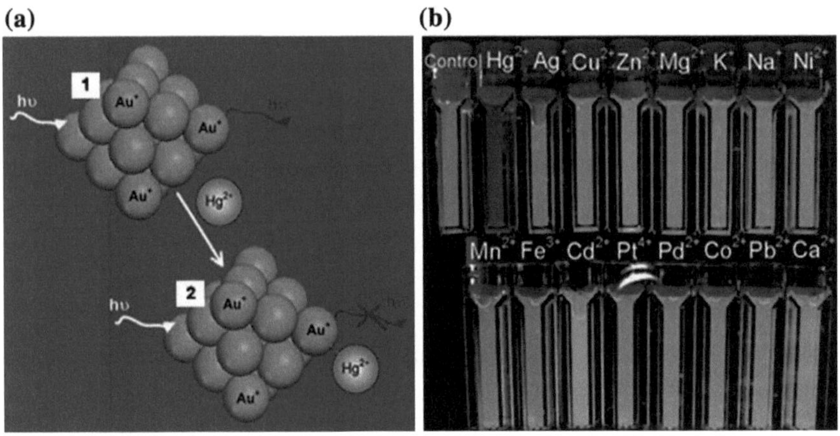

Fig. 9 **a** Schematic representation of Hg^{2+} sensing based on fluorescence quenching of Au NCs resulting from the high-affinity metallophilic Hg^{2+}–Au^+ bonds. **b** *Photographs* of aqueous BSA–Au NCs solutions (20 μM) in the presence of 50 μM of various metal ions under UV light. Reprinted with permission from Ref. [67] © 2010 the Royal Society of Chemistry

Cu^{2+} is another significant environmental pollutant and an essential trace element in biological systems. The detection of Cu^{2+} ion is recently been carried out using fluorescent metal NCs. A simple and sensitive fluorescent sensor for Cu^{2+} detection using fluorescent PMAA–Ag NCs was developed by Shang et al. [68]. The LOD of Cu^{2+} was measured to be 8 nM, which is much lower than that the U.S. EPA limit for Cu^{2+} in drinking water. The mechanism of the fluorescence quenching was explained by the interaction of Cu^{2+} with the carboxylic acid groups of polymers surrounding the emissive Ag NCs where the ligand shell act as the recognition component. Glutathione-capped Au NCs were also employed in a Cu^{2+} sensor application based on aggregation-induced fluorescence quenching [69]. Chang and coworkers [70] developed a novel, turn-on fluorescent assay for Cu^{2+} using DNA-templated Ag NCs. The introduction of Cu^{2+} resulted in the formation of DNA–Cu/Ag NCs and thereby enhanced the fluorescence.

The luminescence enhancement can also be based on the metal deposition on the NC surface. The deposition of Ag on the surface of Au NCs can be readily used as the recognition chemistry for the Ag^+ detection and lead to enhancement in the luminescence intensity. Wu and coworkers [71] used the well-defined GSH–Au_{25} NCs for Ag^+ detection based on the increased luminescence intensity of GSH–Au_{25} NCs due to the deposition of Ag^+ (or Ag^0). The LOD of this method was about 200 nM, which is lower than the maximum level of Ag^+ in drinking water permitted by EPA (460 nM).

It is well known that gold is a kind of chemically inert metal and a few anions can react with it. Cyanide belongs to a class of substance that is toxic to environment and human health. Au and Ag NCs with strong fluorescence show promising application in the cyanide detection. Liu et al. [72] demonstrated the successful application of Au NCs as a fluorescent sensor in detection of cyanide ion. The formation of a stable $Au(CN)_2^-$ complex will lead to the fluorescence quenching of clusters according to the Elsner reaction: [73]

$$4\ Au + 8\ CN^- + 2H_2O + O_2 \rightarrow\ 4\ Au(CN)_2^- + 4OH^-.$$

Upon the introduction of cyanide to the aqueous BSA–Au NC solution, the solution became colorless within 20 min, and the red emission was completely quenched.

6.1.2 Detection of Biomolecules

Bi-thiols such as cysteine (Cys) play a critical role in important cellular functions including detoxification and metabolism. The analysis of their levels in human body fluids plays an important role in early diagnosis of many diseases. Shang et al. [74] developed a simple fluorescent method for Cys detection based on PMAA–Ag NCs. The fluorescence of Ag NCs can be efficiently quenched by Cys which allowed the selective determination of Cys with a LOD of 20 nM.

The mechanism of fluorescence quenching was suggested as the thiol-adsorption-accelerated oxidation of the Ag clusters.

Luminescence quenching based on the enzymatic reaction was demonstrated by Wen and coworkers [75] by fabricating luminescent Au NCs protected by horse-radish peroxidase (HRP). The luminescent HRP-Au NCs showed excellent selectivity for hydrogen peroxide (H_2O_2). The HRP enzymes on the NC surface were still active and could catalyze the reduction of their substrate (H_2O_2). This catalytic reaction generated reactive oxygen species (ROS) could break down the thiolate-Au bond between HRP and Au core by oxidizing the thiolate to disulfide. As a result, some HRP on the surface on Au NC lost their protecting capability, which in turn destabilized the NCs and induced aggregation, thus leading to the luminescence quenching of NCs. Wang and coworkers successfully utilized the luminescent Au–BSA NCs to detect proteases [76]. The red emission of the Au–BSA NCs was completely quenched after the addition of proteinase K. The luminescence quenching of the NCs is due to the catalytic degradation of BSA by the protease and the subsequent exposure of Au core to the dissolved O_2. A variety of other biomolecules (e.g., lysozyme, insulin and transferring etc.) can serve as protecting ligands for highly luminescent Au/Ag NCs and thereby provides this type of luminescence probes for the construction of biosensors for a variety of analytes.

6.1.3 Biological Imaging

The biological imaging relies mainly on stable, biocompatible, highly specific and sensitive markers. The commonly used fluorophores for imaging application include organic dyes, engineered fluorescent proteins. But their limited photostability made them an inefficient tool for long-term experiments in live cells with high sensitivity. An alternative for this was the introduction of semiconductor quantum dots with good photostability and high fluorescence brightness. However these quantum dots have prompted potential safety concerns for in vivo use. Moreover their large dimensions, typically comparable to or larger than the size of proteins affect the function of attached ligands. In contrast, fluorescent metal NCs are smaller in size and exhibit bright emission with good biocompatibility, making them attractive alternatives as fluorescent probes for bio-imaging.

In vivo fluorescence imaging of tumors may offer a possibility for the direct bio-imaging of tumors for precise diagnosis of cancer and monitoring of the treatment process. In situ fluorescent bio-imaging is also of great significance for visualizing the activity of particular molecules, cells, and biological processes that influence the behaviour of tumors. In this perspective, Au NCs are a promising biocompatible fluorescent probes, offering surfaces and core exhibiting physiochemical properties. Wang and coworkers [77] proved that Au NCs could readily be prepared by in situ biosynthesis in live tumor cells, offering a promising opportunity for in vivo bio-imaging application (Fig. 10).

The bright green fluorescence of Au NCs biosynthesized in situ inside HepG2 and K562 cancer cells appear to be adequate for their use for in vivo bio-imaging of

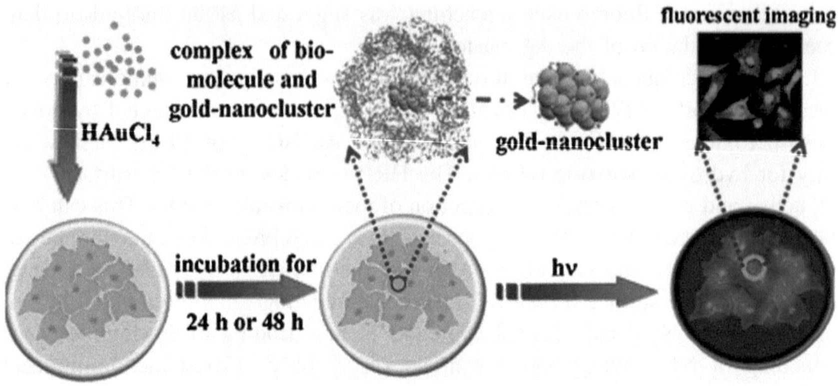

Fig. 10 Schematic illustration of in situ biosynthesis of gold nanoclusters in cells and tumor imaging (Reprinted with permission from Ref. [67] © 2013, Nature Publishing Group)

relevant live tumor cells. Importantly this does not occur in non-cancerous cells, as evidenced with human embryo liver cells. Thus Wang and coworkers proposed a different novel strategy that relies on the fact that cancer cells have a completely different redox homeostasis from normal cells.

In particular, for in vivo imaging and photodynamic therapy, two-photon absorption (TPA) in the NIR region increases the penetration depth, spatial resolution due to lower scattering, and minimizes auto fluorescence. The outstanding TPA cross sections of metal NCs make them good candidates for application in two-photon cellular imaging. Using dextran-encapsulated 11-MUA-Au NCs as the fluorophore, Chou and coworkers [78] investigated their potential for two-photon imaging of human mesenchymal stem cells (hMSCs). Bright luminescence from Au NCs could be observed in fixed cell samples via two-photon excitation in a confocal microscope using 800 nm laser pulses, thus validating the applicability of fluorescent Au NCs in two-photon imaging.

References

1. Kreibig U, Vollmer M (1995) Optical properties of metal clusters. Springer, Berlin
2. Ashcroft NW, Mermin ND (1976) Solid state physics. Holt/Rinehart & Winston, New York
3. Sardar R, Funston AM, Mulvaney P, Murray RW (2009) Gold nanoparticles: past, present, and future. Langmuir 25:13840
4. Jin R, Zhu Y, Qian H (2011) Quantum-sized gold nanoclusters: bridging the gap between organometallics and nanocrystals. Chem Eur J 17:6584 (and references therein)
5. Haberland H (1994) Clusters of atoms and molecules: Theory experiment, and clusters of atoms. Springer, Berlin
6. Kubo R (1962) Electronic properties of metallic fine particles. J Phys Soc Jpn 17:975
7. Schaaff TG, Knight G, Shafigullin MN, Borkman RF, Whetten RL (1998) Isolation and selected properties of 10.4 kDa gold: glutathione cluster compound. J Phys Chem B 102:10643
8. Wallace WT, Whetten RL (2002) Coadsorption of CO and O_2 on selected gold clusters: evidence for efficient room-temperature CO_2 generation. J Am Chem Soc 124:7499

9. Campbell CT, Parker SC, Starr DE (2002) The effect of size-dependent nanoparticle energetics on catalyst sintering. Science 298:811
10. Sanchez A, Abbet S, Heiz U, Schneider W-D, Häkkinen H et al (1999) When gold is not noble: nanoscale gold catalysts. J Phys Chem A 103:9573
11. Link S, Beeby A, FitzGerald S, El-Sayed MA, Schaaff TG, Whetten RL (2002) Visible to infrared luminescence from a 28-atom gold cluster. J Phys Chem B 106:3410
12. Felix C, Sieber C, Harbich W, Buttet J, Rabin I et al (2001) Ag_8 fluorescence in Argon. Phys Rev Lett 86:2992
13. Zheng J, Nicovich PR, Dickson RM (2007) Highly fluorescent noble-metal quantum dots. Annu Rev Phys Chem 58:409
14. Brust M, Walker M, Bethell D, Schiffrin DJ, Whyman R (1994) Synthesis of thiol-derivatised gold nanoparticles in a two-phase liquid-liquid system. Chem Commun 7:801–802
15. Tracy JB, Kalyuzhny G, Crowe MC, Balasubramanian R, Choi JP, Murray RW (2007) Poly (ethylene glycol) ligands for high-resolution nanoparticle mass spectrometry. J Am Chem Soc 129:6706–6707
16. Castro EG, Salvatierra RV, Schreiner WH, Oliveira MM, Zarbin AJG (2010) Dodecanethiol-stabilized platinum nanoparticles obtained by a two-phase method: synthesis, characterization, mechanism of formation, and electrocatalytic properties. Chem Mater 22:360–370
17. Wu ZK, Lanni E, Chen WQ, Bier ME, Ly D, Jin RC (2009) High yield, large scale synthesis of thiolate-protected Ag_7 clusters. J Am Chem Soc 131:16672–16674
18. Ang TP, Wee TSA, Chin WS (2004) Three-dimensional self-assembled monolayer (3D SAM) of n-alkanethiols on copper nanoclusters. J Phys Chem B 108:11001–11010
19. Zhao MQ, Sun L, Crooks RM (1998) Preparation of Cu nanoclusters within dendrimer templates. J Am Chem Soc 120:4877–4878
20. Jin RC, Qian HF, Wu ZK, Zhu Y, Zhu MZ, Mohanty A, Garg N (2010) Size focusing: a methodology for synthesizing atomically precise gold nanoclusters. J Phys Chem Lett 1:2903–2910
21. Lin CAJ, Yang TY, Lee CH, Huang SH, Sperling RA, Zanella M, Li JK, Shen JL, Wang HH, Yeh HI, Parak WJ, Chang WH (2009) Synthesis, characterization, and bioconjugation of fluorescent gold nanoclusters toward biological labeling applications. ACS Nano 3:395–401
22. Lopez-Quintela MA, Tojo C, Blanco MC, Rio LG, Leis JR (2004) Microemulsion dynamics and reactions in microemulsions. Curr Opin Colloid Interface Sci 9:264–278
23. Reetz MT, Helbig W (1994) Size-selective synthesis of nanostructured transition metal clusters. J Am Chem Soc 116:1401–1402
24. Rao TUB, Nataraju B, Pradeep T (2010) Ag_9 quantum cluster through a solid–state route. J Am Chem Soc 132:16304–16307
25. Xie J, Zheng Y, Ying JY (2009) Protein-directed synthesis of highly fluorescent gold nanoclusters. J Am Chem Soc 131:888–889
26. Xavier PL, Chaudhari K, Verma PK, Pal SK, Pradeep T (2010) Luminescent quantum clusters of gold in transferrin family protein, lactoferrin exhibiting FRET. Nanoscale 2:2769–2776
27. Yan L, Cai Y, Zheng B, Yuan H, Guo Y, Xiao D, Choi MMF (2012) Microwave-assisted synthesis of BSA-stabilized and HSA-protected gold nanoclusters with red emission. J Mater Chem 22:1000–1005
28. Yang X, Shi M, Zhou R, Chen X, Chen H (2011) Blending of HAuCl4 and histidine in aqueous solution: a simple approach to the Au_{10} cluster. Nanoscale 3:2596–2601
29. Huang T, Murray R (2001) Visible luminescence of water-soluble monolayer-protected gold clusters. J Phys Chem B 105:12498–12502
30. Lee D, Donkers RL, Wang G, Harper AS, Murray RW (2004) Electrochemistry and optical absorbance and luminescence of molecule-like Au_{38} nanoparticles. J Am Chem Soc 126:6193–6199
31. Paau M, Lo C, Yang X, Choi M (2010) Synthesis of 1.4 nm α-cyclodextrin-protected gold nanoparticles for luminescence sensing of mercury(II) with picomolar detection limit. J Phys Chem C 114:15995–16003

32. Wang Z, Cai W, Sui J (2009) Blue luminescence emitted from monodisperse thiolate-capped Au_{11}clusters. Chem Phys Chem 10:2012–2015
33. Shang L, Dorlich RM, Brandholt S, Schneider R, Trouillet V, Bruns M, Gerthsen D, Nienhaus GU (2011) Facile preparation of water-soluble fluorescent gold nanoclusters for cellular imaging applications. Nanoscale 5:2009–2014
34. Muhammed MAH, Ramesh S, Sinha SS, Pal SK, Pradeep T (2008) Two distinct fluorescent quantum clusters of gold starting from metallic nanoparticles by pH-dependent ligand etching. Nano Res 1:333–340
35. Zhu M, Aikens CM, Hollander FJ, Schatz GC, Jin R (2008) Correlating the crystal structure of a thiol-protected Au_{25} cluster and optical properties. J Am Chem Soc 130:5883–5885
36. Zeng C, Liu C, Pei Y, Jin R (2013) Thiol ligand-induced transformation of $Au_{38}(SC2H4Ph)_{24}$ to $Au_{36}(SPh\text{-}t\text{-}Bu)_{24}$. ACS Nano 7:6138–6145
37. Gao Y (2013) Ligand effects of thiolate-protected Au_{102} nanoclusters. J Phys Chem C 117:8983–8988
38. Huang CC, Liao HY, Shiang YC, Lin ZH, Yang Z, Chang HT (2009) Synthesis of wavelength-tunable luminescent gold and gold/silver nanodots. J Mater Chem 19:755–759
39. Duan H, Nie S (2007) Etching colloidal gold nanocrystals with hyperbranched and multivalent polymers: a new route to fluorescent and water-soluble atomic clusters. J Am Chem Soc 129:2412–2413
40. Li L, Li Z, Zhang H, Zhang S, Majeed I, Tan B (2013) Effect of polymer ligand structures on fluorescence of gold clusters prepared by photoreduction. Nanoscale 5:1986–1992
41. Zheng J, Petty JT, Dickson RM (2003) High quantum yield blue emission from water-soluble Au_8 nanodots. J Am Chem Soc 125:7780–7781
42. Jao Y-C, Chen M-K, Lin S-Y (2010) Enhanced quantum yield of dendrimer-entrapped gold nanodots by a specific ion-pair association and microwave irradiation for bioimaging. Chem Commun 46:2626–2628
43. Bao Y, Zhong C, Vu DM, Temirov JP, Dyer RB, Martinez JS (2007) Nanoparticle-free synthesis of fluorescent gold nanoclusters at physiological temperature. J Phys Chem C 111:12194–12198
44. Liu G, Shao Y, Ma K, Cui Q, Wu F, Xu S (2012) Synthesis of DNA-templated fluorescent gold nanoclusters. Gold Bull 45:69–74
45. Liu G, Shao Y, Wu F, Xu S, Peng J, Liu L (2013) DNA-hosted fluorescent gold nanoclusters: sequence-dependent formation. Nanotechnology 24:1–7
46. Choi S, Dickson RM, Yu J (2012) Developing luminescent silver nanodots for biological applications. Chem Soc Rev 41:1867–1891
47. Diez I, Ras RHA (2011) Fluorescent silver nanoclusters. Nanoscale 3:1963–1970
48. Li T, Zhang L, Ai J, Dong S, Wang E (2011) Ion-tuned DNA/Ag fluorescent nanoclusters as versatile logic device. ACS Nano 5:6334–6338
49. Richards CI, Choi S, Hsiang JC, Antoku Y, Vosch T, Bongiorno A, Tzeng YL, Dickson RM (2008) Oligonucleotide-stabilized Ag nanocluster fluorophores. J Am Chem Soc 130:5038–5039
50. Petty JT, Zheng J, Hud NV, Dickson RM (2004) DNA-templated Ag nanocluster formation. J Am Chem Soc 126:5207–5212
51. Yu J, Choi S, Richards CI, Antoku Y, Dickson RM (2008) Live cell surface labeling with fluorescent Ag nanocluster conjugates. Photochem Photobiol 84:1435–1439
52. Gwinn EG, O'Neill P, Guerrero AJ, Bouwmeester D, Fygenson DK (2008) Sequence-dependent fluorescence of DNA-hosted silver nanoclusters. Adv Mater 20:279–283
53. Díez I, Pusa M, Kulmala S, Jiang H, Walther AA, Goldmann AS, Müller AHE, Ikkala O, Ras RHA (2009) Color tunability and electrochemiluminescence of silver nanoclusters. Angew Chem Int Ed 48:2122–2125
54. Shang L, Dong S (2008) Facile preparation of water-soluble fluorescent silver nanoclusters using a polyelectrolyte template. Chem Commun 9:1088–1090
55. Xu H, Suslick KS (2010) Sonochemical synthesis of highly fluorescent Ag nanoclusters. ACS Nano 4:3209–3214

56. Díez I, Jiang H, Ras RHA (2010) Enhanced emission of silver nanoclusters through quantitative phase transfer. Chem Phys Chem 11:3100–3104
57. Zhou T, Rong M, Cai Z, Yanga CJ, Chen X (2012) Sonochemical synthesis of highly fluorescent glutathione-stabilized Ag nanoclusters and S^{2-} sensing. Nanoscale 4:4103–4106
58. Rao TUB, Pradeep T (2010) Luminescent Ag_7 and Ag_8 clusters by interfacial synthesis. Angew Chem Int Ed 49:3925–3929
59. Mrudula KV, Bhaskara Rao TU, Pradeep T (2009) Interfacial synthesis of luminescent 7 kDa silver clusters. J Mater Chem 19:4335–4342
60. Zheng J, Dickson RM (2002) Individual water-soluble dendrimer-encapsulated silver nanodot fluorescence. J Am Chem Soc 124:13982–13983
61. Zeng L, Miller EW, Pralle A, Isacoff EY, Chang CJ (2006) A selective turn-on fluorescent sensor for imaging copper in living cells. J Am Chem Soc 128:10–11
62. Basabe-Desmonts L, Reinhoudt DN, Crego-Calama M (2007) Design of fluorescent materials for chemical sensing. Chem Soc Rev 36:993–1017
63. Jans H, Huo Q (2012) Gold nanoparticle-enabled biological and chemical detection and analysis. Chem Soc Rev 41:2849–2866
64. Xu Z, Chen X, Kim HN, Yoon J (2010) Sensors for optical detection of cyanide ion. Chem Soc Rev 39:127–137
65. Holmes P, James KAF (2009) Levy LS is low-level environmental mercury exposure of concern to human health. Sci Total Environ 408:171–182
66. Pyykkç P (2004) Theoretical chemistry of the gold. Angew Chem Int Ed 43:4412–4456
67. Xie J, Zheng Y, Ying JY (2010) Highly selective and ultrasensitive detection of Hg^{2+} based on fluorescence quenching of Au nanoclusters by Hg^{2+}–Au^{+} interactions. Chem Commun 46:961–963
68. Shang L, Dong SJ (2008) Silver nanocluster-based fluorescent sensors for sensitive detection of Cu(II). J Mater Chem 18:4636–4640
69. Chen W, Tu X, Guo X (2009) Fluorescent gold nanoparticles-based fluorescence sensor for Cu^{2+} ions. Chem Commun 13:1736–1738
70. Lan G-Y, Huang C-C, Chang H-T (2010) Silver nanoclusters as fluorescent probes for selective and sensitive detection of copper ions. Chem Commun 46:1257–1259
71. Yue Y, Liu TY, Li HW, Liu Z, Wu Y (2012) Microwave-assisted synthesis of BSA-protected small gold nanoclusters and their fluorescence-enhanced sensing of silver(I) ions. Nanoscale 4:2251–2254
72. Liu YL, Ai KL, Cheng XL, Huo LH, Lu LH (2010) Gold-nanocluster-based fluorescent sensors for highly sensitive and selective detection of cyanide 2043 in water. Adv Funct Mater 20:951–956
73. Wang XB, Wang YL, Yang J, Xing XP, Li J, Wang LS (2009) Evidence of significant covalent bonding in $Au(CN)_2^-$. J Am Chem Soc 131:16368–16370
74. Shang L, Dong S (2009) Sensitive detection of cysteine based on fluorescent silver clusters. Biosens Bioelectron 24:1569–1573
75. Wen F, Dong Y, Feng L, Wang S, Zhang S, Zhang X (2011) Horseradish peroxidase functionalized fluorescent gold nanoclusters for hydrogen peroxide sensing. Anal Chem 83:1193–1196
76. Wang Y, Wang Y, Zhou F, Kim P, Xia Y (2012) Protein-protected Au clusters as a new class of nanoscale biosensor for label-free fluorescence detection of proteases. Small 8:3769–3773
77. Wang J, Zhang G, Li Q, Jiang H, Liu C, Amatore C, Wang X (2013) In vivo self-bio-imaging of tumors through in situ biosynthesized fluorescent gold nanoclusters. Sci Rep 3:1157
78. Liu C, Ho M, Chen Y, Hsieh C, Lin Y, Wang Y, Yang M, Duan H, Chen B, Lee J (2009) Thiol-functionalized gold nanodots: two-photon absorption property and imaging in vitro. J Phys Chem C 113:21082–21089

56. Liu L, Jiang D, Xia N (2016) Enhanced emission of silver nanoclusters through ... phase transfer. Chem Phys Chem [illegible]

Plasmonic Properties of Metallic Nanostructures, Two Dimensional Materials, and Their Composites

Lauren Rast

Abstract The intense and highly tunable optical field enhancement provided by nanomaterials supporting plasmon resonances has diverse applications including biophotonics, terahertz spectroscopy, and subwavelength microscopy. This chapter compares plasmon resonance behavior and tunability in noble metal nanostructures with that of two dimensional and quasi-two dimensional materials including graphene, silicene, germanene, and the transition metal dichalcogenides. Plasmonic optical behavior and related advancements in two-dimensional materials functionalized by metallic nanostructures are discussed. Finally, possibilities for new directions for work on similar composite plasmonic systems are outlined.

Keywords Plasmonics · Plasmon resonance · Metallic nanostructures · Two-dimensional materials · Quasi-two dimensional materials · Graphene · Silicene · Germanene · Dichalcogenides · Composites

1 Overview

The quantized units of collective plasma oscillation are known as "plasmons". When the real part of the dielectric function in a medium changes sign across an interface, for example, where a metal and dielectric materials share a boundary, a surface-plasmon resonant (SPR) mode is supported [1]. These collective oscillations can couple strongly with a variety of particles and quasiparticles to form other quasiparticles and collective modes. When a plasmon couples with a photon, an electromagnetic wave known as a surface plasmon polariton (SPP) is formed. These waves are constrained in one dimension—that is, they are confined to propagate along the interface.

The strong optical confinement of plasmons relative to the wavelength of light leads to intense local electromagnetic field enhancement. The field enhancement

L. Rast (✉)
University of Alabama at Birmingham, Department of Physics, Birmingham, AL 35294-1170, USA
e-mail: laurenr@uab.edu

P. Misra (ed.), *Applied Spectroscopy and the Science of Nanomaterials*,
Progress in Optical Science and Photonics 2, DOI 10.1007/978-981-287-242-5_8

induced by SPR in metallic nanoclusters can provide up to a 10^{15} intrinsic enhancement factor, allowing for impressive feats such as single molecule detection via surface-enhanced Raman spectroscopy (SERS) [2].

Normally, when an optical component has dimensions near the wavelength of light, light propagation is limited by optical diffraction. In plasmonic-photonic circuits, the size of components such as lenses and fibers need not be limited by the classical Abbe diffraction limit, a consequence of their propagation in two dimensions [3, 4]. Additionally, the strong coupling of plasmons with light allows control over optical signals at length scales on the order of electronic wavelengths [5]. This allows for extreme miniaturization of photonic circuit components.

Given the enormous electromagnetic field enhancement, unparalleled control over signal propagation, and ability to overcome classical diffraction limits provided by plasmonic materials, it's not surprising that the field of plasmonics has seen remarkable growth in recent years. Nanoplasmonic materials have found a multitude of applications in fields as diverse as spectroscopy, photonics, biomedicine, and telecommunications. Practical application of these materials, however, still presents significant challenges. Traditional plasmonic materials such as noble metals provide the large negative refractive index necessary for generation of intense SPPs. Yet, these materials are notoriously plagued by losses. SPPs travelling along a metallic surface are rapidly damped through a variety of loss pathways, which will be discussed further in the next section of this chapter. In recent years, there has been significant debate over whether or not loss minimization and negative refractive index were simultaneously possible or feasible under realistic conditions [6]. This controversy was associated with overlooked subtleties in the application of causality-related arguments towards limiting behavior of the dielectric response [7]. Although it has been determined that negative index materials may be fabricated with low losses, it is clear new materials and careful structural design will be necessary for overcoming the loss barrier [8, 9].

Recent advances in the fabrication and characterization of atomically thin materials including materials including graphene and beyond, such as the transition-metal dichalcogenides and other semiconductors, present exciting possibilities for progress in low-loss plasmonics. This chapter highlights advantages and disadvantages of a variety of platforms for plasmonics, with a focus on guidance for low-loss applications. These include traditional noble metallic nanostructures, novel two-dimensional and quasi-two dimensional materials, and more complex functionalized and composite devices.

2 Plasmon Resonances and Their Transport Properties in Metallic Nanostructures

The goals of this section are to: (1) To describe some of the advantages of traditional plasmonic materials (2) To present some of their shortcomings as stand-alone plasmonic media, and (3) Most importantly, describe quantitatively the characteristics of an ideal plasmonic material, in order to determine a basis for identifying better materials.

The extreme adjustability of SPR features in noble metal nanoparticle systems make them useful for a myriad of functional devices for biomedical and optical applications, including biosensing, nanoscale optics, catalysis, and drug delivery. Optical properties of metal nanoparticle systems display great sensitivity to the morphology and geometrical configurations of the individual nanoparticles that comprise them [10, 11]. Particle size and size distribution, shape, and dielectric environment are known to significantly affect the position and intensity of SPR peaks [12, 13]. For instance, when metal nanoparticles are in close proximity to one another, interaction of the particles' local electric field can result in significant changes to the properties of the field, which can result in additional resonance peaks for the coupled system.

In order to demonstrate the acute shape sensitivity of metal nanoparticle spectra, I've used the convenient discrete dipole approximation (DDA) method, as implemented in code DDSCAT by Draine [14, 15], to calculate absorption spectra for single silver nano-disks and nano-triangular prisms. These data are displayed in Fig. 1, with inset TEM images of silver nanoparticles of geometry approximately that of the calculated particles, fabricated by methods similar to those detailed in my work with Stanishevsky [10]. The calculated particles are slightly simplified as compared to the fabricated particles. In particular, the fabricated triangles have slightly truncated edges, whereas the calculated triangles have sharp edges. The general method of fabrication of the particles displayed in the TEM images is as follows: A silver nanoparticle precursor solution was prepared by the reduction reaction of 30 mM $AgNO_3$ solution in 2-methoxyethanol in the presence of poly (vinyl)-N-pyrrolidone (PVP). The PVP molecular weight was 8000 g/mol, and mass ratio of PVP to $AgNO_3$ varied from 0.2:1 to 10:1. The solution was kept at

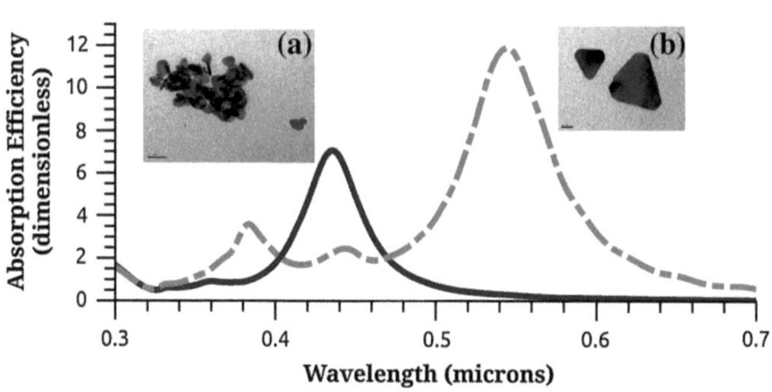

Fig. 1 DDA simulated absorption spectra for non-interacting silver (**a**) nano-disks and (**b**) nano-triangles. Simulated nano-disks have 30 nm radius and nano-triangles have 52 nm edge length (corresponding to a 30 nm effective radius). Both geometries have a thickness of 10 nm. Wave propagation is taken to be normal to the particle surface with in-plane polarization

90 °C for 30 min while stirring (using a magnetic stirrer). Once the $AgNO_3$ was completely reduced, the solution was then removed from the hotplate and transferred to cool in a bath of ice. The PVP matrix, solution PH, and thermal processing then shape the nanoparticle morphology [10, 11].

The dielectric function used for the nanoparticles in the DDA calculation is that of Johnson and Christy [16], with a size correction factor from Doyle for the surface damping of small particles [17]. The results compare very well with experimental results for both single silver nano-triangular prisms and single silver nano-disks of similar size and aspect ratio [18, 19]. Simulated nano-disks have 30 nm radius, nano-triangles have 52 nm edge lengths, which corresponds to a 30 nm effective radius (the radius of the triangle's circumsphere), and both geometries have a thickness of 10 nm.

The large difference in spectral profile and dominant peak position displayed in Fig. 1 are due entirely to shape, rather than differences in particle size. This is an extremely convenient finding from a fabrication standpoint, as metal nanoparticle shape can easily be tuned through slight adjustments in thermal-processing parameters (such as processing time and temperature) or changes in the pH of the reaction environment [20, 21]. Nanoparticle spectra show similar sensitivity to other changes in morphology, such as the pattern of clustering or aggregation, which are also adjustable via changes in the thermal processing parameters and the relative concentration of reactants [10, 20, 22].

Traditional metallic nanomaterials, with all of the advantages they provide, have yet to reach their full potential for industrial applications. The chief hindrance towards more broad application of these materials is the significant losses they incur due to a variety of inherent energy dissipation mechanisms, which occur during plasmon propagation. In metals, primary plasmon loss mechanisms include ohmic losses, scattering, and radiative losses. Different mechanisms dominate depending upon frequency as well as nanomaterial size, morphology, and makeup. Radiative damping is smaller for small metal particles, but even in the small size regime ohmic losses can remain quite significant, and are largest for wavelengths near the surface plasmon resonance [9]. Plasmonic efficiency may be quantified through the use of parameters such as propagation length or a quality factor. Let us first define the propagation length. The behavior of a plasmon excitation in a medium can be described by the frequency (ω)—dependent dielectric function of the medium,

$$\varepsilon(\omega) = \varepsilon'(\omega) + i\varepsilon''(\omega), \tag{1}$$

where both ε' and ε'' are real-valued. Losses in a conducting medium due to the various mechanisms are all represented by ε''. The complex in-plane plasmon wavenumber k is expressed as

$$k_x = k_x'(\omega) + ik_x''(\omega). \tag{2}$$

and is given by

$$k_x = \frac{\varepsilon_1 \varepsilon_2}{\varepsilon_1 + \varepsilon_2} \frac{\omega^2}{c^2}. \tag{3}$$

where ε_1 and ε_2 are the complex dielectric functions in two adjacent media.

Propagation loss mechanisms may be represented in general through the imaginary part of the plasmon wave number, k_x'', which determines the propagation length of the plasmon. The propagation length, L, of a plasmon is typically defined as the length at which the intensity of the SPP has decayed by a factor of 1/e; this occurs when:

$$L = \frac{1}{2k_x''}. \tag{4}$$

The 1/e decay length of the electric field is 1/k″ or 1/(2 k″) for the intensity. So, small plasmon losses and longer propagation length are inextricably related. Larger L values are pursued particularly in applications such as photonic circuits, waveguides, and photovoltaics.

Another useful, and related, parameter for quantifying plasmonic material desirability is plasmon oscillation quality factor. Both the propagation length and quality factor are determined by plasmon damping due to energy loss. The quality factor, Q, of a plasmonic material is then given by [23]:

$$Q = \frac{E}{\left(\sqrt{\omega_p^2 - \gamma^2}\right)^{-1} \frac{dE}{dt}} = \frac{\left(\sqrt{\omega_p^2 - \gamma^2}\right)}{2\gamma}, \tag{5}$$

where E is the stored in the plasmonic oscillator. Multiplying by Planck's constant $\hbar$, we then find:

$$Q \approx \frac{\omega_p}{2\gamma} = \frac{\hbar\omega_p}{\hbar 2\gamma} \equiv \frac{E_p}{\Gamma}. \tag{6}$$

A conceptual illustration of the Q-factor for plasmonic materials is given in Fig. 2. Materials with a narrow and intense plasmon resonance peak are represented by a higher Q-factor.

In principle, one way of creating a material with a higher Q-factor is to mitigate bulk losses through the fabrication of very thin metallic films, so that plasmon resonances couple more strongly to surface modes. Figure 3 demonstrates this effect, known as the *Begrenzung effect*, (or boundary effect) [1, 24, 25] in the electron energy loss function for an increasingly thin silver film on a silicon substrate. These calculated spectra were determined by first obtaining individual-layer dielectric functions. The silver layer from the work of Johnson and Christy [16], and silicon layer dielectric function is 11.68 [26]. Dielectric function values are used as input to a composite dielectric function first obtained by Lambin [27]. Further information on

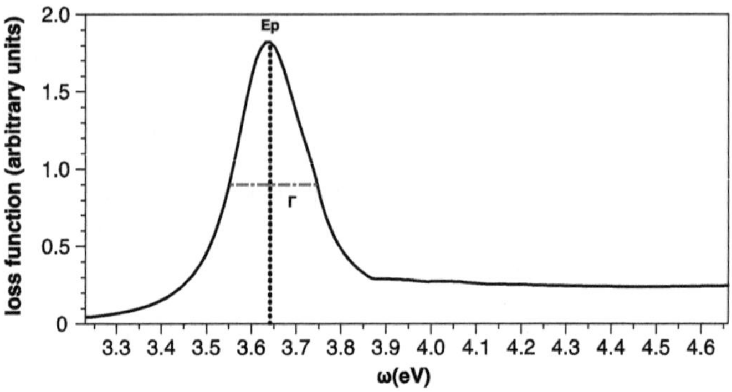

Fig. 2 Conceptual illustration for a plasmon resonance quality factor, given by Ep/Γ, where Ep is the maximum peak height, and Γ is the FWHM value for the of the plasmon resonance

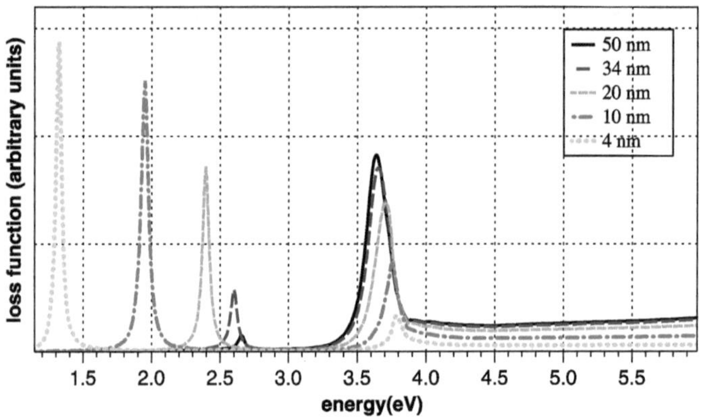

Fig. 3 Electron energy loss spectra for a composite system comprised of a thin silver film on a silicon substrate. The silver film thicknesses shown are 50 nm (*solid line*), 34 nm (*long dashes*), 20 nm (*short dashes*), 10 nm (*dash-dot*), and 4 nm (*dotted*)

this technique is included in Sect. 4.1 of this chapter, and a more detailed analysis may be found in [28]. Fabrication of thin and uniform metal films is very challenging. Thin metallic layers tend to form small islands and may require adhesion layers, which alter the optical and electronic properties of the device [29, 30]. Additionally, even thin noble-metal films are subject to significant ohmic losses at visible wavelengths [31]. Thin noble metallic layers may be more advantageous for low-loss plasmonics as part of a composite structure. Section 4.1 details a numerical example of such a system, similar to the system whose spectra are displayed in Fig. 3, but with the addition of a graphene top coating.

Another important feature of an ideal plasmonic material is plasmon localization. This is a feature that is often times, *though not necessarily* [32], at odds with a large L value or Q-factor. Plasmon localization may also be quantified, though different approaches can lead to very different values. A detailed analysis of this is outside the scope of this chapter, but an excellent and thorough derivation of four different measures (including both statistically and phenomenologically derived values) for confinement by Oulton et al. may be found in the recommended reading, Sect. 6.1.

Materials with simultaneously large values for L (or Q) and significant localization are the supreme goal for most, though not all, plasmonics applications. Much debate has occurred in the literature over whether or not materials with highly confined SPR features could also provide the desirable low-loss characteristics, because in systems such as a noble metal-dielectric interface, high confinement at the metallic interface can lead to large Ohmic losses. There are a few ways of circumventing the common notion that low-loss and a high degree of localization are mutually exclusive. Even in the case of simple silver/silicon systems, it has been found that at system resonances (corresponding to anomalous dispersion) high electromagnetic field confinement with increased propagation length is possible [32]. In addition, low-loss systems can be created that have high local confinement, but low global confinement. Hybrid plasmonic materials may be able to provide the best of both worlds, allowing exhibit longer propagation distances without a reduction in confinement.

3 Current and Prospective Atomically Thin Materials for Plasmonics

In this section, I will present some of the new atomically thin materials, which have been fabricated or proposed theoretically for plasmonics applications. Applications-related advantages provided by some of these materials are also discussed. Lastly, in order to facilitate materials by design based on these materials, a table is presented (Table 1). This table includes both π plasmon peak and $\pi + \sigma$ plasmon peak locations for 2D and quasi-2D materials currently found in the literature.

In a recent paper, Khurgin and Boltasseva [33] made a leap forward for plasmonics, by identifying the fundamental loss mechanisms in plasmonics and suggesting specific means for overcoming these challenges. Beneath the various plasmonic loss mechanisms, there are two fundamental processes. These are intraband or interband transitions, i.e. from occupied to unoccupied states within the s-p band or the d-band to states in the s-p band. Then, the primary way to minimize loss would be to minimize the density of states available for scattering. Khurgin and Boltasseva therefore suggest materials such as highly doped semiconductors or graphene as low-loss plasmonic materials, as these allow for tuning the density of states (DoS). Atomically thin materials such as zero-gap/semi-metal semiconductors and non-zero-gap semiconductors have in general been found to have highly tunable electronic properties. In contrast to noble metals, the dielectric function and DoS of these materials may be tuned via a variety of mechanisms including chemical doping, an applied bias voltage, or strain engineering [34–36].

Two dimensional materials have a clear advantage over other materials in terms of field enhancement due to confinement. Metallic films embedded in a dielectric

Table 1 Plasmonic peak locations for a variety of 2D and quasi-2D materials

Material	π Plasmon Peak (eV) (π − π[a] excitations)	π + σ Plasmon Peak (eV) (π − σ[a] and σ − σ* excitations)
Graphene (monolayer)	4.7[a] [69], 4.8† [69], 4.55† [70]	14.6[a] [69], 14.5† [69]
Graphene (bilayer)	~5[a] [69], ~5.5† [69]	~15[a] [69], ~15.5† [69]
Graphane	3.80† [71]	
Silicene	1.6† [72], 1.74† [73], 1.23† [48]	3.94† [73], 3.9† [48]
Germanene	1.7† [72], 1.10† [48]	3.0† [48]
Silicane	3.0† [48]	3.94† [73], 4.0† [48]
Germanane	1.45† [48]	~3.5† [48]
Hexagonal BN	6.5[a] [74], 6.5 † [75] 6.5† [76]	15.5[a] [74], 15.4 † [75], 16.0† [76]
Hexagonal BN (bilayer)	7.0† [76]	19.0† [76]
MoS_2	6.5* [74], 7.6† [75]	20[a] [74], 15.6† [75]
$MoSe_2$	7.0† [75]	14.9† [75]
$MoTe_2$	6.3† [75]	14.3† [75]
TaS_2		14.4† [75]
$TaSe_2$		13.8† [75]
$TaTe_2$		11.9† [75]
WS_2		15[a] [74], 15.2† [75]
WSe_2		14.1† [75]
WTe_2		12.4† [75]
NbS_2	7.7† [75]	14.0† [75]
$NbSe_2$	7.0† [75]	13.8† [75]
$NbTe_2$	6.0† [75]	11.7† [75]
$NiTe_2$		10.2† [75]
Bi_2Te_3		
Bi_2Se_3		
SiC (monolayer)		3.3 [77]
SiC (bilayer)	2.4† [77]	3.3† [77], 3.4† [77]
SiN		
GaS (monolayer)	6.25† [76]	12.0† [76]
GaS (bilayer)	6.25† [76]	13.5† [76]
GaSe (monolayer)	6.0† [76]	11.0† [76]
GaSe(bilayer)	6.0† [76]	13.5† [76]
Bi2Sr2CaCu2Ox		
Mica		

[a] Denotes experimental value, and † corresponds to a calculated value

are known to produce plasmonic modes with smaller volume as the metallic film thickness is reduced [37]. Two dimensional materials represent the extreme end of the spectrum in terms of spatial confinement. Graphene, for instance, have a plasmon mode volume 10^6 smaller than the diffraction limit [38].

It's clear that alternative plasmonic materials beyond noble metals must be sought out in order to find low-loss, highly tunable plasmonic materials with strong resonances. I've compiled Table 1 includes a list of two-dimensional materials for which data on the plasmonic features is thus far available in the literature. Materials that are predicted to exist, or have been fabricated in 2D form, but for which neither theoretical nor experimental plasmon peak values were available at the time of writing, are listed in the table with the field for plasmon peak values left blank. These missing values are a reminder that 2D material-based plasmonics is a field still in its infancy, with much room for new work.

In recent years, great strides have been made in large-scale fabrication of monolayer materials as well as mechanical isolation of atomically thin layers of two dimensional and quasi-two-dimensional materials [39]. These advances in nano-material fabrication along with recent leaps forward in highly spatially resolved and time-resolved characterization techniques have opened up the possibility of creating nanoelectronic and optoelectronic devices with a wide variety of 2-D material components. The lexicon of monolayer materials continues to grow, and thus far includes graphene, hexagonal boron nitride (h-BN), silicene, germanene, and many materials based on transition metal dichalcogenides (TMDCs), such as MoS_2, $MoSe_2$, WS_2, TaS_2, $NbSe_2$, and WSe_2.

Table 1 includes both π plasmon peak ($\pi - \pi^*$ excitations) and $\pi + \sigma$ plasmon peak ($\pi - \sigma^*$ and $\sigma - \sigma^*$ excitations) locations for these 2D and quasi-2D materials, as well as some bi-layer materials. Plasmon peak values for each of these materials can be used for roughly estimating the peak values that would be found in composites based on these materials (by simply averaging the locations of nearby peaks for the different materials).

Figure 4 displays the crystal structures for some representative two-dimensional and quasi-two dimensional plasmonic materials. Represented materials include graphene and h-BN, silicene and germanene, and TMDCs of trigonal coordination (MoS2, for example). Each of these materials provides distinctive advantages, both individually and in composite. For example, consider the case of graphene and h-BN. Graphene provides the advantages of unparalleled electronic and thermal transport properties, while h-BN provides exceptional insulating properties with thermal transport properties similar to those of graphene, an ultra-flat morphology, and extreme uniformity. When h-BN is used as a substrate for graphene, the flat morphology and uniformity, along with nearly isomorphic lattice properties of the two materials, lead to electronic transport properties for the graphene layer that are nearly equal to those of suspended samples [40]. Recent ab initio calculations have found similar results for Moiré superstructures of silicene on hexagonal boron nitride [41].

Silicene and germanene both provide electron transport properties similar to graphene along with compatibility with existing silicon-based technology [36, 41].

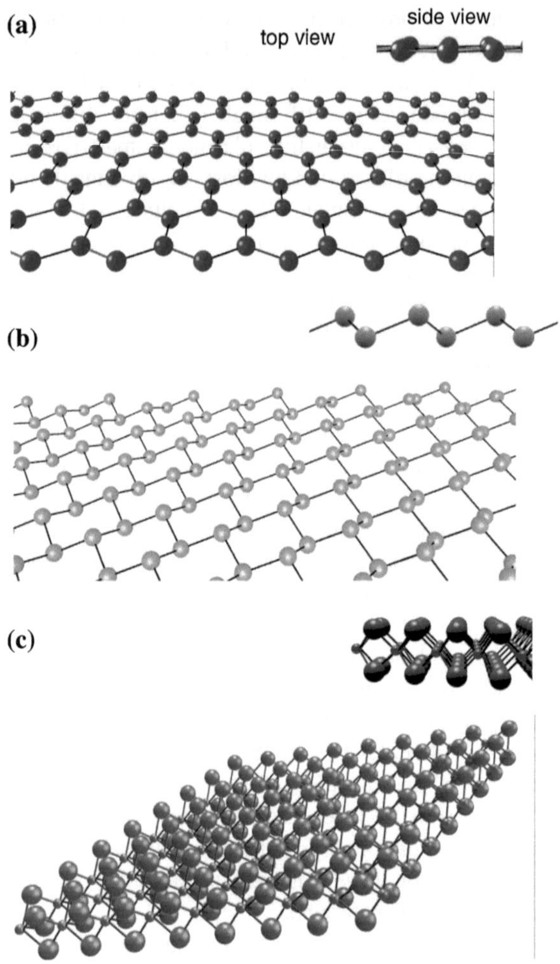

Fig. 4 Crystal structures of representative two dimensional and quasi-two dimensional materials. Surface and side view of lattice structures of (**a**) graphene (**b**) silicene and germanene (**c**) transition metal dichalcogenides of trigonal coordination

Another major advantage provided by these materials is that the inversion symmetry breaking imparted by the buckled lattice structure of both materials may be taken advantage of, through the application of an external electrical field perpendicular to the plane, for extremely controllable bandgap tunability.

As an example of the plasmon peaks seen in 2D material spectra, the plasmon peak positions for graphene, silicene and germanene calculated via DFT methods are displayed in Fig. 5. These ab initio calculations were performed using the highly efficient GLLBSC exchange correlation functional [42]. They are implemented in the code GPAW[1] [43–46]. GPAW is an electronic structure python code based on

---

[1] Certain commercial equipment, instruments, or materials are identified in this paper in order to specify the experimental procedure adequately. Such identification is not intended to imply recommendation or endorsement by the National Institute of Standards and Technology, nor is it

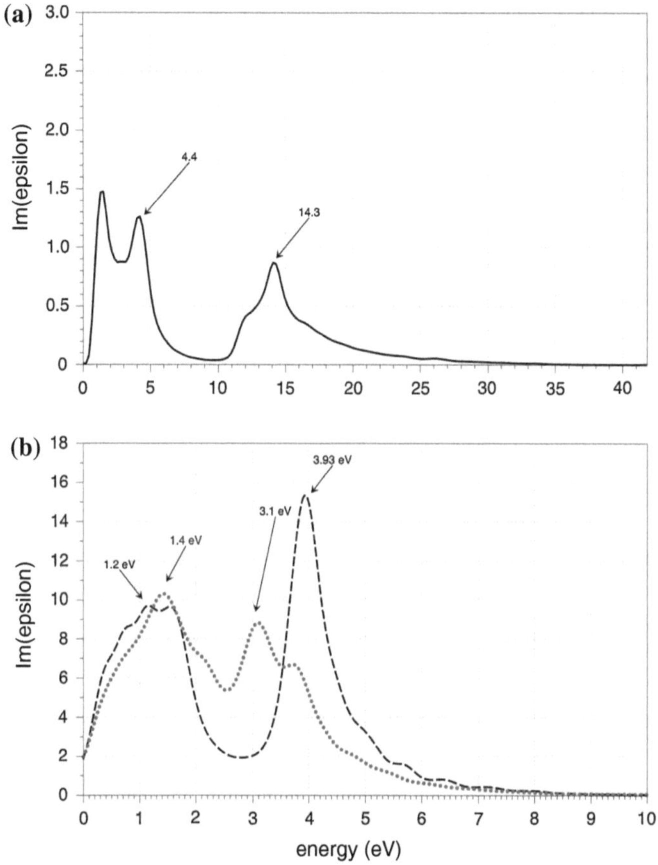

Fig. 5 Imaginary dielectric function for representative freestanding two-dimensional materials (**a**) graphene and (**b**) silicene (*dashed*) and germanene (*dotted*) calculated via the Density Functional Theory (DFT) code GPAW

the projector augmented wave method. For all materials, dielectric functions are calculated in the optical limit. The momentum transfer used is 0.005° A^{-1}, along the surface Brillouin zone. The k-point sampling with 20 × 20 × 1 Monkhorst–Pack grid was chosen to compute the band-structure and dielectric function.

The method combining the use of the GLLBSC functional and the Bethe Salpeter Equation (BSE) was selected for the germanene and silicene dielectric function calculations due to the extreme accuracy of this method in predicting experimental values of dielectric functions and bandgaps for similar materials, such as a variety of

(Footnote 1 continued)
intended to imply that the materials or equipment identified are necessarily the best available for the purpose.

bulk semiconductors including silicon, as well as two-dimensional materials graphene and hexagonal boron nitride [47]. The graphene calculation was performed as part of work that was earlier than the silicene and germanene calculations; thus, the less advanced (though still reasonably accurate) Random Phase Approximation method was used in the optical part of the calculation. This method neglects excitonic effects. Similarly obtained results may also be seen for graphene in our paper [28].

The GLLBSC potential includes the derivative discontinuity of the exchange correlation potential, critical for obtaining physically meaningful band structure via a DFT calculation. This functional has also been shown to have computational cost similar to the Local Density Approximation (LDA) with accuracy comparable to that of methods such as the LDA-GW method. The use of the BSE in the optical calculation is important due to the inclusion of excitonic effects, which is a prominent spectral feature for silicene and germanene [48]. A two dimensional Coulomb cutoff is employed so that the BSE may be used in the calculation of the dielectric function of these monolayer materials [49].

In the graphene calculation, a low energy peak may be observed at $\sim$1 eV. This feature, which is due to the low-energy $\pi \rightarrow \pi^*$ single-particle excitation, would appear as a broad shoulder in the electron energy loss function. The π plasmon peak for graphene appears at $\approx$4.4 eV, and the $\pi + \sigma$ plasmon peak at 14.3 eV. The results compare well with those obtained by others, which are displayed in Table 1. Silicene peaks are found to be 1.2 (π plasmon peak) and 3.93 eV (π + σ plasmon peak). Finally, the values calculated for germanene are 1.4 (π plasmon peak) and 3.1 (π + σ plasmon peak). The values for silicene and germanene are also in good agreement with those reported in the literature and displayed in Table 1.

The TMDCs are also very interesting materials for plasmonics. Like graphene, these materials have strong in-plane bonding and weak-interlayer interaction and may therefore be prepared by exfoliation in order to obtain single (quasi-two-dimensional) layers. TMDCs such as MoS_2 possess high carrier mobilities (density functional theory predicts $\sim$400 $cm^2\ V^{-1}\ s^{-1}$ at room temperature and a carrier density of $10^{11}cm^{-2}$), but unlike graphene, these materials possess sizeable bandgaps of 1–2 eV [50]. This on-off feature is compulsory for devices such as field-effect transistors or all-optical switches. Lastly, these materials provide the interesting advantage of tuning of band structure properties via stoichiometry, as the choice of the chalcogenide determines the band gap for the semiconducting TMDCs, and the choice of the transition metal determines whether or not the material is fully metallic versus semiconducting.

4 Functionalized and Hybrid Plasmonic Materials

The goal of this section is to illustrate some of the most advanced types of composite nanomaterial structures that are currently being explored for low-loss plasmonics applications. Superlattice hybrid structures made of 2D materials, 2D antidot lattices, and 2D materials functionalized by metallic nanodots are examples of materials that allow one to combine the advantages of a variety of components, as well as to take advantage of emergent properties (properties that arise due to interactions within the composite structure). It is then possible to design metamaterial structures with both low-loss characteristics and a high degree of confinement.

4.1 2D Material-Based Heterostructures

As progress continues forward towards ever-more convenient and rapid production of 2D materials, the road is being paved for the fabrication of tailored heterostructures on demand. 2D crystal-based metamaterial composites could, for example, combine the sizable bandgaps of TMDCs, which change from indirect to direct in single layers, with the unrivaled strength of graphene [51], and the atomically flat morphology of h-BN [40].

Metamaterial heterostructures have physical properties that are derived from the properties of their constituent materials, as well as properties that arise due to collective interaction between component materials and various physical processes in each material layer, such as interfacial phenomena. Rational design of 2D-based heterostructures therefore hinges upon reliable and comprehensive data on material properties for the individual layers, efficient methods for multiscale calculation of composite properties, and advanced software frameworks for optimal design and model inversion. Further reading on computational frameworks for materials by design may be found in Sect. 6.3 of the Recommended Reading section of this chapter.

As mentioned in Sect. 2, hybrid plasmonic materials (such as the layered heterostructure displayed in Fig. 6) are good candidates for higher Q-factor plasmonic

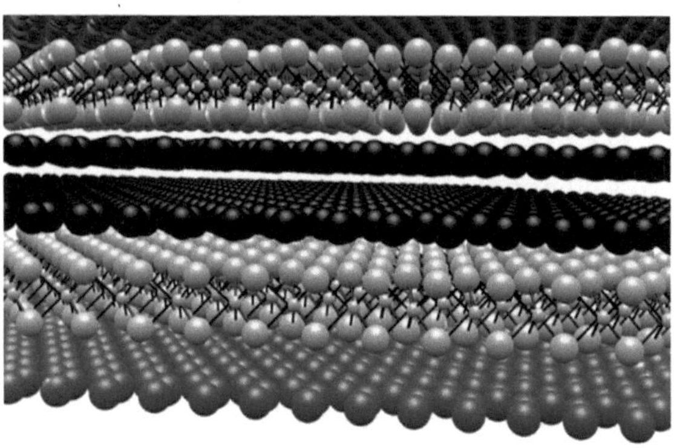

Fig. 6 Conceptual illustration of a heterostructure based on two-dimensional and quasi-two dimensional crystals. Heterostructures combine advantageous optoelectronic and mechanical properties of the individual materials and exhibit novel properties that may be attributed to cooperative effects

materials. For example, those that contain metal and a combination of low and high refractive index dielectric materials, exhibit longer propagation distances without a reduction in confinement. The effective dielectric function, $\xi(\omega, k, z)$, of such stratified structures may be found using the convenient and efficient expression first derived by Lambin et al. [27]. ω, k, and z are the plasmon frequency, wavenumber, and the spatial coordinate normal to the surface of the stratified structure, respectively. $z = 0$ is considered to be the upper surface of the top layer.

The expression by Lambin et al. for ξ was found based on electron energy loss spectroscopic (EELS) theory in a reflection geometry. At the z = 0 surface, ξ is:

$$\xi_0 = a_1 - \cfrac{b_1^2}{a_1 + a_2 - \cfrac{b_2^2}{a_2+a_3-\cfrac{b_3^2}{a_3+a_4-\ldots}}}, \tag{7}$$

where

$$a_1 = \varepsilon_i \coth(kd_i), \tag{8}$$

and

$$b_1 = \frac{\varepsilon_i}{\sinh(kd_i)}. \tag{9}$$

The dielectric function of an individual layer of thickness d_i is given by ε_i.

The expression has been shown to be applicable to calculation of EELS spectra for both phonons and plasmon polaritons [52] in multilayer heterostructures with interacting interfaces and histogram-like dielectric functions (continuous within each layer). Lambin's initial paper in which the expression was derived focused on semiconductor applications. However, the expression and the formalism from which the expression is derived are also applicable to modeling of surface plasmon resonances in alternating of metal-insulator layers [52, 53]. For small values of the wavevector, the Lambin model has been shown to agree with the spectroscopic predictions of the Bloch hydrodynamic model. [24]. Tewary and Sullivan, and I employed Lambin's continued fraction expression in a semi-classical approach for calculating electron energy loss properties of graphene/noble metal/insulating substrate layered structures [28]. Our semi-classical method was as follows: Dielectric functions for the individual material layers were initially obtained. Ab initio methods were used to obtain dielectric functions for the 2-D materials of interest (graphene and h-BN). The ab initio calculations were performed in the following way: Time-dependent DFT was employed (using the DFT code GPAW) in the LDA. These optical calculations were performed with materials in the armchair configuration. The momentum transfer value was 0.005 Å^{-1} along the Γ-M direction.

The optical properties of other materials, such as the noble metals, were empirical values from the literature. Once individual dielectric functions were obtained, Eq. (7)

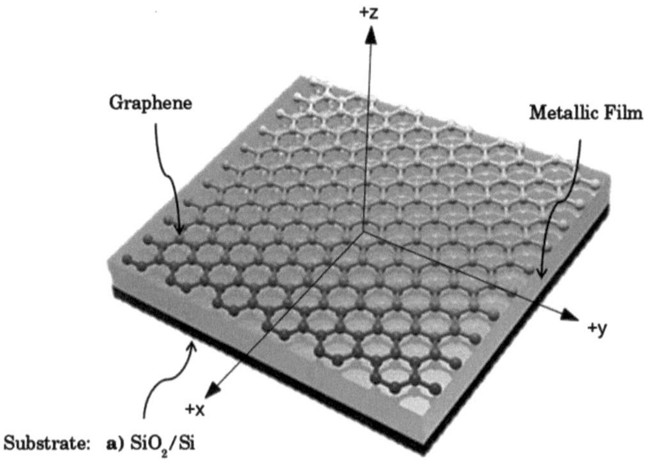

Fig. 7 Schematic representation of a graphene layered structure consisting of a semi-infinite Si bottom layer, SiO_2 layer above the Si layer, followed by a 34 nm Ag layer, single graphene top layer coating, and semi-infinite vacuum

was used to obtain the composite dielectric function for the layered system, and finally, the electron energy loss function (EELF) was found by:

$$\mathrm{EELF} = \mathrm{Im}\left[\frac{-1}{\xi(\omega, \mathrm{k}) + 1}\right]. \tag{10}$$

This procedure allows for multiscale EELF calculations for a extensive variety of layered systems. Layer thickness and material can easily be replaced, with each EELF calculation running in under a second. The speed of these calculations is nearly independent of the frequency range.

Here, I present an example based on data obtained in a similar matter as the data presented in [28]. The model for the multilayer structure is presented in Fig. 7. The calculated structure consists of a semi-infinite Si bottom layer, SiO_2 layer above the Si layer, followed by a 34 nm Ag layer, a single graphene top-layer coating, and semi-infinite vacuum, which begins at $z = 0$.

In this case, the graphene dielectric function was obtained in the optical limit, and the LDA calculation was performed with a Monkhorst-Pack grid of 20x20x1 k-points. As in previous sections of this chapter, the silver dielectric function is that of Johnson and Christy [16]. Static relative permittivities of 11.68 and 3.9 (both broadly-used values obtained from the literature [54, 55]) are used for the Si and SiO_2 dielectric constants, respectively.

The graphene lattice constant used here is 2.46 Å. The composite dielectric function is derived in a specular reflection geometry, and thus, the model employs dielectric functions exclusively associated with surface parallel excitations. This approximation is reasonable, since out-of-plane resonances are small in graphene at energies less than $\approx$10 eV, an energy well outside of the regime considered in these calculations.

Figure 8 shows the effect of the addition of a graphene top coating to a thin noble metallic (Ag, in this case) film. For reference, the results may be compared with those displayed in Fig. 3 for a silver SiO_2/Si system. The bulk peak (at approximately 3.7 eV) amplitude is reduced by approximately 20 % with the addition of a graphene top coating. Additionally, there is no change to the lower energy surface peak (seen at approximately 2.6 eV). In our work published in reference [28], we found that for up to three graphene layers, the surface peak remains nearly unchanged, while the bulk peak amplitude is diminished by roughly 50 %. We attribute this effect to the introduction of the thin boundary layer (graphene), leading to concurrent weakened coupling to bulk modes and strengthened coupling to surface modes. This behavior bears physical similarity to the *begrenzung* effect discussed in Sect. 2 and as seen in Fig. 3. The *begrenzung* effect arises due to a reduction of the degrees of freedom for electronic excitations. Additional surface confinement is at the cost of the bulk excitation, leading to a reduction in losses. In our work in reference [28], we find no

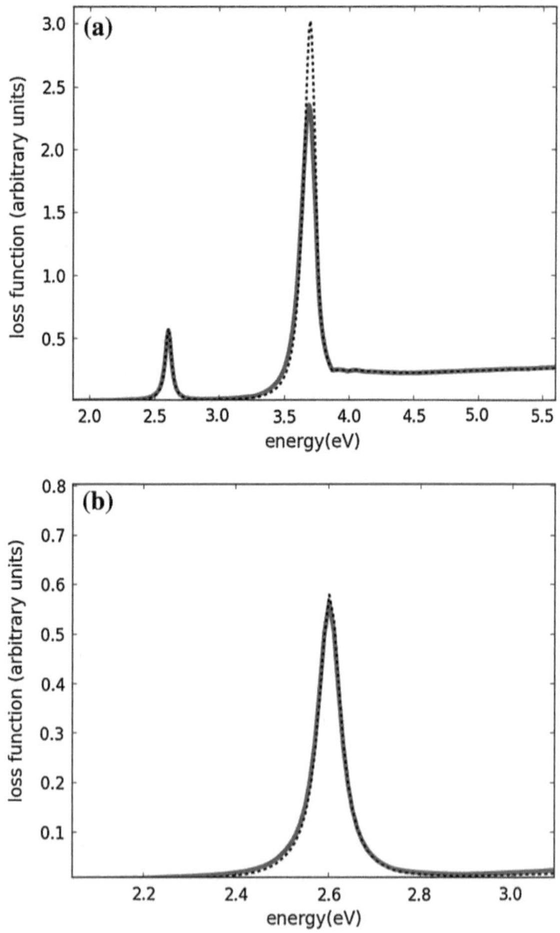

Fig. 8 **a** Electron energy loss function for a layered structure consisting of a semi-infinite Si bottom layer, SiO_2 layer above the Si layer, followed by a 34 nm Ag layer and, then, vacuum (*dotted line*) and semi-infinite Si bottom layer, SiO_2 layer above the Si layer, followed by a 34 nm Ag layer and, then, single graphene top layer coating (*solid line*). **b** close up of lower energy surface peak at ≈2.6 eV

significant broadening of the bulk peak or reduction in the surface peak until the number of graphene layers exceeds 10 layers. Bulk-like behavior is observed at 20 layers. In this case, the bulk peak broadens significantly, and the system has an optical spectrum similar to that of graphite on silver. The reduction in intensity for the lossy bulk mode, along with no reduction in intensity for the surface mode, means that for a few-layer graphene coating, losses are reduced as compared to the Ag/SiO_2/Si system. Broadening corresponds to increased losses (recall Fig. 2 and the discussion of plasmonic quality factor in Sect. 2), our results indicate that loss is minimized in the case of a few layer graphene coating without compromising the strength of the surface plasmon. We obtained qualitatively similar results for other noble metallic films in our previous work [28], including Au and Cu, as well as further enhancement of the surface coupling through the addition of a hBN substrate, rather than SiO_2/Si. The highly insulating hBN substrate reduced bulk losses and enhanced surface confinement. This is due in particular to a reduction in scattering. The results for the multilayer structures presented here indicate that multilayer plasmonic devices based on 2D materials are promising structures for low-loss plasmonics. Highly insulating, wide-bangap materials other than hBN, such as SiC for example, are interesting materials also worth consideration for future work in low-loss plasmonics.

4.2 Two Dimensional Materials with Antidots

One very interesting and novel possibility for tuning plasmons in atomically thin materials is the tuning of antidot arrays (such as that pictured in Fig. 9). Graphene with arrays of micro- or nano-size antidots is known to possess strong optical

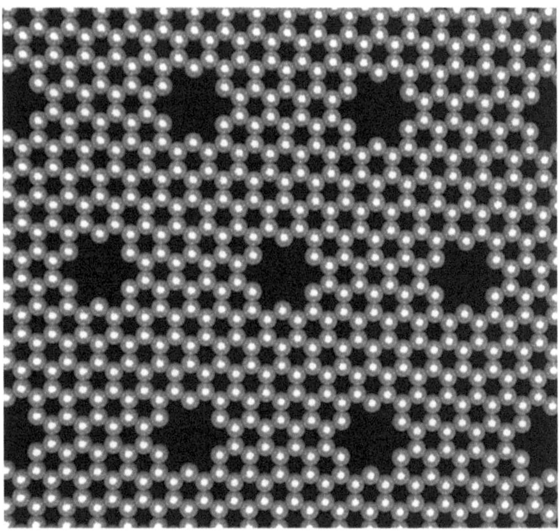

Fig. 9 Two-dimensional material with an ordered array of antidot perforations

resonances in the microwave and terahertz regions [56]. The resonances correspond to surface plasmons of a continuous graphene sheet, with a perturbation due to the antidot lattice. It's reasonable that such a model could be extended to other two-dimensional and quasi-two-dimensional materials with antidot arrays (referred to from here on as 2DAAs). Such materials provide a variety of advantages for plasmonics. Graphene antidot layers have been shown to provide enhanced absorption relative to a continuous graphene film, which is of use for applications such as transformation optics. One of the most exciting possibilities for 2DAAs, however, is the inherent tunability of the plasmon resonance peaks. Plasmon resonances can be tuned by tuning the perturbation due to the antidot lattice, and one of the most convenient and predictable ways of doing this is by tuning the geometric properties of the antidot pattern, such as antidot size and spacing. Schultz et al. [57] have given the mathematical formulation of the shift in plasmon dispersion due to the graphene antidot lattice. The plasmon dispersion for the unperturbed lattice is given by:

$$\omega_c = \mu C \sqrt{\frac{q}{\mathrm{k}_F}}, \tag{11}$$

where ω_c is the plasmon frequency of the continuous 2D lattice, μ is the chemical potential, and

$$C = \sqrt{\frac{g_s g_\vartheta e^2}{8\pi\varepsilon_r\varepsilon_0\vartheta_f}}, \tag{12}$$

where g_s and g_ϑ are the spin and valley degeneracies, ε_r is the relative dielectric constant for the 2D material (graphene in Schultz et al.), q is the wave vector (weak azimuthal dependence is neglected), and k_F and ϑ_f are the Fermi wave number and Fermi velocity, respectively. Zhang et al. [58] found that for graphene, the antidot lattice allowed for the formation of a bandgap, and that adjusting the separation of the antidots allowed for continuous tuning of the band gap.

If the chemical potential is close to the band edge, then the low-energy regime of the electronic structure dominates the behavior. Then, a "gapped" model is a reasonable approximation of the perturbation due to the antidot lattice. This gapped model represents a small perturbation on the pristine lattice, and is valid for very small antidots compared to the unit cell of the array. The plasmon dispersion of the 2DAA is then given by:

$$\omega_{2DAA} = \mu C \sqrt{\frac{q}{\mathrm{k}_F}} \sqrt{1 - \frac{E_{gap}^2}{4\mu^2}}. \tag{13}$$

The perturbation term, $\sqrt{1 - \frac{E_{gap}^2}{4\mu^2}}$, is of course zero when the bandgap E_{gap} vanishes, in the case of pristine graphene, for example.

In addition to the ability to fine-tune plasmon dispersion through antidot lattice patterning, some fascinating phenomena have very recently been observed in graphene antidot structures, including the so called *Hoftstadter butterfly* [59]. The ability to smoothly tune plasmon characteristics in a controlled manner, by surface patterning of antidot lattices, make graphene and other 2DAA materials are attractive prospects for future work in the area of materials by design.

4.3 Two Dimensional Materials Functionalized by Metallic Nanoparticles

It has been possible for well over a decade to pattern regular arrays of metal nanoparticles with controllable size and morphology by use of techniques including electron beam lithography [60], colloidal lithography [61], and block copolymer self-assembly [62]. As mentioned earlier in this chapter, SPR due to noble metallic nanoclusters can provide up to a 10^{15} intrinsic Raman enhancement factor, allowing for single molecule detection via SERS. Variation of particle shape and interparticle distance allow for highly controllable tuning of the optical absorption maximum wavelength and bandwidth. For example, adjusting the shape of Ag particles in regular arrays and adjusting the interparticle distance between 44 and 95 nm, has been shown to allow for independent tuning of the extinction peak wavelength between 460 and 520 nm and the bandwidth between 50 and 100 nm [56]. More recently, carbon nanotubes [63], graphene [64], and single layer MoS2 [65] have all been functionalized by metallic nanoparticles. (A conceptual illustration of a two-dimensional material functionalized by a regular array of metallic nanoparticles is given in Fig. 10).

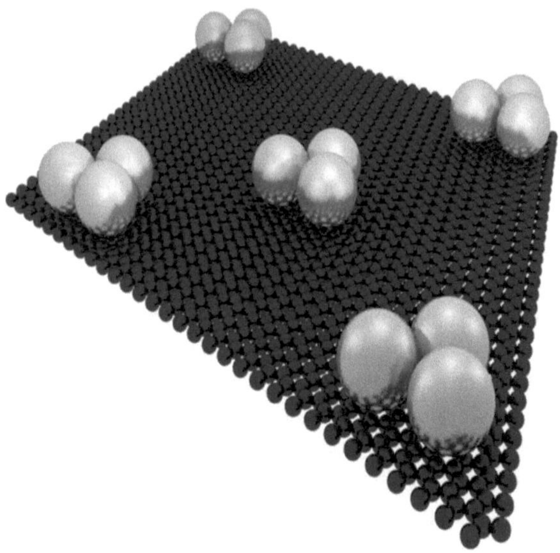

Fig. 10 Conceptual illustration of a two dimensional material functionalized by metallic nanoparticles. As plasmon resonances are very sensitive to changes in geometric configuration and local environment, the spectral profile of such a composite structure may be tuned via changes in nanoparticle surface patterning

When graphene is doped with other plasmonic materials or included in a multilayer structure, the result is extremely enhanced electromagnetic response and highly tunable opto-electronic properties [66]. Tunability of plasmon resonances to the infrared regime is an important feature for ultra-fast communications and biosensing applications.

It's been shown that Au nanoparticles impose extraordinary p-doping effects to the MoS_2 surface without significantly degrading the MoS_2 electron transport properties. More specifically, CVD-MoS2 electric double layer transistors demonstrated a 50-fold increase in on/off ratio [65]. Au nanoparticle surface decoration has also led to a three-fold improvement in the photodetection efficiency of MoS_2 [67].

Clearly, two-dimensional materials decorated with plasmonic nanoparticles show promise for biosensing techniques such as SERS, as well as a variety of other optoelectronics applications where enhanced sensitivity, tunable optoelectronic characteristics, and the low-loss capabilities of two-dimensional materials such as MoS_2 provide advantages. Much work has already been done in the area of fabrication techniques for these types of systems. There is much work yet to be done, however, in terms of predictability of the electromagnetic response of these composite materials with adjustable composition, morphology, and patterning. Standardized material properties for two-dimensional and quasi-two-dimensional materials are needed, as are well-validated multiscale models for these types of composite systems (including quasiparticle and inter-layer interactions, strain effects, and defects), and the incorporation of predictive modeling frameworks with these multiscale models. Predictive modeling frameworks (with capabilities for uncertainty quantification as well as parameter sensitivity analysis) should be used in tandem with experimental analysis, in order to be useful in predicting optoelectronic characteristics of these materials. More information on such frameworks is given in the Recommended Reading section of this chapter, under the heading Rational Design.

5 New Directions

In this chapter, I've discussed some of the most current work on low-loss plasmonic materials and composites, including metallic nanoparticles, two-dimensional and quasi-two-dimensional materials, 2D material-based heterostructures, two-dimensional antidot lattices, and two-dimensional materials functionalized by metallic nanoparticle arrays. Now, a final comment on the future of low-loss plasmonic materials: Fabrication techniques for producing advanced structures such as functionalized two-dimensional materials, hybridized plasmonic structures for applications such as integrated plasmonic circuits, and heterostructures based on atomically thin materials are continuing to progress at a fast clip.

Simultaneously, our ability to do rational design of materials is moving forward, owing to innovations in multiscale modeling, model-inversion techniques, uncertainty quantification, and a rapidly growing database of material properties arising from recent initiatives such as the Materials Genome Initiative [68].

These innovations have enabled high throughput in materials design-coupling of experiment and simulation in this framework can drastically reduce the time for material development. They also have led, and will very likely continue to lead, towards discovery of materials with emergent properties that would not have been possible without the advent of materials informatics. There is a current drive toward the development of a "standard model" for two-dimensional material-based heterostructures. The thermo-opto-electronic properties in this novel class of materials are not yet well understood.

Fundamental work on transport properties, doping effects, quantum confinement, and interfacial coupling effects in these materials is needed for realization of their full potential. This is particularly true for applications where the tolerances for material properties would be very narrow, such as low-loss plasmonic circuitry. The burgeoning capability for creation of efficient and predictive models for quasiparticle interaction in these materials has stunning implications for future work, which will likely enable unprecedented tuning of plasmonic properties as well as many other material properties of two-dimensional and quasi-two-dimensional heterostructures at the atomic scale.

6 Recommended Reading

6.1 Theory and General Introduction to Plasmonics

Maier (2007) Plasmonics: fundamentals and applications. Springer, New York.

Oulton RF, Bartal G, Pile DFP, Zhang X (2008) Confinement and propagation characteristics of subwavelength plasmonic modes. New J Phys 10:105018.

6.2 Computational Modeling of Plasmonic Systems

Loke, VLY, Huda GM, Donev EU, Schmidt V, Hastings JT, Mengüç M, Pinar Wriedt T (2013) Comparison between discrete dipole approximation and other modelling methods for the plasmonic response of gold nanospheres. Appl Phys B. doi:10.1007/s00340-013-5594-z

Jensen LL, Jensen L (2009) Atomistic electrodynamics model for optical properties of silver nanoclusters. J Phys Chem C 113 (34):15182–15190.

Chen H, McMahon JM, Ratner MA, Schatz GC (2010) Classical electrodynamics coupled to quantum mechanics for calculation of molecular optical properties: a RT-TDDFT/FDTD approach. J Phys Chem C 114 (34):14384–14392.

Esteban R, Borisov AG, Nordlander P, Aizpurua J (2012) Bridging quantum and classical plasmonics with a quantum-corrected model. Nat Commun 3:825.

Sholl DS, Steckel JA (2009) Density functional theory: a practical introduction. Wiley, New Jersey.

6.3 *Rational Design*

Owhadi H, Scovel C, Sullivan TJ, McKerns M, Ortiz M (2013) Optimal uncertainty quantification. SIAM Rev 55(2):271–345.

Dienstsfrey A, Boisvert RF (2011) Uncertainty quantification in scientific computing. Springer, Heidelberg, Dordrecht, London, New York.

6.4 *Functionalized Plasmonic Heterostructures*

Grigorenko AN, Polini M, Novoselov KS (2012) Graphene plasmonics. Nature Photon 6:749–758.

Novoselov KS and Castro Neto AH (2012) Two-dimensional crystals-based heterostructures: materials with tailored properties. Phys Scrs T146:014006.

Acknowledgments The author thanks Vinod K Tewary, Timothy J. Sullivan, Katie Rice, and Ann Chiaramonti Debay for enlightening and helpful discussions. This research was performed while the author held a National Research Council Research Associateship Award at the National Institute of Standards and Technology. This work represents an official contribution of the National Institute of Standards and Technology and is not subject to copyright in the USA.

References

1. Ritchie RH (1957) Plasma losses by fast electrons in thin films. Phys Rev 106:874–881
2. Nie S, Emory SR (1997) Probing single molecules and single nanoparticles by surface-enhanced Raman scattering. Science 275(5303):1102–1106
3. Ozbay E (2006) Plasmonics: merging photonics and electronics at nanoscale dimensions. Science 311(5758):189–193
4. Liu N, Wei H, Li J, Wang Z, Tian X, Pan A, Xu H (2013) Plasmonic amplification with ultra-high optical gain at room temperature. Sci Rep 3:1967
5. Dionne JA, Atwater HA (2012) Plasmonics: Metal-worthy methods and materials in nanophotonics. MRS Bull 37(8):717–724
6. Stockman M (2007) Criterion for negative refraction with low optical losses from a fundamental principle of causality. Phys Rev Lett 98:177404
7. Kinsler P, McCall MW (2008) Causality-based criteria for a negative refractive index must be used with care. Phys Rev Lett 101(16):167401
8. Dirdal C, Skaar J (2013) Negative refraction in causal media by evaluating polar paths for rational functions. J Opt Soc Am B: Opt Phys 30(2):370–376
9. Boltasseva A, Atwater H (2011) Low-loss plasmonic metamaterials. Science 331 (6015):290–291
10. Rast L, Stanishevsky A (2005) Aggregated nanoparticle structures prepared by thermal decomposition of poly(vinyl)-N-pyrrolidone/Ag nanoparticle composite films. Appl Phys Lett 87
11. Kamat P (2002) J Phys Chem B. Photophysical, photochemical and photocatalytic aspects of metal nanoparticles. 106(32):7729–7744

12. Klar T, Perner M, Grosse S, Von Plessen G, Sprikl W, Feldmann J (1998) Surface-plasmon resonances in single metallic nanoparticles. Phys Rev Lett 80:4249
13. Kelly KL, Coronado E, Zhao LL, Schatz GC (2003) The optical properties of metal nanoparticles: the influence of size, shape, and dielectric environment. J Phys Chem B 107 (3):668–677
14. Draine BT, Flatau PJ (1994) Discrete dipole approximation for scattering calculations. J Opt Soc Am A: 11:1491–1499
15. Draine, BT, Flatau, PJ (2012) User Guide to the Discrete Dipole Approximation Code DDSCAT 7.2. arXiv preprint arXiv: 1202.3424
16. Johnson PB, Christy RW (1972) Optical constants of the noble metals. Phys Rev B 6:4370–4379
17. Doyle WT, Agarwal A (1965) Optical Extinction of Metal Spheres. J Opt Soc Am 55:305–308
18. Sherry LJ, Chang SH, Schatz GC, Van Duyne RP (2006) Localized surface plasmon resonance spectroscopy of single silver triangular nanoprisms. Nano Lett 6(9):2060–2065
19. Chen S, Fan Z, Carroll DL (2002) Silver nanodisks: synthesis, characterization, and self-assembly. J Phys Chem B 106(42):10777:10781
20. Stanishevsky A, Williamson H, Yockell-Leviere H, Rast L, Ritcey AM (2005) Synthesis of complex shape gold nanoparticles in water-methanol mixtures. J Nanosci Nanotechnol 6 (7):2013–2017
21. Barnard AS (2012) Mapping the shape and phase of palladium nanocatalysts. Catal Sci Technol 2:1485–1492
22. Schaefers S, Rast L, Stanishevsky A (2006) Electroless silver plating on spin-coated silver nanoparticle seed layers. Mater Lett 60(5):706–709
23. Bosman M, Ye E, Tan SF, Nijhuis CA, Yang JKW, Marty R, Mlayah A, Arbouet A, Girard C, Han MY (2013) Surface plasmon damping quantified with an electron nanoprobe. Nat Sci Rep 3:1312
24. Ritchie RH, Marusak AL (1966) The surface plasmon dispersion relation for an electron gas. Surf Sci 4(3):234–240
25. Osma J, Garcia de Abajo FJ (1997) Surface effects in the energy loss of ions passing through a thin foil. Phys Rev A 56:2032–2040
26. Yi GC (2012) Semiconductor nanostructures for optoelectronic devices: processing, characterization, and applications. Springer, Heidelberg
27. Lambin Ph, Vigneron JP, Lucas AA (1985) Electron-energy-loss spectroscopy of multilayered materials: theoretical aspects and study of interface optical phonons in semiconductor superlattices. Phys Rev B 32:8203–8215
28. Rast L, Tewary VK (2013) Stratified graphene/noble metal systems for low-loss plasmonics applications. Phys Rev B 87(4):045428
29. Boragno C, Buatier de Mongeot F, Felici R, Robinson IK (2009) Critical thickness for the agglomeration of thin metal films. Phys Rev B 79:155443
30. Rha JJ, Park JK (1997) Stability of the grain configurations of thin films—A model for agglomeration. J Appl Phys 82:1608
31. Atwater HA, Polman A (2010) Plasmonics for improved photovoltaic devices. Nat Mater 9:205
32. Dionne JA, Sweatlock LA, Atwater HA, Polman A (2005) Planar metal plasmon waveguides: frequency-dependent dispersion, propagation, localization, and loss beyond the free electron model. Phys Rev B 72:075405
33. Khurgin JB, Boltassevaa A (2012) Reflecting upon the losses in plasmonics and metamaterials. MRS Bull 37(8):768–779
34. Messinal R, Hugonin JP, Greffet JJ, Marquier F, De Wilde Y, Belarouci A, Luc Frechette L, Cordier Y, Ben-Abdallah P (2013) Tuning the electromagnetic local density of states in graphene-covered systems via strong coupling with graphene plasmons. Phys Rev B 87:085421
35. Ghatak S, Nath Pal A, Ghosh A (2011) Nature of electronic states in atomically thin MoS2 field-effect transistors. ACS Nano 5(10):7707–7712

36. Ni Z, Liu Q, Tang K, Zheng J, Zhou J, Qin R, Gao Z, Yu D, Lu J (2012) Tunable Bandgap in Silicene and Germanene. Nano Lett 12(1):113–118
37. Berini P (1999) Plasmon polariton modes guided by a metal film of finite width. Opt Lett 24 (15):1011–1013
38. Koppens FHL, Chang DE, Garcia de Abajo FJ (2011) Graphene plasmonics: A platform for strong light-matter interaction. Nano Lett 11(8):3370–3377
39. Novoselov KS, Jiang D, Schedin F, Booth TJ, Khotkevich VV, Morozov SV, Geim AK (2005) Two-dimensional atomic crystals. PNAS 102(30):10451–10453
40. Xue J, Sanchez-Yamagishi J, Bulmash D, Jacquod P, Deshpande A, Watanabe K, Taniguchi T, Jarillo-Herrero P, LeRoy BJ (2011) Scanning tunnelling microscopy and spectroscopy of ultra-flat graphene on hexagonal boron nitride. Nat Mater 10:282–285
41. Li L, Wang X, Zhao X, Zhao M (2013) Moiré superstructures of silicene on hexagonal boron nitride: a first-principles study. Phys Lett A 377(38):2628–2632
42. Kuisma M, Ojanen J, Enkovaara J, Rantala TT (2010) Kohn-Sham potential with discontinuity for band gap materials. Phys Rev B 82:115106
43. Mortensen JJ, Hansen LB, Jacobsen KW (2005) Real-space grid implementation of the projector augmented wave method. Phys Rev B 71:035109
44. Enkovaara J, Rostgaard C, Mortensen JJ et al (2010) Electronic structure calculations with GPAW: a real-space implementation of the projector augmented-wave method. J Phys: Condens Mater 22:253202
45. Walter M, Hakkinen H, Lehtovaara L, Puska M, Enkovaara J, Rostgaard C, Mortensen JJ (2008) Time-dependent density-functional theory in the projector augmented-wave method. J Chem Phys 128:244101
46. Yan J, Mortensen JJ, Jacobsen KW, Thygesen KS (2011) Linear density response function in the projector augmented wave method: applications to solids, surfaces, and interfaces. Phys Rev B 83:245122
47. Yan J, Jacobsen KW, Thygesen KS (2012) Optical properties of bulk semiconductors and graphene/boron nitride: The Bethe-Salpeter equation with derivative discontinuity-corrected density functional energies. Phys Rev B 86:045208
48. Wei W, Dai Y, Huang B, Jacob B (2013) Many-body effects in silicene, silicane, germanene and germanane. Phys Chem Chem Phys 15:8789–8794
49. Rozzi CA, Varsano D, Marini A, Gross EKU, Rubio A (2006) An exact Coulomb cutoff technique for supercell calculations. Phys Rev B 73:205119
50. Wang QH, Kalantar-Zadeh K, Kis A, Coleman JN, Strano MS (2012) Electronics and optoelectronics of two-dimensional transition metal dichalcogenides. Nature Nanotech 7:699–712
51. Lee C, Wei X, Kysar JW, Hone J (2008) Measurement of the elastic properties and intrinsic strength of monolayer graphene. Science 321(5887):385–388
52. Dereux A, Vigneron JP, Lambin Ph, Lucas AA (1988) Polaritons in semiconductor multilayered materials. Phys Rev B 38:5438–5452
53. Economou EN (1969) Surface plasmons in thin films. Phys Rev 182(2):539–554
54. Murarka SP, Eizenberg M, Sinha AK (2003) Interlayer Dielectrics for Semiconductor Technologies. Elsevier, Amsterdam
55. Yi GC (2012) Semiconductor Nanostructures for Optoelectronic Devices: Processing, Characterization, and Applications. Springer, Heidelberg, Dordrecht, London, New York
56. Nikitin AY, Guinea F, Martin-Moreno L (2012) Resonant plasmonic effects in periodic graphene antidot arrays. Appl Phys Lett 101(15):151119
57. Schultz MH, Jauho AP, Pedersen TG (2011) Screening in graphene antidot lattices. Phys Rev B 84:045428
58. Zhang YT, Li QM, Li YC, Zhang YY, Zhai F (2010) Band structures and transport properties of zigzag graphene nanoribbons with antidot arrays. J Phys: Condens Matter 22:315304
59. Pedersen JG, Pederson TG (2013) Hofstadter butterflies and magnetically induced band-gap quenching in graphene antidot lattices. Phys Rev B 87:235404

60. Gotschy W, Vonmetz K, Leitner A, Aussenegg FR (1996) Thin films by regular patterns of metal nanoparticles: Tailoring the optical properties by nanodesign. Appl Phys B 63 (4):381–384
61. Hulteen JC, Treichel DA, Smith MT, Duval ML, Jensen TR, Van Duyne RP (1999) Nanosphere lithography: size-tunable silver nanoparticle and surface cluster arrays. J Phys Chem B 103:3854–3863
62. Hamley IW (2003) Nanostructure fabrication using block copolymers. Nanotech 14(10):R39
63. Chu H, Wang J, Ding L, Yuan D, Zhang Y, Liu J, Li Y (2009) Decoration of gold nanoparticles on surface-grown single-walled carbon nanotubes for detection of every nanotube by surface-enhanced raman spectroscopy. J Am Chem Soc 131(40):14310–14316
64. Zedan AF, Moussa S, Terner J, Atkinson G, El-Shall MS (2012) Ultrasmall gold nanoparticles anchored to graphene and enhanced photothermal effects by laser irradiation of gold nanostructures in graphene oxide solutions. ACS Nano 7(1):627–636
65. Shi Y, Huang JK, Jin L, Hsu YT, Yu SF, Li LJ, Yang HY (2013) Selective decoration of au nanoparticles on monolayer mos2 single crystals. Sci Rep 3:1839
66. Zaniewski AM, Schriver M, Lee JG, Crommie MF, Zettl A (2013) Electronic and optical properties of metal-nanoparticle filled graphene sandwiches. Appl Phys Lett 102:023108
67. Lin J, Li H, Zhang H, Chen W (2013) Plasmonic enhancement of photocurrent in MoS_2 field-effect-transistor. Appl Phys Lett 102(20):203109
68. Materials Genome Initiative for Global Competitiveness http://www.whitehouse.gov/sites/default/files/microsites/ostp/materials_genome_initiative-final.pdf
69. Eberlein T, Bangert U, Nair RR, Jones R, Gass M, Bleloch AL, Novoselov KS (2008) Plasmon spectroscopy of free-standing graphene films. Phys Rev B 77(23):223406
70. Yang L, Deslippe J, Park CH, Cohen ML, Louie SG (2009) Excitonic effects on the optical response of graphene and bilayer graphene. Phys Rev Lett 103(18):186802
71. Cudazzo P, Attaccalite C, Tokatly IV, Rubio A (2010) Strong charge-transfer excitonic effects and the Bose-Einstein exciton condensate in graphene. Phys Rev Lett 104(22):226804
72. Bechstedt F, Matthes L, Gori P, Pulci O (2012) Infrared absorbance of silicene and germanene. Appl Phys Lett 100(26):261906
73. Chinnathambi K, Chakrabarti A, Banerjee A, Deb SK (2012) Optical properties of graphene-like two dimensional silicene. arXiv preprint arXiv:1205.5099
74. Coleman J, Lotya M, O'Neill A et al (2011) Two-dimensional nanosheets produced by liquid exfoliation of layered materials. Science 331:568–571
75. Johari P, Shenoy VB (2011) Tunable dielectric properties of transition metal dichalcogenides. ACS Nano 5(7):5903–5908
76. Ma Y, Dai Y, Guo M, Yu L, Huang B (2013) Tunable electronic and dielectric behavior of GaS and GaSe monolayers. Phys Chem Chem Phys 15:7098–7105
77. Lin X, Lin S, Xu Y et al (2013) Ab initio study of electronic and optical behavior of two-dimensional silicon carbide. J Mater Chem C 1:2131–2135

Gotschy W, Vonmetz K, Leitner A, Aussenegg FR (1996) Thin films by regular patterns of metal nanoparticles: Tailoring the optical properties by nanodesign. Appl Phys B 63 [illegible]

[illegible]

Application of Graphene Within Optoelectronic Devices and Transistors

F.V. Kusmartsev, W.M. Wu, M.P. Pierpoint and K.C. Yung

Abstract Scientists are always yearning for new and exciting ways to unlock graphene's true potential. However, recent reports suggest this two-dimensional material may harbor some unique properties, making it a viable candidate for use in optoelectronic and semiconducting devices. Whereas on one hand, graphene is highly transparent due to its atomic thickness, the material does exhibit a strong interaction with photons. This has clear advantages over existing materials used in photonic devices such as Indium-based compounds. Moreover, the material can be used to 'trap' light and alter the incident wavelength, forming the basis of the plasmonic devices. We also highlight upon graphene's nonlinear optical response to an applied electric field, and the phenomenon of saturable absorption. Within the context of logical devices, graphene has no discernible band-gap. Therefore, generating one will be of utmost importance. Amongst many others, some existing methods to open this band-gap include chemical doping, deformation of the honeycomb structure, or the use of carbon nanotubes (CNTs). We shall also discuss various designs of transistors, including those which incorporate CNTs, and others which exploit the idea of quantum tunneling. A key advantage of the CNT transistor is that ballistic transport occurs throughout the CNT channel, with short channel effects being minimized. We shall also discuss recent developments of the graphene tunneling transistor, with emphasis being placed upon its operational mechanism. Finally, we provide perspective for incorporating graphene within high frequency devices, which do not require a pre-defined band-gap.

Keywords Graphene · Indium-based compounds · Plasmonic devices · Nonlinear optical response · Saturable absorption · Zero band-gap · Pre-defined band gap · Chemical doping · Deformation · Carbon nanotubes (CNTs) · Honeycomb

F.V. Kusmartsev (✉) · W.M. Wu · M.P. Pierpoint
Department of Physics, Loughborough University, Loughborough LE11 3TU, UK
e-mail: f.kusmartsev@lboro.ac.uk

K.C. Yung
Department of Industrial and Systems Engineering, The Hong Kong Polytechnic University, Hung Hom, Kowloon, Hong Kong, China

P. Misra (ed.), *Applied Spectroscopy and the Science of Nanomaterials*,
Progress in Optical Science and Photonics 2, DOI 10.1007/978-981-287-242-5_9

structure · Quantum tunneling · Nanomaterials · Nanotransistors · Optoelectronics · High frequency devices

1 Introductom

Two-dimensional materials have always been considered unstable due to their thermal fluctuations [1, 2], in what were famously referred to as the Landau-Peierls arguments. However, many scientists have not given up hope that such two-dimensional structures exist. In 2004, a research team based in Manchester successfully segregated graphene flakes from a graphite sample via 'mechanical exfoliation' (more commonly referred to as the scotch-tape method) [1, 3–6]. They witnessed a full preservation of graphene's hexagonal honeycomb structure, with astounding electrical, thermal and optical characteristics.

Graphene is an allotrope of carbon—other examples include diamond, fullerene and charcoal, all with their own unique properties. Usually graphene will be found in the form of highly ordered pyrolytic graphite (HOPG), whereby individual graphene layers stack on top of one another to form a crystalline lattice. Its stability is due to a tightly packed, periodic array of carbon atoms [7] (cf. Fig. 1), and an sp^2 orbital hybridization—a combination of orbitals p_x and p_y that constitute the σ-bond. The final p_z electron makes up the π-bond, and is key to the half-filled band which permits free-moving electrons [8]. In total, graphene has three σ-bonds and one π-bond. The right-hand portion of Fig. 1, emphasizes how small displacements of the sub-lattices A and B can be shifted in the z-direction [9].

Moreover, graphene's mode of preparation will have a strong influence upon its overall quality and characteristics. As conducted by Geim et al. [10], mechanical exfoliation consists of gradually stripping more and more layers from a graphite sheet, until what remains are a few layers of graphene. In terms of overall mobility and the absence of structural defects, this method will produce the highest quality material. Other methods such as vacuum epitaxial growth or chemical vapour deposition (CVD), each have their own merits, but will generally lead to inferior quality. For a more in-depth discussion of the available manufacturing methods, one can refer to [1, 11–21].

On the other hand, graphene is highly impermeable [7]—the mobility can become severely compromised upon molecular attachment. Yet, this apparent flaw has immediate applications for molecular sensors. By monitoring the deviation of electrical resistivity [22, 23], one could, for example, envision novel smoke detection systems. So too is graphene more than 100 times stronger than steel [7], possessing a Young's modulus as large as 1 TPa [24]. Together with its outstanding electrical [8], thermal [25–27] and in particular, optical properties [11, 24, 28, 29], graphene has thus become a widely sought after material for use in future semiconducting and optoelectronic devices [12, 30].

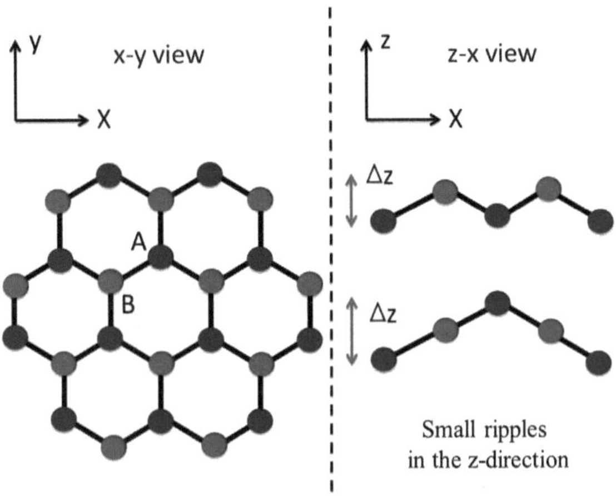

Fig. 1 The honeycomb structure of graphene is presented in the *left-hand* figure. The *right-hand* figure depicts small quantum corrugations of the sub-lattices A and B, which are shifted in the transverse direction by a small fraction of the inter-atomic spacing 'a'

Electrical Mobility—As a material, graphene harbors some remarkable qualities; highly elastic due to its monolayer structure, and more conductive than copper with mobilities reaching up to 2,00,000 $cm^2\ V^{-1}\ s^{-1}$ for perfect structures [24, 31, 32]. Charge carriers in graphene travel with a Fermi velocity $v_F = \sqrt{3}\gamma_0 a/2 \approx 10^6\ ms^{-1}$. Here, $\gamma_0 \approx 3$ eV is the energy required to 'hop' from one carbon atom to the nearest neighbor, $a \approx 1.42$ Å is the inter-atomic spacing between two neighboring carbon atoms, and ħ is Planck's constant [8, 24, 33]. This Fermi velocity is approximately 1/300 the speed of light, thus presenting a miniaturized platform upon which to test many features of quantum electrodynamics (QED) [1]. Theoretical studies with graphene show that the density of states (DoS) of electrons approaches zero at the Dirac point. However, a minimum conductivity $\sigma_0 \sim 4e^2/h$ has been displayed [1], which is approximately double that for the conductance quantum [34, 35]. Even at room temperature, electrons can undergo long range transport with minimal scattering [1, 32, 36].

Thermal Conductivity—Heat flow in suspended graphene was recently shown to be mediated by ballistic phonons, and has been verified by Pumarol et al. [25] with the use of high resolution vacuum scanning thermal microscopy. However, when considering multiple layers of graphene, this transport will be reduced due to an increase of inelastic scattering. The same is observed for graphene coated upon a substrate—the mean free path of thermal phonons degrading to less than 100 nm. Nevertheless, graphene on a silicon substrate can still retain a thermal conductivity of around 600 $Wm^{-1}\ K^{-1}$ [26]—even higher than copper. Whilst the mechanism of heat transport across the graphene-substrate interface remains unknown [37], it is possible this may be linked to the in-plane thermal conductivity [25, 27].

Optical Response—Graphene's atomic thickness makes it almost perfectly transparent to visible light [9, 38], allowing such a material to become widely accessible to a number of applications. These cover everything from photovoltaic cells, to graphene photonic transistors [30, 33, 39, 40]. Being a single layer of carbon atoms, graphene also exhibits many interesting photonic properties. As such, our focus will be directed mainly upon those which are associated with applications to optoelectronic devices. The transmittance between multiple graphene layers, how optical frequency relates to conductivity, nonlinear optical response, saturable absorption and plasmonics will all be discussed in later sections.

Most semiconducting photonic devices will be governed in some way by electron excitation and electron-hole recombination. Excitation refers to an electron absorbing photon energy of a very specific wavelength within the allowed energy bands. On the other hand, recombination is a process which leads to the emission of photons (cf. electro-luminescence) [38]. Gallium arsenide (GaAs), indium functional compounds and silicon are all common semiconductors for use in photonic devices [41, 42]. However, graphene exhibits a strong interaction with photons with the potential for direct band-gap creation, thus being a good candidate for optoelectronic and nanophotonic devices [43]. Its strong interaction with light arises due to the Van Hove singularity [44]. Graphene also possesses different time scales in response to photon interaction, ranging from femtoseconds (ultra-fast) to picoseconds [43, 45]. Overall, graphene could easily be an ideal candidate for transparent films, touch screens and light emitting cells. It may even be used as a plasmonic device which confines light, and altering the incident wavelength. We shall elaborate upon this in later sections.

2 Energy Spectrum, Band-Gap and Quantum Effects

Theoretical studies of monolayer graphite (i.e., graphene) first began in 1947 by Wallace [8], who considered a simple tight-binding model with a single hopping integral. This model takes into account the hopping of an electron from one carbon atom to its first and second nearest neighbors only. Wallace's conclusions were stark; an electrical conductivity should *theoretically* exist for two-dimensional graphene. To elaborate; at six positions of the Brillouin zone, Dirac points (K and K′) exist. These are points in momentum space for which the energy $E(\mathbf{p}_0) = 0$, where $\mathbf{p}_0 = \hbar\,\mathbf{K}$ (or $\hbar\,\mathbf{K}'$). Here, we have denoted the momentum as a vector $\mathbf{p} = (p_x, p_y) = \hbar\,\mathbf{k}$, where $\mathbf{k} = (k_x, k_y)$ is the wave vector [1]. The energy eigenvalues were found to take a gapless form [8],

$$E^{\pm}(k_x, k_y) = \pm\gamma_0\sqrt{1 + 4\cos\left(\frac{\sqrt{3}}{2}k_x a\right)\cos\left(\frac{1}{2}k_y a\right) + 4\cos^2\left(\frac{\sqrt{3}}{2}k_x a\right)} \quad (1)$$

where the plus and minus signs refer to the upper and lower half-filled bands respectively [24, 34]. By expanding the above equation in the vicinity of the K or K′ points, one can thus obtain a linear dispersion relation that is given by $E^{\pm} = \pm v_F\,\hbar$

$|\delta\mathbf{k}|$, where $\mathbf{k} = \mathbf{K} + \delta\mathbf{k}$. These constitute what are known as Dirac cones, and are clearly emphasized by Fig. 2. Here, a direct contact of the conduction and valence bands is found [5, 8, 10, 11], thus pertaining to a zero energy band-gap E_g [1, 8, 12]. Therefore, generating a band-gap in graphene will be essential for its application within semiconducting devices (e.g., transistors). On the other hand, graphene may secure its place in high-frequency devices, which do not require a logical OFF state [38].

A. Dirac Energy-Momentum Dispersion

Supposing we consider the Hamiltonian $\hat{H}$ as given by Wallace [8]—in the low energy limit, spinless carriers in graphene possess a zero effective mass, and are well approximated by the relativistic Dirac Hamiltonian $\hat{H}$ [34],

$$\hat{H} = v_F \mathrm{h} \hat{\sigma} \cdot \delta\mathbf{k} \tag{2}$$

where $\hat{\sigma} \cdot \delta\mathbf{k} = \sigma_x \delta k_x + \sigma_y\, \delta k_y$. Here, $\hat{\sigma} = (\sigma_x, \sigma_y)$ is the vector of 2 × 2 Pauli matrices:

$$\sigma_x = \begin{pmatrix} 0 & 1 \\ 1 & 0 \end{pmatrix} \qquad \sigma_y = \begin{pmatrix} 0 & -i \\ i & 0 \end{pmatrix} \tag{3}$$

The spinor wave function ψ of graphene can be obtained from,

$$\hat{H}\psi = E\psi, \tag{4}$$

where E denotes the energy eigenvalues of $\hat{H}$ [24]. Here, $\psi = (\psi_A, \psi_B)^T$ is a vector containing the two component wave function. These components represent the sub-lattices A and B accordingly [34].

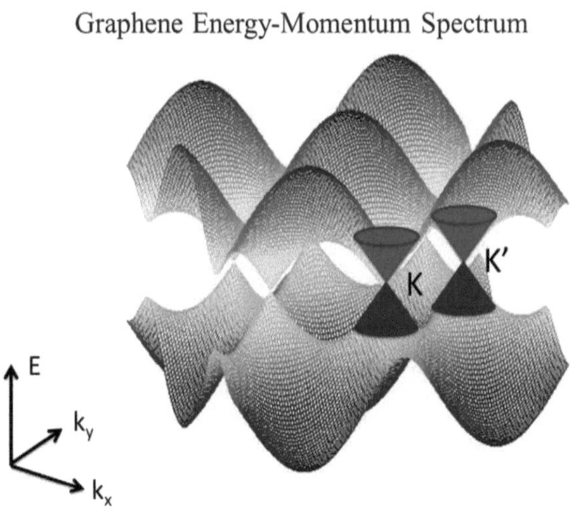

Fig. 2 The energy-dispersion spectrum as given by Eq (1). Here, the z-axis represents the energy $E(\mathbf{k})$, with the x–y plane corresponding to the momentum $\mathbf{k} = (\mathbf{k}_x, \mathbf{k}_y)$. Dirac cones are located at both the **K** and **K′** points of the Brillouin zone

B. Band-Gap Creation

Generally speaking, the electrical conductivity of a material can fall into one of three groups: conductors, insulators, semiconductors [46–48]. For a conductor, electrons are able to move freely in the conduction band since electron states are not fully occupied. However, the conduction and valence band may sometimes be separated by an energy band-gap E_g (e.g., for insulators and semiconductors), thus preventing the free movement of electrons in the conduction band. For an insulator, an electron requires a huge energy in order to excite from the valence to conduction band. A small band-gap is present for semiconductors, with an electronic band structure that is parabolic in shape [12, 31]. Doped semiconductors will make the band-gap even smaller, and hence more easy to control (cf. Fig. 3).

Graphene's high mobility makes it a particularly enticing material for use in electronic devices. However, we have already mentioned that in the vicinity of the Dirac point, graphene possesses a conical band structure which is *gapless* (i.e., $E_g = 0$) [1, 10]. Thus our main concern with regards to logical devices, is the absence of this well-defined OFF state pertaining to zero current flow. To rectify this, we must open up an energy band-gap such that $E_g \neq 0$. With regards to optoelectronic devices, a tunable band-gap can specify the range of wavelengths which can be absorbed. The energy bands for pure graphene, and graphene with a small band-gap E_g are displayed in Fig. 3. The Fermi energy level E_F is situated at the Dirac point for pure graphene [3]. For graphene that has been modified to

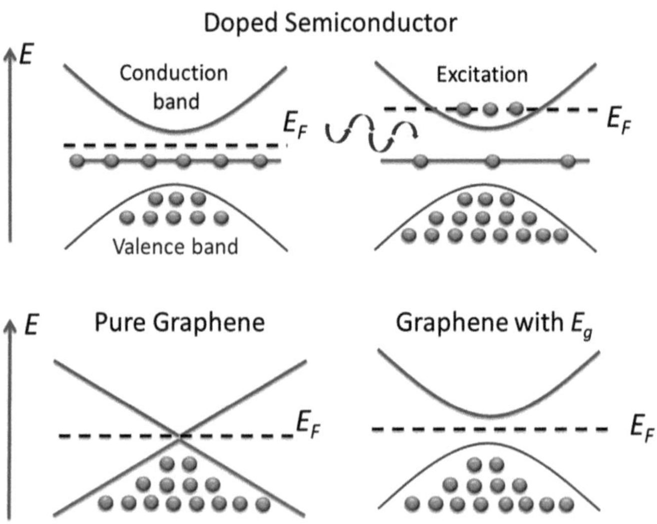

Fig. 3 The *upper half* of this figure depicts the electronic band structure of a doped semiconductor. Typically, the band-gap for a doped semiconductor is very small, with only a small energy being required to excite an electron from the valence to conduction band. The *lower* figure shows the electronic band structure for graphene. For pure samples, no energy band-gap E_g exists. In principle, an energy band-gap E_g can be created via many methods

include a band-gap, an energy is required to excite electrons from the valence to conduction band, and hence an ON/OFF state regime is established [10]. Amongst many others, existing methods include the use of carbon nanotubes (CNTs), graphene nanoribbons (GNRs) or even bilayer graphene [1, 31, 38, 49]. However, it is important to note that although bilayer graphene does possess a zero energy band-gap, an applied electric field can be used to create one [9, 12, 31]. Other methods include deformed structures, graphene oxide (GO) [12, 22, 23], and also the use of chemical doping via compounds such as Boron Nitride (BN) [50, 51]. The idea here, is that the doped atoms alter graphene's honeycomb structure, similar to deformation or localized defects [52, 53]. All in all, one has to note that the aforementioned methods are not well-developed enough to maintain a high mobility. Much more exotic concepts are required, which we shall now discuss.

C. Quantum Phenomena

Of its more surprising attributes, graphene has also displayed signs of anomalous quantum behaviors, even at room temperature [5, 54]. We shall briefly discuss two key phenomena in particular.

Quantum Hall Effect: QHE has been observed for both single and bilayer graphene [11, 55, 56], in the presence of a magnetic field B. The Landau levels for graphene are given by,

$$E_{Landau} = \sqrt{|2e\text{h}v_F^2 \text{B}j|} \tag{5}$$

where e is the electric charge, and $j \in \mathbb{Z}$ is the Landau index [5, 54]. In conventional 2-D semiconductors, the Landau levels are $E = \hbar\omega_c\ (j + 1/2)$, where ω_c is the cyclotron frequency [5, 54]. The anomalous energy spectrum for graphene subject to a B field leads to a one half shift of the minimum conductivity at the zero energy Landau level, whereas traditional QHE semiconductors give an integer one [5, 54]. The Hall conductivity σ_{H} is therefore givenby [24, 34],

$$\sigma_H = g\left(j + \frac{1}{2}\right)\frac{e^2}{h} \tag{6}$$

where g is the degeneracy. For graphene, a fourfold degeneracy exists—two spins, and the valley degeneracy of the K and K′ Dirac points [5]. Additionally, the fractional QHE has been observed for both monolayer and bilayer graphene (cf. for details [5, 55, 56]).

Klein Tunneling: Intuition states that if a particle's kinetic energy *KE* is less than some value *U*, then it will be physically incapable of surpassing a potential barrier of the same energy *U*. However, quantum mechanics states that a particle is able to tunnel the potential barrier *U* with a certain decay probability [5]. Furthemore, *relativistic* quantum mechanics permits a remarkable phenomenon called Klein tunneling. Much like a freight train instead taking a tunnel from one side of a mountain to the other, an electron can perform a similar process [54, 57].

This occurs when an electron experiences a strong repulsive force from the barrier U, and hence induces a hole inside the barrier [5, 55, 56]. This leads to a matching of the energy spectrum inside and outside the barrier, with the transmission probability becoming very close to one [54]. A perfect transmission is demonstrated for square potentials only, and is dependent upon the energy KE, and the angle of incidence θ relative to the barrier [58]. Confined bound states will arise for energies close to the Dirac point [58]. Further details regarding how this confinement effect may relate to the special waveguide geometry has been discussed in references [59–63].

3 Photonic Properties

Optical communication networks are ubiquitous nowadays, affecting our everyday lives. A fiber-optic cable provides a much wider bandwidth, and less energy loss than some traditional copper wiring [41, 42]. According to the Shannon-Hartley theorem [64], the maximum capacity of a channel is given by

$$\max\{C\} = B\log_2\left(1 + \frac{P_s}{P_n}\right),$$

where B is the channel bandwidth, and P_s and P_n are the average signal and noise powers respectively. It is therefore obvious that optical cable provides a much larger channel capacity, where $P_s / P_n \gg 1$.

When optical and electronic devices work together (e.g., a modulator), light signals are converted into an electrical equivalent. Generally speaking, the term 'optoelectronic' refers to an optical (photonic) electronic device, which transmits signals via light waves, or electron- photon interaction [41, 42]. A photonic device can be made of semiconductors, either being integrated into electronic circuits or transistors. Optoelectronics also play an important role as the mediator of optical communication. Devices will typically operate with an optical frequency ranging from ultraviolet to infrared (400–700 nm) [41, 42], although graphene photonic devices can possess an even wider spectrum than this [65].

A. Transmittance Properties

As emphasized by Fig. 4, a single layer of graphene absorbs a mere 2.3 % of incident light, allowing around 97.7 % to pass through. Wavelengths typically range from the infrared to ultraviolet regions [33]. The transmittance T of single-layer graphene (SLG) can be approximated by the following Talyor expansion [33, 39, 40, 66, 67]

$$T = \frac{1}{(1 + \alpha\pi/2)^2} \approx (1 - \alpha\pi) \approx 97.7\,\% \tag{8}$$

where $\alpha = e^2/c\hbar \approx 1/137$ is the fine structure constant. For multiple layers of graphene, this can be roughly estimated by

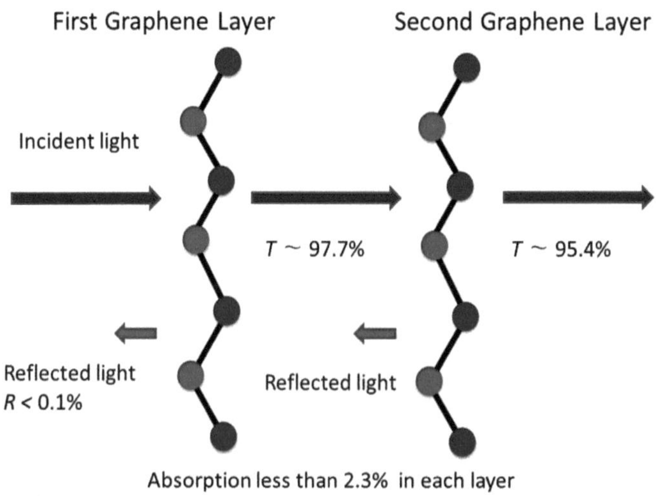

Fig. 4 Incident light passes through the to layers of graphene. The transmission, absorption and reflection coefficients are all shown. Each layer of graphene only absorbs 2.3 % of incident light, transmitting around 97.7 %, and reflecting less than 0.1 %

$$T \approx (1 - N\alpha\pi), \tag{9}$$

where N is number of layers (cf. Bao et al. [39]). For example, the transmittance of bilayer graphene ($N = 2$) is around 95.4 % (cf. Fig. 4). Indium Tin Oxide (ITO) is a semiconductor which is typically used in photonic devices, with a transmittance of around 80 % [33]. It is therefore obvious that graphene film has a clear advantage over ITO. Bonaccorso et al. [33] also point out that the resistance per unit area for ITO is much smaller than for graphene. However, this value can be minimized by increasing the concentration of charge carriers via methods such as doping.

The degree of reflection from SLG is almost negligible, just less than 0.1 % [33]. Avouris et al. [40, 66, 67] also mention that graphene shows a strong interaction of photons, much stronger than some traditional photonic materials per unit depth. It is also surprising that absorption can rise from 2.3 % to around 40 % with high concentration doping [40, 66, 67]. Unquestionably, these properties present graphene as an excellent candidate for use in photonic applications.

B. Optical Conductivity

As mentioned by Avouris et al. [40, 66, 67], graphene possesses a universal optical conductance $G_{op} = e^2/4\hbar$. In general, the optical conductivity $\sigma_{\rm op}$ depends upon the frequency ω, Fermi energy E_F (via chemical doping or an applied gate voltage), and the transition rate Γ. Moreover, the optical conductivity can be divided into real and imaginary components, $\Re(\sigma_{\rm op})$ and $\Im(\sigma_{\rm op})$ [33, 39, 40, 66, 67],

$$\sigma_{\text{op}}(\omega, E_F, \Gamma) = \Re(\sigma_{\text{op}}) + i\Im(\sigma_{\text{op}}), \tag{10}$$

with energy loss originating from the imaginary part [40, 66, 67].

Bao et al. [39] further explain that the interband and intraband carriers' transitions are the major factors governing the optical conductivity σ_{op} (cf. Fig. 5). Interband transitions refer to an exchange of charge carriers between the conduction and valence bands, whereas intraband transitions refer to a 'jump' between quantized energy levels. For carriers performing an interband transition (at high frequency), the energy of a photon $\hbar\omega$ should be satisfying the relationship $\hbar\omega \geq 2E_F$ [39]. For the low frequency THz range ($\hbar\omega < 2E_F$), the intraband transition would be a significant contribution to the optical conductivity, while interband transitions are prohibited in this range due to the Pauli exclusion principle (Pauli block) [67]. It is important to note that a change in doping concentration would alter the Fermi energy E_F, and hence the optical conductivity. Bao et al. [39] state that one can tune the optical conductivity by controlling the chemical doping (shift of E_F) and the frequency response. However, one must remain aware that a high doping concentration may deteriorate the transmittance T of graphene itself.

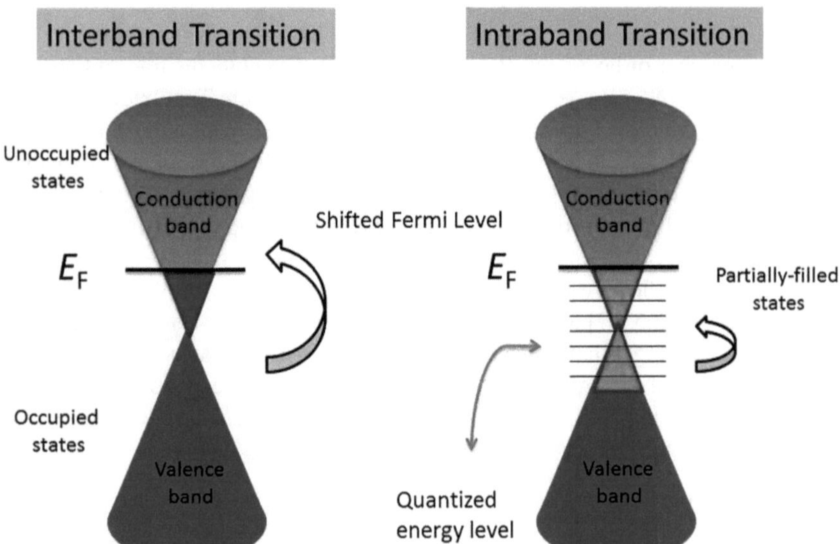

Fig. 5 The Fermi energy level E_F can be shifted upward due to either chemical doping or an applied electric field. Interband transitions refer to an electron 'jumping' from the valence to conduction bands, satisfying the relationship $\hbar\omega \geq 2E_F$. Intraband transitions refer to an electron moving through quantized energy levels, and requires less energy ($\hbar\omega < 2E_F$), provided the states are not already occupied

C. Linear and Nonlinear Optical Response

Graphene also exhibits a strong nonlinear optical response to an electric field, and is an important factor in modifying the shape of the wavefront for incident light [40, 66, 67]. The displacement field D_z is given by the dielectric response of an applied electric field E_z along the 'z' direction, with polarization $P(E_z)$ (cf. Fig. 6);

$$D_z = \epsilon_0 \epsilon_r E_z = \epsilon_0 E_z + P(E_z). \tag{11}$$

Here, ϵ_0 is the electric permittivity of free space, and ϵ_r is the relative permittivity. The polarization response can be written in terms of a power series (cf. for details [39, 68])

$$P(E_z) = C_0 + \epsilon_0 \sum_{j=1}^{\infty} \chi_j (E_Z)^j, \tag{12}$$

where C_0 is a constant associated with the hysteresis (typically $C_0 = 0$), χ_j refers to the dielectric susceptibility of the jth order correction, and $(E_z)^j$ is the j-th power of E_z. The linear dielectric susceptibility χ_1 can again be divided into a real part χ_{R1} and an imaginary part χ_{I1} [39]. The relative dielectric constant can then be expressed in terms of $\epsilon_r = \chi_{R1} + 1$, with an optical refractive index n_{op} given by,

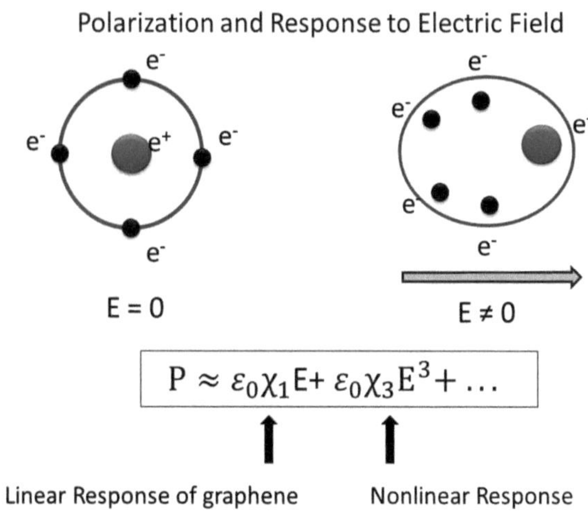

Fig. 6 This figure shows the linear and nonlinear responses to an electric field E. The *left-hand* figure schematically represents an atom without electric field, whilst the *right-hand* figure has a non-zero electric field. The equation is related to the polarization response P. The linear susceptibility χ_1 is usually associated with the refractive index, whereas the nonlinear susceptibility χ_3 provides a unique contribution to the optical properties of graphene. Due to the symmetry of graphene's honeycomb structure, the χ_2 component is very small, and is therefore neglected here

$$n_{op} \approx \sqrt{\epsilon_r} = \sqrt{\chi_{R1} + 1} \tag{13}$$

Thus, the refractive index is determined by the real part of the linear susceptibility χ_{R1}, as mentioned by Bao et al. [39]. Meanwhile, the imaginary part of the linear susceptibility χ_{I1} corresponds to the tangent loss arising at optical frequencies. Bao et al. [39] also come to the conclusion that the 2nd order susceptibility χ_2 is generally small, provided that the symmetry of the graphene honeycomb structure is not broken (i.e., flat). The major contribution to the nonlinear response of graphene originates via the 3rd order term $\epsilon_0\chi_3 E_z^3$, which modifies the current density in graphene (cf. for details [39]).

D. Surface Plasmons

Surface plasmons describe a set of quantized charge oscillations of electrons and holes, acting upon the graphene-substrate interface [39] (cf. Fig. 7). Plasmons, in general, interact with photons or phonons to form the surface plasmon polariton (SPP). At present, aluminium, silver and gold are all ideal materials for plasmonic platforms [39, 40, 66, 67]. The basic idea is as follows—a dielectric material can be coated upon a graphene layer. Electrons then oscillate on the graphene-substrate interface, excited by the phonon or photon interactions of electromagnetic (EM) fields [39]. The SPP wavelength λ_{SPP} is normally suppressed, and much smaller than the incident wavelength λ_{in}—the ratio of these wavelengths typically being around $\lambda_{in}/\lambda_{SPP} \approx 10$–$100$ [39, 40, 66, 67]. The plasmonic frequency ω_{SPP} on the graphene surface is proportional to the square root of the Fermi energy, as given by [33, 39, 40]

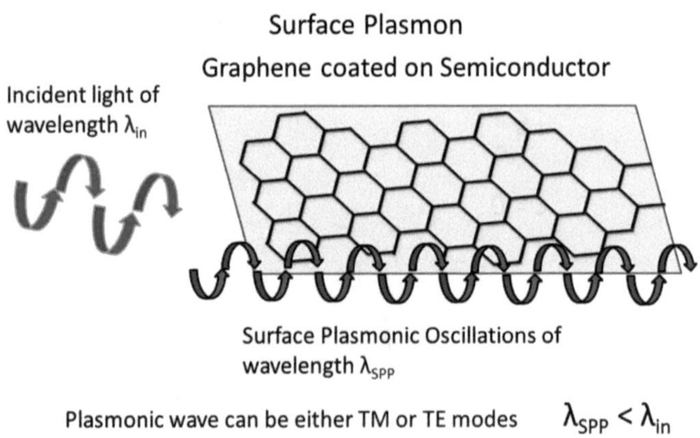

Fig. 7 This figure shows the surface plasmonic wave for graphene coated on a semiconductor. Plasmonic waves are trapped, and oscillate along the graphene and semiconductor interface. Typically, the surface plasmonic wavelength λ_{SPP} will be suppressed, and is much smaller than the incident wave λ_{in}. Either TM or TE wave modes can propagate along the plasmonic surface, depending upon the imaginary component of the optical conductivity

$$\omega_{\mathrm{SPP}} \propto \sqrt{E_F} \propto n^{\frac{1}{4}} \quad (14)$$

where n is a carrier density. In practice, graphene can trap incident light, and an EM wave can propagate along the graphene surface in the THz to infrared range [40, 66, 67]. As mentioned by Avouris et al. [40, 66, 67], the distance traveled for a plasmonic wave in graphene is around $d_{\mathrm{SPP}} \approx$ 10–100 λ_{SPP}. Graphene is thus a suitable material for a waveguide. Bao et al. [39] further remark that graphene is suitable for guiding transverse magnetic (TM) waves when the imaginary part of the optical conductivity $\Im(\sigma_{\mathrm{op}}) > 0$, and suitable for guiding transverse electric (TE) waves when $\Im(\sigma_{\mathrm{op}}) < 0$.

E. Saturable Absorption and Optical Excitation

There is an interesting property which prevents graphene from absorbing photons at high intensity, and can be used to adjust the wavefront of the light [40, 66, 67]. This is referred to as 'saturable absorption', and is dependent upon the wavelength and incident light intensity. This will be elaborated upon in later sections when we discuss the saturable absorber and photonic (optical) limiter.

The timescale of graphene's response to the interaction of photons, phonons and electron hole recombination can be divided into three regimes [33, 39, 40, 66, 67] (cf. Fig. 8). Graphene has a very quick response to incident photons, around 10–100 fs, whereby 'hot' electrons are excited from the valence to conduction band [33, 39]. This also links to an excitation of the non-equilibrium state. Electrons may then cool down via the intraband phonon emission, with timescales of 0.1–1 ps [40, 66, 67]. Finally, an electron-hole pair may recombine, thus emitting photons, and an equilibrium state being reached. This process takes a mere 1−10 ps. It is important

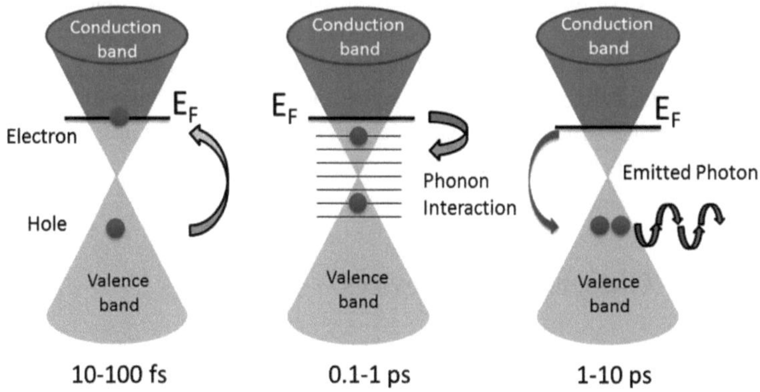

Fig. 8 There are three time scales associated with optical response. The *left-hand* figure represents the inter-band non-equilibrium excitation, and lasts around 10–100 fs. The *middle* figure relates to phonon cooling via the intra-band interaction (0.1–1 ps). Finally, the *right-hand* figure is the process of electron-hole recombination (1–10 ps)

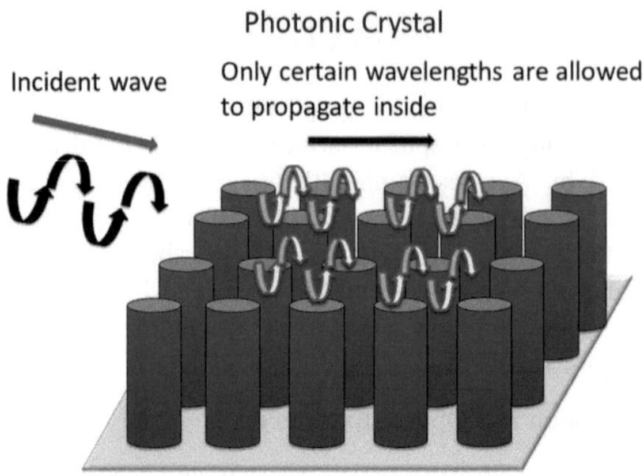

Fig. 9 Here, we present a periodic photonic crystal lattice on the substrate. The lattice forms a band-gap, allowing only certain wavelengths to propagate inside. The photonic properties can therefore be controlled

to note that these excitations and scattering processes are influenced by both topological defects of the lattices (e.g., dislocation and disclination) and boundary characteristics [40, 66, 67].

F. Graphene Photonic Crystal

The photonic crystal is a kind of optical device, whereby a lattice can be periodically allocated upon or within a semiconductor [69, 70] (cf. Fig. 9). A band-gap can be obtained in these periodic structures, and only a certain range of photon energies (i.e., frequencies) are allowed to propagate within. The basic idea is that the periodic dielectric behaves as a superlattice, with restriction being placed upon the wave properties of the electrons [70]. Moktadir et al. [71] find that the graphene photonic crystal provides a wide transmission range, which can be tuned via an applied gate voltage. It has also been reported by Majumdar et al. [72] that the resonance reflectivity can be increased fourfold via a slight 2 nm shifting of the graphene crystalline structure (i.e. dislocations). Graphene's flexible nature therefore offers numerous applications.

4 Graphene Optoelectronic Devices

In this section, we will present ideas for optical devices which incorporate graphene, with emphasis being placed upon their photonic properties. The various photodetectors, optical modulator, and the photonic limiter (mode-locked laser) will all be discussed.

A. Photodiode and Graphene Photodetector

The n-p or p-n junctions are comprised of two different semiconductors (n-type and p-type). Electrons from the n-type semiconductor will flow across the p-type, whereas holes in the p-type will move to the n-type [73]. In any case, a depletion layer is formed at the junction interface. In principle, n-p or p-n junctions can be forward or reverse biased [73]. Since a band-gap can be created in graphene (cf. Sect. 2), it is therefore feasible to conceive of a graphene-semiconductor junction.

Photodiodes are a key component for use in logical devices. It is a current generating device that is sensitive to incoming light (cf. Fig. 10 for the p-n junction configuration). In the absence of light, the device carries a high resistance. However, incoming photons can break down some of the bonding within the compounds at the depletion layer (cf. Fig. 10). Electrons and holes will then be created, and hence a drift current I_d flows across the diode [73]

$$I_d \approx c_1(1 - \exp(-c_2 W_{DL})). \tag{15}$$

Here, c_1 is a constant associated with electric charge and photonic flux, W_{DL} is the width of the depletion layer, and c_2 is a constant associated with the photon energy and band-gap [45].

The working principle of the photodetector is similar to that of the photodiode, transforming photons into an observable current [33, 74] (cf. Fig. 11). More specifically, photons transfer energy to electrons, causing them to 'jump' from the valence band to conduction band (cf. interband transition). This has a typical timescale of ~1 ps [39].

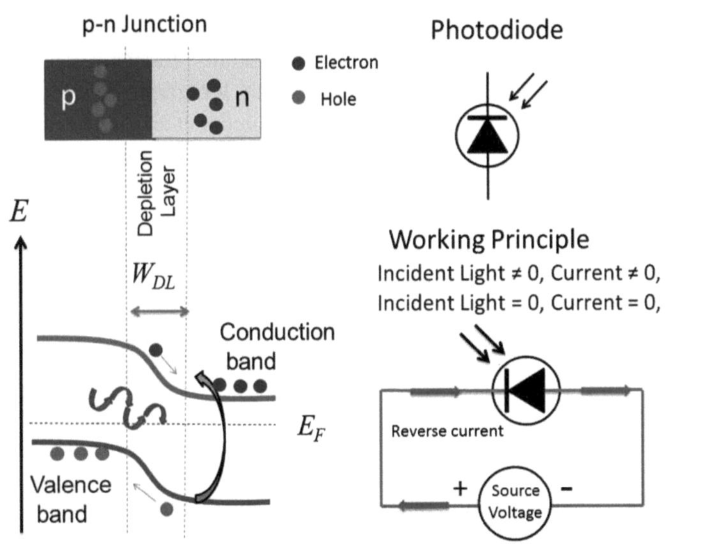

Fig. 10 A schematic of the photodiode is shown. The basic idea is that a reverse current flows upon illumination of the photodiode. The figure also emphasizes that a depletion layer is formed at the interface of the p-n junction

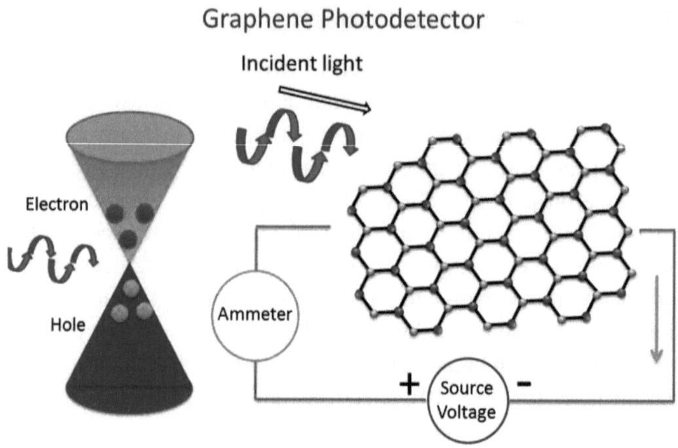

Fig. 11 This figure shows how electrons in the valence band can be excited to conduction band by incident photons. The conductivity of graphene increases, and a measurable current can flow around the circuit. In practice, the idea can be used to measure incident light intensity, for example

$$\gamma(\text{photon}) + e^-_{\text{val}} \rightarrow e^-_{\text{con}} \tag{16}$$

Here, e^-_{val} and e^-_{con} refer to electrons in the valence and conduction bands respectively. Bonac-corso et al. [33] point out that the absorption bandwidth of light spectra depends upon the choice of semiconductor. As we mentioned before, graphene interacts with an EM range covering the majority of the visible spectrum [33, 39]. Xia et al. [75] have also reported that the frequency response of graphene can be upwards of 40 GHz, with a theoretical limit reaching even 500 GHz. This response generally depends upon the electrical mobility, resistance and capacitance of the materials [39]. An appropriate bandwidth for graphene can therefore be adjusted via doping or an applied electric field. Mueller et al. [76] further reveal that their results for graphene display a strong photonic response at a wavelength of 1.55 μm, when applying the graphene photodetector on fast data communication links.

Bao et al. [39] have summarized that current from a photodetector can also be generated just by the contact of graphene and a semiconductor, due to the differing work functions and thermal gradient. Current leakage is one of the major drawbacks of the graphene photodetector, although this can be optimized by reducing the band-gap, or coating some dielectric material on the graphene surface [33, 39]. Echtermeyer et al. [77] show that a number of metallic nanoparticles can be allocated on the graphene substrate, vastly improving the efficiency of the devices. The basic idea is that a metallic nanoparticle touches the graphene film, and forms a 'junction like' contact. Metallic nanoparticles on the graphene layer would thus act as small photodetectors at the same time, and thus enhance the sensitivity [77, 78].

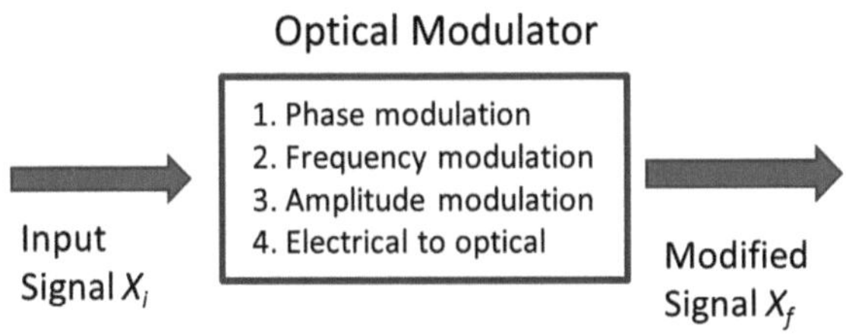

$$M.I. = \frac{Maximum\ of\ Modified\ Signal}{Input\ Signal}$$

Fig. 12 The optical modulator is an important device for converting electrical signals into an optical equivalent, and therefore an ideal bridge between electronic and optical devices. This device can also change the properties of the incident wave, such as the phase, frequency, and amplitude. The Modulation Index (*M.I*) is defined as the maximum of the modified signal X_f, divided by the input signal X_i

Some other applications such as the measurement of refractive index [79], and the analysis of metamaterials via the graphene sensor [80], are all being studied on the graphene photonic detector platform.

B. Optical Modulator

The optical modulator is a photonic device which transforms electrical signals into an optical equivalent [74, 81, 82] (cf. Fig. 12 for a schematic overview). It is an essential communication link within many electronic devices, and can also alter the properties of light via doping or an applied electric field [39]. For example, assume a plane wave propagates as

$$A = |A|\exp(i\theta), \tag{17}$$

where A can be either electric or magnetic in origin, and θ is the phase of the wave. A modulator changes the amplitude $|A|$ and phase θ of the input wave [39]. Graphene is a suitable material for a modulator since it has a strong response to a wide range of light spectra (i.e., bandwidth) [67, 81, 82]. Typically, graphene will be coated upon the silicon substrate to enhance the absorption rate [67].

Optical modulators can generally be divided into two types [39]. The first is an absorptive modulator, converting photons into some other form of energy. Normally, an absorption modulator can tune the transmitted light intensity via adjustment of the Fermi energy level E_F [39, 67]. The second type is a refractive modulator which can change the dielectric constant according to variation of the electric field. Graphene is a promising material for an absorption modulator due to its wide bandwidth and tunable Fermi energy level [33, 39]. Bao et al. [39] further reveal that the interband transition can be tuned to a logical ON/OFF state, dependent upon E_F. Regardless, graphene provides a high optical Modulation Index (*M.I*), making it an ideal material for any modulator [39]. This index is given by

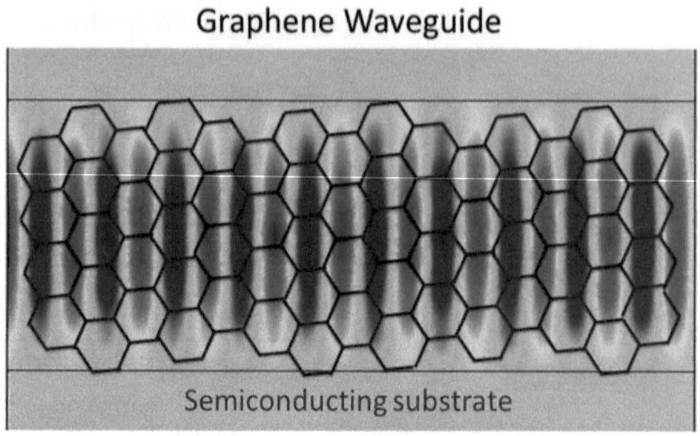

Fig. 13 Graphene is coated upon the semiconducting substrate, and either the TE or TM wave-modes can be transmitted along the graphene thin film

$$M.I. = \frac{\max(X_f)}{X_i}, \tag{18}$$

where X_i and X_f refer to the variable before and after modulation respectively. The graphene modulator can also be applied to the optical resonator, allowing the wavelength to be altered (cf. for details [39]).

Recently, the dielectric sandwich—two layers of graphene with dielectric filling—has been used as an optical signaling modulator [81, 82]. Gosciniak et al. [81] estimate that this graphene optical modulator can reach speeds of up to 850 GHz, with 3 dB modulation and small losses. Liu et al. [82] have also reported a wide absorption range of 1.35–1.60 μm in wavelength.

C. Graphene Waveguide

A waveguide is a physical channel which traps light, guiding it through a designated path [61, 83]. For example, fiber-optic cable is a common waveguide for the communication of light signals—its high refractive index n_{op} trapping light inside the fiber [61]. As we have already seen, the refractive index depends upon the linear dielectric susceptibility χ_{R1} [39]. Zhang et al. [83] have studied the wave-modes of the graphene quantum well, identifying energy dispersion relations associated with Klein tunneling and classical wave-modes [83]. Zhang et al. [83] further note an absence of the 3rd order classical, and 1st order tunneling wave-modes.

Graphene plasmonic waveguides (Fig. 13) have become an essential component for integration with logical devices [84, 85]. Kim et al. [84] have studed the plasmonic waveguide for a dielectric substrate coated on graphene, discovering little optical loss and very fast operating speeds. They show that at the peak wavelength $\lambda = 1.31$ μm, the transmission ratio is around 19 dB for the TM mode [84].

D. Saturable Absorber

As we have already highlighted upon, saturable absorption refers to an absorption of photons decreasing as the light intensity increases [66, 86] (cg. Fig. 14). It is usually applied via the mode-locked laser [39, 87]. Many semiconductors exhibit

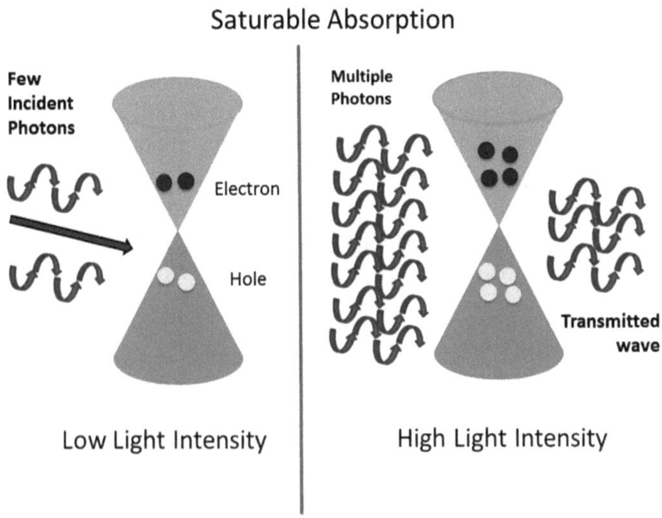

Fig. 14 The idea of saturable absorption for graphene is shown. Graphene can absorb photons, and create electron-hole pairs at low incident light intensity. However, electrons are incapable of occupying the conduction band at high light intensity, since most of the states have already been occupied (cf. Pauli exclusion principle)

saturable absorption, but are not as sensitive as graphene [39, 66]. The basic idea is as follows—a number of excited electrons occupy the conduction band during high intensity exposure, and electrons in the valence band are no longer able to absorb photons due to the Pauli exclusion principle [39, 87]. This property originates from the nonlinear susceptibility of graphene for a short response time [33]. In application, a saturable absorber can be used to transform a continuous wave to a very short wave pulse [33]. Generally speaking, monolayer graphene provides a high saturable absorption coefficient, and recently, some research has uncovered that CNTs may also be suitable candidate for a saturable absorber [86, 88]. Bao et al. [87] also report that a single layer graphene (SLG) saturable absorber can provide around 66 % modulation depth, and produce picosecond wave pulses.

E. Photonic Limiter

A photonic limiter is used to reduce the intensity of light that is emitted from the source [33, 39, 89, 90]. The mechanism is to permit the passage of low intensity light, and to filter out light of higher intensity [89, 90]. Dispersed graphene-oxide solutions are generally used for studying the optical limiter [86, 87]. In particular, graphene, has a strong response to a change of light intensity [39], with a transmittance $T(I)$ that is inversely dependent upon the light intensity I. Such a device can therefore, for example, be implemented to protect the human eye when working with laser apparatus [33]. Wang et al. [89] also note how graphene's nonlinear response is the working principle behind the reduction of light transmitted at high intensity, and also show that graphene can limit a wide range of the visible spectrum [89, 90].

According to Bao et al. [39], the *reverse* saturable absorption (opposite to saturable absorption) is the key nonlinear response that filters high intensity light, and subject to certain conditions. This relates to an optical limiter absorbing more high-energy photons than low-energy photons [39]. Lim et al. [86] have reported that, in practice, the property will change from saturable absorption to reverse saturable absorption, only when microplasmas or microbubbles appear. These lead to a nonlinear thermal scattering, which is also an important factor in limiting high intensity light [39]. Nevertheless, the graphene photonic limiter is still in the early stages of development, with more drastic efforts being required in the near future.

5 Transistors

Nowadays, field effect transistors (FETs) are a key component of most integrated circuitry, commonly acting as a simple logic gate. These devices can be of either n-p-n or p-n-p type, depending upon the desired operation. In this day and age, when the speed and size of devices are becoming all important, scientists are having to seek revolutionary new materials to replace the likes of Silicon (Si), Germanium (Ge) and Gallium Arsenide (GaAs). With outstanding electrical mobility, graphene-based materials are becoming evermore prominent as candidates within future transistors (cf. Fig. 15).

A. Carbon Nanotube Transistor

Before we commence any in-depth discussion of this particular design [22, 23, 91], we must first discuss the physical properties of graphene nanoribbons (GNRs) [92] and carbon nanotubes (CNTs) [38, 93–96]. A GNR is considered to be a piece of graphene of exceptionally narrow width [92]. The electrical attributes of GNRs are determined by their boundary conditions (BCs) (cf. Fig. 16). The 'armchair' BC can cause either metallic or semi-metallic behavior to be exhibited, whereas the zig-zag BC yields only metallic characteristics [38, 93, 94]. Therefore, GNRs are another

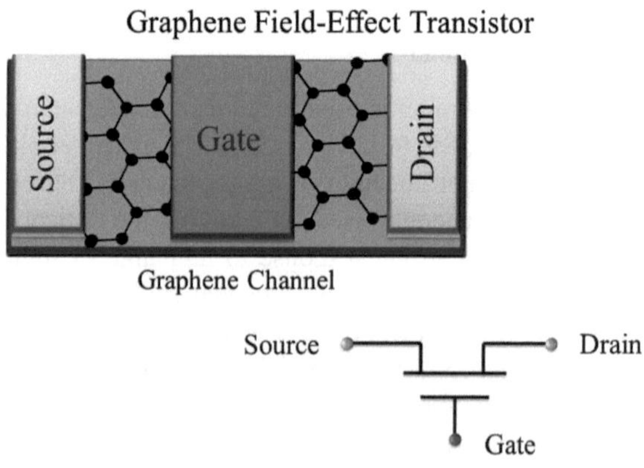

Fig. 15 The idea of the graphene FET is shown. The channel of the transistor is made of graphene, and a gate voltage V_G controls the current flow I_{DS} from drain to source

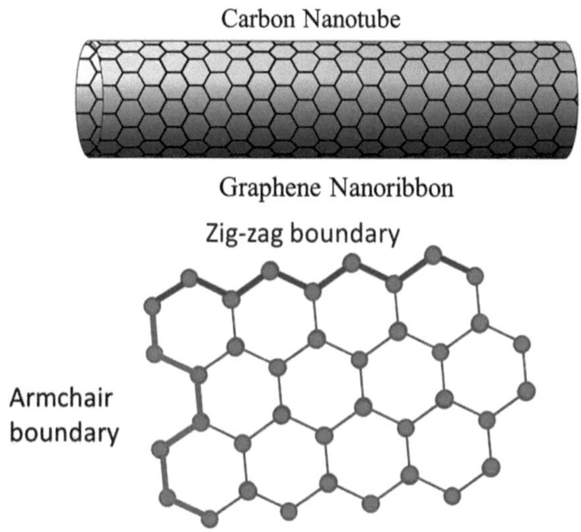

Fig. 16 The lower portion of this figure highlights the two possible boundary conditions (BCs) that a graphene nanoribbon (GNR) can satisfy. The zig-zag BC yields only a conducting state, whereas the armchair BC can either imply a conducting or semiconducting state (dependent upon the width of the nanoribbon). The GNR can also be curled to form a carbon nanotube (CNT), with a band-gap that is inversely proportional to its radius

means of generating an energy band-gap. In this case, the gap size is inversely proportional to the nanoribbon width.

Carbon nanotubes are often considered to be one-dimensional structures, and can be formed by curling a GNR (typically 10–100 nm in width [38, 92]) into a cylindrical configuration (for further details concerning their fabrication, one can refer to [38, 95]). The nanotubes can either be single-walled or multi-walled, although this must be taken into account when considering the CNT radius r_{CNT}. Since this process leads to structural deformation of graphene's honeycomb lattice, there is an overall modification of the electronic band structure [38, 93, 94]. Quantum equivalents of the capacitance, inductance and resistance have all been exhibited within the electrical properties of CNTs, and an energy band-gap E_g is found to be inversely proportional to this CNT radius r_{CNT} [38],

$$E_g \sim 1/r_{CNT}. \tag{19}$$

Thus, together with graphene's capability for long-range ballistic transport (even at room temperature), this presents many useful applications. For example, CNTs would be an apt source-drain channel within semiconducting devices such as transistors [93].

For decades now, the CNT transistor has been subject to intense study [31, 38, 93, 97–99], with a recently reported high switching ratio [99]. They consume much less power, and can possess shorter channel lengths than their silicon-based counterparts. They can also exist in many forms, the most popular being the

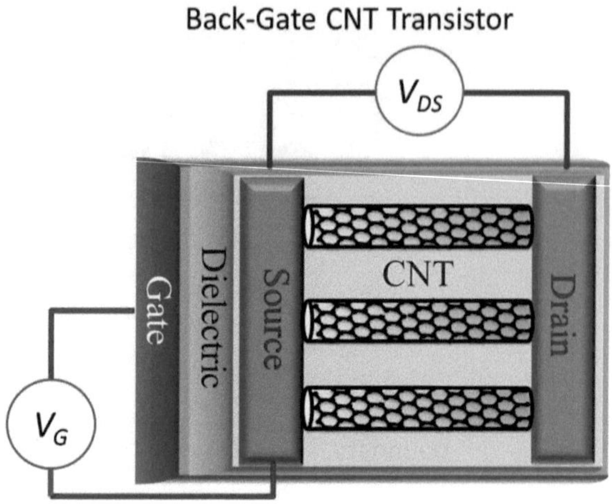

Fig. 17 The figure shows the back-gated CNT transistor. It contains multiple CNTs as the intermediary channels, providing stable performance and current flow. Since the device is back-gated, it can easily be applied within integrated circuitry

top-gate, back-gate and wrap-around gate designs [93, 100–102]. Recently, Shulaker et al. [100] have developed a simple computer from the CNT-based transistors, which can perform more than 20 different instructions. Figure 17 provides a diagrammatic representation of how a back-gated multi-CNT transistor may look—the CNTs themselves acting as the intermediary channels. Currently, one can produce a purified CNT having less than 0.0001 % impurity—which can minimize any inelastic scattering in the channel [101]. The ON/OFF drain-source current I_{DS} can be tuned by using an applied electric field (i.e., the gate voltage V_G) to act upon the CNT channel [101]. Moreover, the ballistic transport of electrons is a result of the one-dimensional CNT structure [38], which again restricts the degree of inelastic scattering. CNT transistors would also appear to alleviate the issue of the short-channel effect in silicon-based devices. In theory, the shorter the channel, the faster the transistor [31, 38]. However, usually when the channel has a length scale in tens of nanometers, the drain-source current I_{DS} tends to become most unstable [38]. A recent study [98, 101] has revealed that the CNT channel can be as short as 9 nm, whilst maintaining a stable current. There are even some predictions that the CNT channel can reach even down to 7 nm in the near future [98, 101].

Schottky barriers at the channel-electrode contacts, are the major obstacle with regards to the CNT's application within transistors [102]. Specifically, they provide a large resistance at the CNT-electrode interface, due to the differing work-functions [38]. The Schottky barrier would generally downgrade the ON/OFF switching ratio [38, 102]. Even worse, this barrier is much larger than for silicon-based devices. A recent study by Javey et al. [102] reveals that the Schottky barrier would be greatly reduced when using the noble metal, Palladium (Pd) as the electrode. They also show that the CNT channel can even then maintain ballistic transport [102]. It is important to emphasize how both classical and quantum equivalents of inductance, resistance and capacitance are exhibited for CNT transistors [38]. In particular, quantum effects become most apparent at the nanoscale. Both quantum inductance and capacitance are determined by the size, BCs and the density of states (DoS) of the CNT [31, 38], whereas the quantum resistance is equal to $h/4e^2$.

B. Tunneling Transistor

1. *Mechanisms of Tunneling*
   As we have already highlighted upon many times now, the absence of a well-defined OFF state in the graphene transistor is a major setback [103, 104]. Assuming a band-gap were to be created, the next hurdle to overcome is the back-current leakage during this OFF state, since this downgrades the power efficiency [103, 104]. Furthermore, opening this band-gap would then reduce the mobility of graphene, with the Dirac fermions being subject to some inelastic scattering [104].
   The tunneling graphene transistor is a revolutionary new concept, and may be capable of alleviating some of the aforementioned drawbacks. It consumes very little power (up to 10^9 times less than silicon-based devices [105]) and possesses a very fast response time (steep sub-threshold slope) [31]. Michetti et al. also report that an ON/OFF switching ratio can reach as high as 10^4, even with a small electric field [106–108]. It is also found that tunneling occurs at exceptional speed [108]. The underlying concept is visualized in Fig. 18. The interband tunneling is tuned via an applied drain-source voltage V_{DS} and gate voltage V_G [104, 109]. Both V_{DS} and V_G are

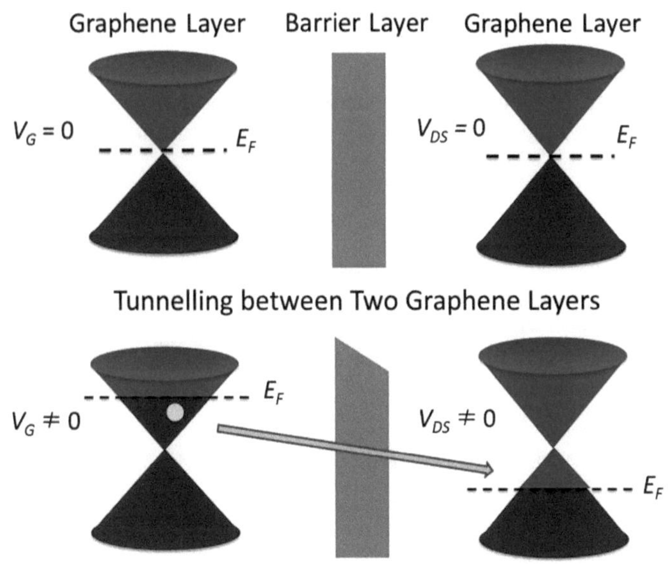

Fig. 18 The *upper* figure shows an intermediate barrier separating the two layers of graphene. In the absence of external electric fields, the Fermi energies are situated at the Dirac points. When the gate voltage V_G and drain-source voltages V_{DS} are applied, electrons begin to accumulate in the conduction band of one graphene layer, and holes in the other. Tunneling can readily occur in this situation, via a fine tuning of both voltages

used to accumulate the electrons and holes in upper and lower graphene layers respectively, and thus altering the shape of the potential barrier [110]. The tunneling is also associated with the channel length, and thickness of the gate oxide layer t_{ox} [106].

2. *Vertical Design*
   A relatively new concept which relies upon *vertical* tunneling has been developed by the Manchester research group [103, 104]. The graphene-based device consists of a few (insulating) layers of hexagonal Boron Nitride (hBN) or molybdenum disulphide (MoS_2) [111–113]. These are positioned between two graphene sheets which then constitute the electrodes (cf. Fig. 19). The key point here, is that the insulating layers act as a barrier, and thus preventing the flow of current. As such, there is no need for a well-defined band gap in graphene [111–113]. This has the added benefit of greatly reducing any current leakage whilst in the OFF state [111–113]. The whole process of current tunneling then acts perpendicular to the layers [111–113]. Electrons in the bottom graphene layer will begin to accumulate once a gate voltage V_G is applied across the lower insulating layer [114, 115]. The drain-source voltage V_{DS} is then added to create holes in the upper graphene layer [103, 111–113]. This has the desired effect of increasing the Fermi energy E_F in bottom graphene layer, and decreasing E_F in the upper layer. Electrons in the bottom graphene layer are then capable of tunneling to the top graphene layer [104]. A recent study by Georgiou et al. [111] reveals that the current modulation can reach a high value of 10^6 (even at room temperature). It is also interesting to note that a resonant tunneling within the vertical transistor occurs in some energy states, and a negative differential conductance exists (i.e., current decreases upon an increase in voltage) [112, 113].

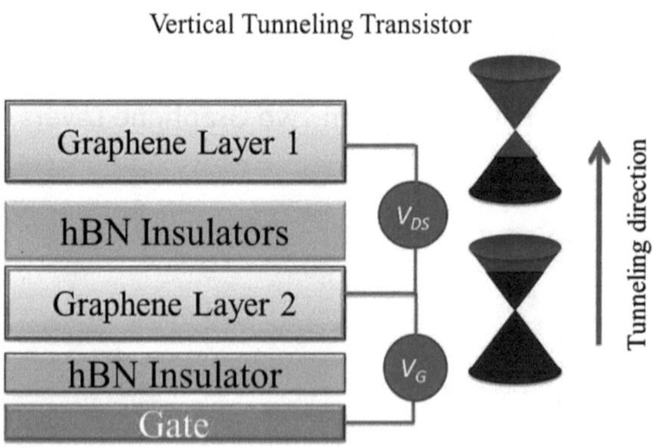

Fig. 19 The vertical tunneling graphene transistor is shown. The hexagonal Boron Nitride (hBN) insulating layers act as an intermediary barrier. The accumulation of holes in the upper graphene layer is controlled by the drain-source voltage V_{DS}, whereas the build-up electrons in the lower graphene layer can be tuned via the gate voltage V_G. Electrons are then capable of tunneling from the bottom to top layer of graphene

C. High Frequency Devices

High frequency transistors do not require an OFF state, and can operate solely through variations of the current or voltage signaling [31, 116]. Graphene may thus be applicable within the realms of high-frequency transistors, inverters, or operational amplifiers [31, 84, 117]. Graphene's response to these signals is incredibly fast, with operating speeds of around a few hundred GHz [84, 116, 118–120].

The performance of high frequency devices is characterized by two important parameters—the cut-off frequency f_{cut} and maximum oscillation frequency f_{max}. The cut-off frequency f_{cut} is given by a current gain G_I equal to unity [116]

$$G_I = 20\log_{10}\left(\frac{I_{out}}{I_{in}}\right) \tag{20}$$

where I_{out} and I_{in} are the output and input currents respectively. Typically, f_{cut} is proportional to the trans-conductance g_{rf} and the thickness of the gate oxide layer t_{ox}, whereas inversely proportional to the transistor gate length L_G and gate width W_G [31, 116]. The whole expression is given by

$$f_{cut} = c_1 \frac{t_{ox} g_{rf}}{L_G W_G} \tag{21}$$

where c_1 is a constant associated with dielectric gate. In experiments, one would only shorten the gate length L_G for simplicity, thereby increasing the cut-off frequency. Wu et al. [116] report that with CVD-prepared graphene, f_{cut} can reach upwards of 155 GHz for a relatively short gate length of 40 nm. Theoretical simulations have indicated that a cut-off frequency of 1 THz can be attained for just a few nanometers gate length [118].

Similar to f_{cut}, the maximum oscillation frequency f_{max} is obtained for a power gain G_P equal to one. Here, we have $G_P = 10\ \log_{10}\ (P_{out}/P_{in})$, where P_{out} and P_{in} are the output and input powers respectively [116]. The value of f_{max} in graphene-based devices is slightly more complicated, and is dependent upon the cut-off frequency, gate resistances, and the trans-conductance g_{rf} [116, 118]. A recent report [116] has mentioned that f_{max} can reach up to 20 GHz. However, it is important to note that a short gate length would not necessarily imply a high value for f_{max} [116, 118]. At present, not much is understood of the *I–V* characteristic curve—this has three regions, firstly linear, then saturating, and finally a second linear region [31, 116, 118]. In addition, a change of gate voltage V_G would alter the shape of the *I–V* curve, and making the saturation region ambiguous. Without a stable saturation region, the value of f_{max} is limited. This problem will require urgent attention in the near future, if the high-frequency transistor is to make any headway [84, 121].

6 Summary

Graphene's outstanding capabilities have drawn the attention of scientists from several interdisciplinary backgrounds—all looking to take advantage. This standalone two-dimensional structure is a playground for Dirac fermions which possess a zero effective mass [4, 122, 123]. Quantum phenomena have been observed even at room temperature; a series of anomalous quantum effects including QHE and Klein tunneling [1, 54]. Graphene's versatility is nigh on endless—in this paper, we have merely focused upon optoelectronic devices and transistors. Optical communications provide a much wider bandwidth, with higher efficiencies than most typical conducting wire. We are thus dawning upon a new golden *photonic* age of higher internet speeds, and entertainment-based devices. Graphene's high transmittance, strong interaction of light with ultra-fast response time, wide absorption spectrum, and tunable optical conductivity [33, 39, 67], present an ideal material for optical devices! Amongst many others, these include the photodetector, optical modulator, plasmonic waveguide and also the saturable absorber. The absence of any discernible band-gap for graphene is an unavoidable issue for logical devices, although one may be created via various methods (e.g., structural deformation or chemical doping) [31]. The CNT transistor is now a well-established technology—developed over more than 30 years. Only now has the dream of a CNT-based computer become a working reality [100]. Carriers in CNT channels can perform ballistic transport, even for very short lengths. However, the graphene vertical tunneling transistor is something rather novel. This device itself does not require band-gap at all, and yet, both operates at exceptional speeds, whilst consuming very little power. Our final talking point was the high frequency transistor, which acts as amplifier in the circuit rather than a typical logical device. The cut-off frequency can reach theoretical estimates of up to 1 THz, for just a few nanometers of gate length [38]. Although we have plainly made the case for graphene's implementation within numerous optical and electronic devices, there are a few obstacles which we must overcome. These are the nonlinear *I–V* characteristic curve, and the emergence of Schottky barriers (although we mention a suitable fix). Fifty years ago, no one would have ever envisaged that optical or silicon-based devices would have their place in everyday life. Graphene may change the world!

References

1. Geim AK, Novoselov KS (2007) The rise of graphene. Nature Mater 6:183–191
2. Meyer JC (2007) The structure of suspended graphene sheets. Nature 446:60–63
3. Geim AK (2009) Graphene: status and prospects. Science 324:1530–1534
4. Geim AK (2012) Graphene prehistory. Phys Scr T146:014003
5. Katsnelson MI (2007) Graphene: carbon in two dimensions. Mater Today 10:20–27
6. Novoselov KS et al (2005) Two-dimensional atomic crystals. PNAS 102:10451–10453
7. Savage N (2012) Materials science: super carbon. Nature 483:S30–S31

8. Wallace PR (1947) The band theory of graphite. Phys Rev 71:622–634
9. O'Hare A, Kusmartsev FV, Kugel KI (2012) A stable flat form of two-dimensional crystals: could graphene, silicene, germanene be minigap semiconductors. Nano Lett 12:1045–1052
10. Novoselov KS et al (2004) Electr Field Eff At Thin Carbon Films Sci 306:666–669
11. Novoselov KS (2005) Two-dimensional gas of massless dirac fermions in graphene. Nature 438:197–200
12. Iyechika Y (2010) Application of graphene to high-speed transistors: expectations and challenge. Sci Techno Trends—Q Rev 37:3776–3792
13. Hlawacek G, Beilstein J et al (2012) Imaging ultra thin layers with helium ion microscopy: utilizing the channeling contrast mechanism. Nanotechnology 3:507–512
14. Robinson JA et al (2009) Correlating Raman spectral signatures with carrier mobility in epitaxial graphene: a guide to achieving high mobility on the wafer scale. Nano Lett 9:2873–2876
15. Gouider Trabelsi AB, Ouerghi A, Kusmartseva OE, Kusmartsev FV, Oueslati M (2013) Raman spectroscopy of four epitaxial graphene layers: macro-island grown on 4H-SiC0001 substrate and an associated strain distribution. Thin Solid Films 539:377–383
16. Das A, Pisana S, Chakraborty B, Piscanec S, Saha SK, Waghmare UV, Novoselov KS, Krishnamurthy HR, Geim AK, Ferrari AC, Sood AK (2008) Monitoring dopants by Raman scattering in an electrochemically top-gated graphene transistor. Nature Nanotechnol 3:210–215
17. Bae S et al (2010) Roll-to-roll production of 30-inch graphene films for transparent electrodes. Nature Nanotechnol 5:574–578
18. Chen JH, Jang C, Xiao S, Ishigami M, Fuhrer MS (2008) Intrinsic and extrinsic performance limits of graphene devices on SiO_2. Nature Nanotechnol 3:206–209
19. Bolotin KI et al (2008) Ultrahigh electron mobility in suspended graphene. Solid State Commun 146:351–355
20. Morozov SV et al (2008) Giant intrinsic carrier mobilities in graphene and its bilayer. Phys Rev Lett 100:016602
21. Lin YM, Farmer DB, Jenkins KA et al (2011) Enhanced performance in epitaxial graphene FETs with optimized channel morphology. IEEE Electron Device Lett 32:1343–1345
22. He QY, Wu SX, Yin ZY, Zhang H (2012) Graphene-based electronic sensors. Chem Sci 3:1764–1772
23. He QY et al (2011) Transparent, flexible, all-reduced graphene oxide thin film transistors. ACS Nano 5:082117
24. Neto AHC, Novoselov KS (2011) New directions in science and technology: two-dimensional crystals. Rep Prog Phys 74:082501
25. Pumarol ME et al (2012) Direct nanoscale imaging of ballistic and diffusive thermal transport in graphene nano-structures. Nano Lett 12:2906–2911
26. Prasher R (2010) Graphene spreads the heat. Science 328:185–186
27. Seol JH et al (2010) Science 328:213–216
28. Chen S (2012) Thermal conductivity of isotopically modified graphene. Nature Mater 11:203–207
29. Rafiee J (2012) Wetting transparency of graphene. Nature Mater 11:217–222
30. Engel M et al (2012) Light matter interaction in a micro-cavity controlled graphene transistor room temperature transistor based on a single carbon nanotube. Nature Commun 3:906–911
31. Schwierz F (2010) Graphene transistors. Nature Technol 5:487–496
32. Kusmartsev FV, Tsvelik AM (1985) Semi-metallic properties of a hetero-junction. JETP Lett 42:257–260
33. Bonaccorso F, Sun Z, Hasan T, Ferrari AC (2010) Graphene photonics and optoelectronics. Nat Photonics 4:611–622
34. Sarma SD, Adam S, Hwang EH, Rossi E (2011) Electronic transport in two-dimensional graphene. Rev Mod Phys 83:407–470
35. Zhou YB, Wu HC, Yu DP, Liao ZM (2013) Magneto-resistance in graphene under quantum limit regime. Appl Phys Lett 102:093116

36. Du X, Skachko I, Barker A, Andrei EY (2008) Approaching ballistic transport in suspended graphene. Nature Nanotechnol 3:491–495
37. Koh YK, Bae MH, Cahill DG, Pop E (2010) Heat conduction across monolayer and few-layer graphenes. Nano Lett 10:4363–4368
38. Avouris P, Chen Z, Perebeinos V (2007) Carbon-based electronics. Nature Nanotech. 2:605–615
39. Bao Q, Loh KP (2012) Graphene photonics, plasmonics, and broadband optoelectronic devices. ACS Nano 6:3677–3694
40. Avouris P (2010) Graphene photonics and optoelectronics. Nano Lett 10:4285–4294
41. Saleh BEA, Teich MC (2007) Fundamentals of photonics, 2nd edn. Wiley Series in Pure and Applied Optics. Wiley, USA
42. Kasap SO (2001) Optoelectronics and photonics: principles and practices, 1st edn. Prentice Hall, New Jersey
43. Koppens FHL, Chang DE, Garca de Aba jo FJ (2011) Graphene plasmonics: a platform for strong light-matter interactions. Nano Lett 11:3370–3377
44. Britnell L et al (2013) Strong light-matter interactions in heterostructures of atomically thin films. Sci Comm 340:1311
45. Rosencher E (2002) Optoelectronics, Cambridge University Press
46. Rudden MN, Wilson J (1993) Element of solid state physics. Wiley, New York (Chapter 4–6)
47. Irwin JD, Kerns DV (1995) Introduction to electrical engineering. Prentice Hall, New Jersey Chapter 8–9
48. Turton R (2000) The physics of solids. Oxford University Press, New York Chapter 4–6
49. Kim KS et al (2009) Large-scale pattern growth of graphene films for stretchable transparent electrodes. Nature 457:706–710
50. Fan X, Shen Z, Liu AQ, Kuo JL (2012) Band gap opening of graphene by doping small boron nitride domains. Nanoscale 4:2157–2165
51. Shinde PP, Kumar Y (2011) Direct band gap opening in graphene by BN doping: Ab initio calculations. Phys Rev B 84:125401
52. Coletti C, Riedl C, Lee DS, Krauss B, Patthey L, von Klitzing K, Smet JH, Starke U (2010) Charge neutrality and band-gap tuning of epitaxial graphene on SiC by molecular doping. Phys Rev B 81:235401
53. Terrones H, Lv R, Terrones M, Dresselhaus MS (2012) The role of defects and doping in 2D graphene sheets and 1D nanoribbons. Rep Prog Phys 75:062501
54. Katsnelson MI, Novoselov KS, Geim AK (2006) Chiral tunneling and the klein paradox in graphene. Nature Phys 2:620–625
55. Dean CR et al (2011) Multicomponent fractional quantum Hall effect in graphene. Nature Phys 7:693–696
56. Novoselov KS et al (2007) Room-temperature quantum hall effect in graphene. Science 315:1379
57. Calogeracos A (2006) Paradox in a pencil. Nature Phys 2:579–580
58. Zalipaev VV, Maksimov DN, Linton CM, Kusmartsev FV (2013) Spectrum of localized states in graphene quantum dots and wires. Phys Lett A 377:216–221
59. Hartmann RR, Robinson NJ, Portnoi ME (2010) Smooth electron waveguides in graphene. Phys Rev B 81:245431
60. Williams JR, Low T, Lundstrom MS, Marcus CM (2011) Gate-controlled guiding of electrons in graphene. Nat Nanotech 6:222–225
61. Wu Q, Turpin JP, Werner DH (2012) Integrated photonic systems based on transformation optics enabled gradient index devices. Light Sci Appl 1:e38. doi:10.1038/lsa.2012.38
62. Downing CA, Stone DA, Portnoi ME (2011) Zero-energy states in graphene quantum dots and rings. Phys Rev B 84:155437
63. Stone DA, Downing CA, Portnoi ME (2012) Searching for confined modes in graphene channels: the variable phase method. Phys. Rev. B 86:075464
64. Shannon CE (1949) Communication in the presence of noise. Proc Inst Radio Engs 37:10–21

65. Gan X et al (2013) High-contrast electrooptic modulation of a photonic crystal nanocavity by electrical gating of graphene. Nano Lett 13:691–696
66. Avouris P, Xia FN (2012) Graphene applications in electronics and photonics. Mater Res Soc 37:1225–1234
67. Avouris P, Freitag M (2014) Graphene photonics, plasmonics and optoelectronics. IEEE J Sel Top Quantum Electron. doi:10.1109/JSTQE.2013.2272315
68. Butcher PN, Cotter D (1991) The elements of nonlinear optics. Cambridge University Press, Cambridge
69. Lu L, Cheong LL, Smith HI, Johnson SG, Joannopoulos JD, Soljacic M (2012) Three-dimensional photonic crystals by large-area membrane stacking. Optics Lett 37:47264728
70. Joannopoulos JD, Johnson SG, Winn JN, Meade RD (2008) Photonic crystals: molding the flow of light, 2nd edn. Princeton University Press, Princeton
71. Moktadir Z, Charlton M, Pollard M, Mizuta H, Rutt H (2011) Tunable transmission in a graphene photonic crystal in mid-infrared. In: Graphene 2011 conference, Bilbao, Spain, 11–14 Apr 2011
72. Majumdar A, Kim J, Vuckovic J, Wang F (2014) Graphene for tunable nanophotonic resonators. IEEE Sel Top Quantum Electron 20:4600204
73. Mohan Kumar D (2003) Optoelectronic devices and their applications. Electronics for You, Oct 2003
74. Furchi M et al (2012) Microcavity-integrated graphene photodetector. Nano Lett 12:2773–2777
75. Xia F, Mueller T, Lin YM, Garcia AV, Avouris P (2009) Ultrafast graphene photodetector. Nature Nanotech 4:839–843
76. Mueller T, Xia F, Avouris P (2010) Graphene photodetectors for high-speed optical communications. Nat Photonics 4:297–301
77. Echtermeyer TJ et al (2011) Strong plasmonic enhancement of photovoltage in graphene. Nature Commun 2:458. doi:10.1038/ncomms1464
78. Nicoletti O et al (2013) Three-dimensional imaging of localized surface plasmon resonances of metal nanoparticles. Nature 502:80–84
79. Xing F et al (2012) Sensitive real-time monitoring of refractive indexes using a novel graphene-based optical sensor. Sci Rep 2:908. doi:10.1038/srep00908
80. Amin M, Farhat M, Bagci H (2013) A dynamically reconfigurable Fano metamaterial through graphene tuning for switching and sensing applications. Sci Rep 3:2105. doi:10. 1038/srep02105
81. Gosciniak J, Tan DTH (2013) Theoretical investigation of graphene-based photonic modulators. Sci Rep 3:1897. doi:10.1038/srep01897
82. Liu M et al (2011) A graphene-based broadband optical modulator. Nature 474:64–67
83. Zhang FM, He Y, Chen X (2009) Guided modes in graphene waveguides. Appl Phys Lett 94 (21):212105
84. Kim JT, Choi SY (2011) Graphene-based plasmonic waveguides for photonic integrated circuits. Optic Express 19:24557–24562
85. Wang X, Cheng Z, Xu K, Tsang HK, Xu JB (2013) High-responsivity graphene/ siliconheterostructure waveguide photodetectors. Nat Photonics 7:888–891
86. Lim GK et al (2011) Giant broadband nonlinear optical absorption reponse in dispersed graphene single sheets. Nat Photonics 5:554–560
87. Bao Q et al (2011) Monolayer graphene as a saturable absorber in a mode-locked laser. Nano Res. 4(3):297–307
88. Hendry E, Hale P, Moger J, Savchenko A, Mikhailov S (2010) Coherent nonlinear optical response of graphene. Phys Rev Lett 105:97401
89. Wang J, Hernandez Y, Lotya M, Coleman JN, Blau WJ (2009) Broadband nonlinear optical response of graphene dispersions. Adv Mater 21:2430–2435
90. Tutt LW, Kost A (1992) Optical limiting performance of C60 and C70 solutions. Nature 356:225–226

91. Park J, Nam S, Lee M, Lieber CM (2011) Synthesis of monolithic graphene-graphite integrated electronics. Nat Mater 98:082117
92. Li X, Wang X, Zhang L, Lee S, Dai H (2008) Chemically derived, ultra-smooth graphene nanoribbon semiconductors. Science 319:1229–1232
93. Tans SJ, Verschueren ARM, Dekker C (1998) Room-temperature transistor based on a single carbon nanotube. Nature 393:49–52
94. Martel R, Schmidt T, Shea HR, Hertel T, Avouris P (1998) Single- and multi-wall carbon nanotube field-effect transistors. Appl Phys Lett 73:2447–2449
95. Zhu HW, Xu CL, Wu DH, Wei BQ, Va jtai R, Ajayan PM (2002) Direct synthesis of long single-walled carbon nanotube strands. Science 296:884–886
96. McCann E, Fal'ko VI (2004) Symmetry of boundary conditions of the Dirac equation for electrons in carbon nanotubes. J Phys Cond Matter 16:2371–2379
97. Kreupl F (2012) Carbon nanotubes finally deliver. Nature 484:321–322
98. Franklin AD et al (2012) Sub-10 nm carbon nanotube transistor. Nano Lett 12:758–762
99. Jang S et al (2010) Flexible, transparent single-walled carbon nanotube transistors with graphene electrodes. Nanotechnology 21:425201
100. Shulaker MM, Hills G, Patil N, Wei H, Chen HY, Wong HSP, Mitra S (2013) Carbon nanotube computer. Nature 501:526
101. Franklin AD (2013) Electronics: the road to carbon nanotube transistors. Nature 498:443
102. Javey A, Guo J, Wang Q, Lundstorm M, Dai H (2003) Ballistic carbon nanotube transistors. Nature 424:654
103. Britnell L et al (2012) Field-effect tunneling transistor based on vertical graphene heterostructures. Science 335:947–950
104. Ponomarenko LA et al (2011) Tunable metal-insulator transition in double-layer graphene heterostructures. Nature Phys. 7:958–961
105. Yang X et al. (2010) Graphene tunnelling FET and its applications in low power circuit design. In: GLSVLSI10 proceedings of the 20th symposium on great lakes symposium on VLSI, pp 263–268
106. Michetti P, Cheli M, Iannaccone G (2010) Model of tunneling transistors based on graphene on SiC. Appl Phys Lett 96:133508
107. Zhao P, Chauhan J, Guo J (2009) Computational study of tunneling transistor based on graphene nanoribbon. Nano Lett 9:684–688
108. Zhang Q, Fang T, Xing H, Seabaugh A, Jena D (2008) Graphene nanoribbon tunnel transistors. IEEE Electron Dev Lett 29:1344–1346
109. Ionescu MA, Reil H (2011) Tunnel field-effect transistors as energy-efficient electronic switches. Nature 479:329–337
110. Nandkishore R, Levitov L (2011) Common-path interference and oscillatory zener tunneling in bilayer graphene p-n junctions. PNAS 108:14021–14025
111. Georgiou T et al (2013) Vertical field effect transistor based on graphene-WS2 Heterostructures for flexible and transparent electronics. Nature Nanotechnol 8:100–103
112. Britnell L et al (2013) Resonant tunnelling and negative differential conductance in graphene transistors. Nature Comm 4:1794
113. Nguyen VH et al (2012) Bandgap nanoengineering of graphene tunnel diodes and tunnel transistors to control the negative differential resistance. J Comput Electron 12:85–93
114. Malec CM, Davidovic D (2011) Transport in graphene tunnel junctions. J Appl Phys 109:064507
115. Cobas E, Friedman AL, Erve OMJ, Robinson JT, Jonker BT (2012) Graphene as a tunnel barrier: graphene-based magnetic tunnel junctions. Nano Lett 12:3000–3004
116. Wu Y et al (2011) High-frequency, scaled graphene transistors on diamond-like carbon. Nature 472:74–78
117. Schall D, Otto M, Neumaier D, Kurz H (2013) Integrated ring oscillators based on high-performance graphene inverters. Sci Rep 3:2592
118. Zheng J et al (2013) Sub-10 nm gate length graphene transistors: operating at terahertz frequencies with current saturation. Sci Rep 3:1314

119. Lin YM et al (2010) 100-GHz transistors from wafer-scale epitaxial graphene. Science 327:662
120. Cheng R et al (2012) High frequency self-aligned graphene transistors with transferred gate stack. PNAS 109:11588–11592
121. Yung KC, Wu WM, Pierpoint MP, Kusmartsev FV (2013) Introduction to graphene electronics—a new era of digital transistors and devices. Cont Phys 54:233–251. doi:10.1080/00107514.2013.833701
122. Pototsky A, Marchesoni F, Kusmartsev FV, Hanggi P, Savel'ev SE (2012) Relativistic Brownian motion on a graphene chip. Eur Phys J B 85:356
123. Yang Y et al (2013) Coherent nonlocal transport in quantum wires with strongly coupled electrodes. Phys Rev B 87:045403

The Versatile Roles of Graphene in Organic Photovoltaic Device Technology

Jayalekshmi Sankaran and Sreekanth J. Varma

The pursuit of truth continues; it is enticing, when each giant leap brings in, new surprises.

Abstract This chapter discusses the potential applications of graphene in the realization of efficient and stable organic optoelectronic devices, especially flexible solar cells. With the introduction of the prospects of graphene and functionalized graphene in modifying the performance characteristics of organic solar cells, the chapter evolves into assessing the prospects of realizing all carbon photovoltaic devices. The combination of unique, yet tunable, electrical and optical properties of graphene, makes it a highly sought after candidate for various technologically important applications in optoelectronics. Graphene has been identified as a suitable replacement for the highly expensive, brittle and less abundant indium tin oxide, as the transparent electrode material for optoelectronic device applications. The best graphene-based transparent conducting films show very low sheet resistance of 20 Ω/sq and high transparency around 90 % in the visible spectrum, making it a better choice compared to the commonly used transparent conductors including indium tin oxide (ITO) and zinc oxide (ZnO). The absence of energy band gap in graphene has originally limited its applications in optoelectronic devices. This problem has since been solved with the advent of graphene nanoribbons (GNRs) and functionalized graphenes. Functionalized graphenes and GNRs have extended the use of graphene as hole and electron transport layers in organic/polymer light emitting diodes and organic solar cells by the suitable tuning of the band gap energy. Blending dispersions of functionalised graphene with the active layers in photovoltaic devices has been found to enhance light absorption and enable carrier transport efficiently. Graphene layers with absorption in the entire visible region can be fine-tuned to be incorporated into the active layers of organic solar cells. Finally

J. Sankaran (✉)
Department of Physics, Cochin University of Science and Technology, 682022 Cochin, Kerala, India
e-mail: lakshminathcusat@gmail.com; jayalekshmi@cusat.ac.in

S.J. Varma
Department of Physics, Sanatana Dharma College, 688003 Alapuzha, Kerala, India
e-mail: sreekanthvarma@gmail.com

P. Misra (ed.), *Applied Spectroscopy and the Science of Nanomaterials*,
Progress in Optical Science and Photonics 2, DOI 10.1007/978-981-287-242-5_10

the synthesis conditions of GNRs and the functionalized graphenes can be optimized to achieve the required structural, optical and electrical characteristics for venturing into developing all carbon-based cost-effective organic solar cells with improved efficiency.

Keywords Graphene · Graphene nanoribbons (GNRs) · Functionalized graphene · Organic solar cells · Photovoltaics · Indium tin oxide · Zinc oxide · Polymer light emitting diodes (LEDs) · Organic LEDs · Electronic devices · All carbon photovoltaic devices

1 Introduction

The past few decades have witnessed unparalleled technological innovations capable of influencing all walks of human activity and enrich the intellectual pursuits in the realization of a sustainable and green living ambience with the added flavors of hitherto unimaginable comfort and luxury. In this context, the possibilities of Nanoscience and Technology have revolutionized the conventional perspectives in the design strategies of the major categories of devices that empower the energy requirements of the present day society. Of the various categories of nanostructures developed and investigated, carbon-based nanostructures have gained an upper hand, by virtue of their intriguing and often exciting characteristics with visionary prospects as the building blocks of new surprises yet to be unveiled. The three important carbon-based nanostructures, the graphene, the carbon nanotubes and the fullerenes, all stem from the three dimensional graphite, invented in 1777 by Scheele.

Graphene, or more precisely monolayer graphene, is a one-atom thick two-dimensional (2D) structure of carbon atoms arranged in a honeycomb lattice with carbon- carbon bond length of about 0.142 nm. The three dimensional parent structure graphite, from where graphene originates, can be considered as the three dimensional stacking of multilayers of graphene with an interplanar spacing of 0.335 nm. The history of graphene is marked with times of approval and disapproval ranging from 1947 to the huge surprise of the 2004 [1] when the possibility of the existence of free standing graphene was experimentally realized. When Wallace [2] in 1947 used the model of the 2D crystal arranged in a honeycomb lattice to solve the problems related to the band structure of graphite, and later in 1990, when the model was utilized for the band structure determination of carbon nanotubes [3], nobody realized that the free standing material realization of this model was waiting to be brought into limelight within a few years. Based on the famous Mermin-Wagner theorem [4], free-standing atomic planes cannot exist naturally because of the thermo dynamical instability on the nanometer scale and, if unsupported, they have a tendency to crumble and form disordered aggregates. In other words, naturally occurring free standing 2D crystals cannot be formed spontaneously because such crystals are thermodynamically unstable [5]. Deeper perception on this theory shows that its validity is confined to infinite systems and hence it is possible to realize finite sized 2D crystals in a perfectly ordered, long lived metastable state. Even though 2D crystals cannot be formed spontaneously, they can exist as stable structures when stacked and held together by van der Waals forces as part of any 3D structure like graphite. This was the breakthrough accomplished by the Manchester team in 2004, under the leadership of Andre Geim, when it was realized that graphene, a one- atom thick, two dimensional crystal of carbon atoms, or more specifically, a single planar sheet of sp^2 bonded carbon atoms, densely packed in a honeycomb crystal lattice, could be isolated from the bonding of its parental three dimensional structure, graphite, by a simple mechanical exfoliation process using Scotch tape [6–8]. The integrity of the 2D layer structure during the exfoliation process is kept well by the robust covalent bonds within the layer which are much stronger than the comparatively weaker van der Waals force between the layers in graphite. The Manchester group also demonstrated that the exfoliated graphene layer can either be supported on any suitable substrate or suspended from a supporting structure [9, 10]. The observation that, through graphene sheets, electrons can travel sub-micrometer distances without scattering, establishes the excitingly high mobility of these electrons and the excellent two dimensional crystal quality of graphene [11]. Here, there is an incompatibility when one considers the theoretical and experimental observation that perfect two-dimensional crystals cannot exist in the free state. Still, one can argue that the graphene structures form either an integral part of larger three-dimensional structures like the embedding graphite matrix or the supporting bulk substrates [12]. Recent transmission electron microscopy studies on suspended one atom thick graphene sheets, exhibiting long range crystalline order, provide experimental footing to the argument that these suspended graphene sheets are not perfectly flat. They exhibit intrinsic microscopic roughening on the scale of 1 nm and these observed corrugations in the third dimension could be quite significant in the stability of these two dimensional crystals [12].

The exciting electronic and structural characteristics of graphene, identified soon after its isolation, released an unparalleled surge of worldwide scientific curiosity, invoking the need for the integration of various scientific disciplines for more dedicated future work to unveil the hidden surprises this new material has yet to offer. When Andre Geim and Konstantin Novoselov of the Manchester University had their memorable achievement of winning the 2010 Nobel Prize in Physics for their innovative experiments regarding the two-dimensional material graphene, the spirit gave wings to the already vibrant scenario of graphene research to venture into the fields of developing the so called 'all carbon-based devices' with prospects to dominate the next generation device technology.

2 Graphene

High quality graphene is very strong, light, quite transparent and an excellent conductor of heat and electricity and is a highly sought after material as a transparent conductor for a variety of optoelectronic device applications. Graphene can be genuinely described as the most perfect 2D electronic material possible in nature [12], by virtue of the fact that the system is one atomic monolayer thick and charge carrier transport is confined in the 2D layer. The miraculous electronic properties of graphene can be attributed to the hexagonal honeycomb lattice network of carbons, in which each carbon atom is bonded to its three neighbors by strong σ bonds and the delocalized π electrons determine the low energy electronic structure. This lattice of graphene with two carbon atoms per unit cell (A and B) and the two triangular sublattices A and B indicated in different colors is shown in Fig. 1. The unit cell of graphene encompasses two π orbitals which form the bonding (π) and anti- bonding (π^*) states and give rise to a rather unique band structure that was first introduced by Walace in 1947 and is illustrated in Fig. 2. In analogy with conventional semiconductors, the π states form the lower energy valence band and the π^* states, the higher energy conduction band. The valence and conduction bands touch at six points, termed as the Dirac or neutrality points. Considering the inherent symmetry properties, these six points get reduced to a pair, K and K′, which are independent of one another. In fact, the Dirac points K and K′ correspond to the inequivalent corners of the Brillouin zone and are of much significance in the transport properties of graphene. The Fermi surface in graphene consists of the two K and K′ points in the Brillouin zone where the π and π^* bands cross. Since the orthogonal π and π^* states do not interact their crossing is allowed, and the π electrons in graphene provide an

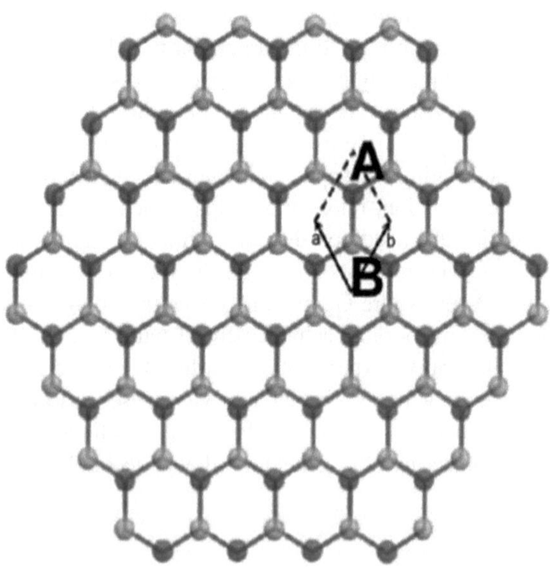

Fig. 1 The hexagonal honeycomb lattice of graphene, with two atoms *A* and *B* per unit cell. The underlying *triangular* Bravais lattice with lattice vectors **a** and **b** is also illustrated. Adapted with permission from [13]. Copyright ©2010 American Chemical Society

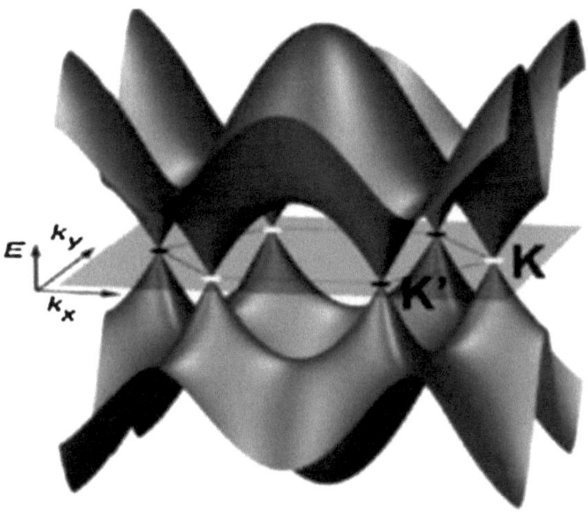

Fig. 2 The energy band structure of graphene showing the six Dirac points and the two inequivalent points K and K′. Adapted with permission from [13]. Copyright ©2010 American Chemical Society

ideal 2D system where one can realize the non-interacting π and π^* states within a single atom thick planar layer [13].

The π (bonding) and π^*(anti-bonding) bands have a linear (conical) dispersion at low energies and the band structure can be visualized as two cones touching at the Dirac point as shown in Fig. 3. Such a band structure has been shown to be quite receptive to modifications by suitable functionalizing and doping procedures. In view of the fact that the valence and conduction bands touch at the Dirac points, and based on the ambipolar nature of charge carriers (both electrons and holes), pristine graphene is considered as a zero band gap semiconductor [14–16].

The most striking aspect of graphene's energy dispersion is its linear energy momentum relationship with the conduction and valence bands intersecting at the Dirac point, corresponding to the wave vector $q = 0$, with zero energy gap. Graphene represents a zero band-gap semiconductor with a linear, energy dispersion for both electrons and holes, quite different from the parabolic energy dispersion

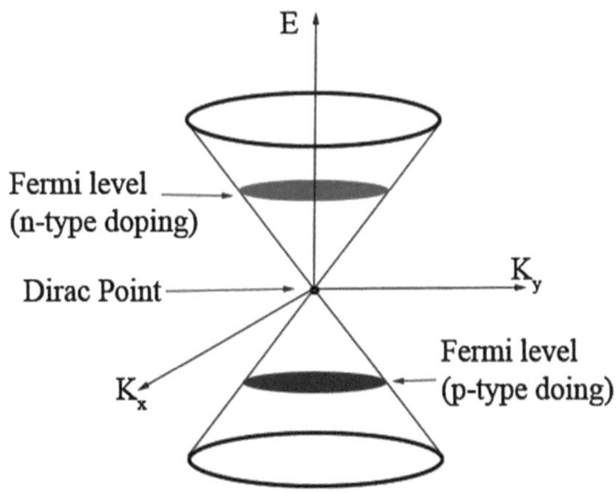

Fig. 3 The conical band structure of graphene, with the two cones touching at the Dirac point. Adapted with permission from [13]. Copyright ©2010 American Chemical Society

characteristic exhibited by other classes of well-studied two dimensional semiconductors including heterostructures, quantum wells and inversion layers. The band structure is symmetric about the Dirac point and hence the electrons and holes in pure, free-standing graphene should have similar properties. In fact, the electron and hole states in graphene are interconnected, exhibiting properties similar to the charge conjugation symmetry in quantum electrodynamics (QED).

Thus, at low energies, the presence of the two equivalent sublattices, A and B, in the structure of graphene allows the Hamiltonian to be written in the form of a relativistic Dirac Hamiltonian,

$$H = vF\sigma\hbar k \tag{1}$$

where v_F is the Fermi velocity of graphene, ħk the momentum vector measured from the Dirac point or K point and σ are Pauli spin matrices acting on the A and B sublattice degrees of freedom. In principle, the sublattice degree of freedom can be considered as an effective spin or a pseudo-spin which is parallel to the momentum vector in the conduction band and antiparallel to that in the valence band. This connection between the pseudo-spin and momentum in graphene sublattice is quite similar to that between real spin and momentum in Dirac equation [17].

The existence of two equivalent, but independent, sublattices A and B (corresponding to the two atoms per unit cell) gives rise to a novel chirality in graphene carrier transport properties. The two linear branches of graphene's conical energy dispersion intersecting at Dirac points, as shown in Fig. 4, are independent of each other and consequently are responsible for the origin of the pseudospin quantum number similar to electron spin, but completely independent of real spin. The carriers in graphene have a pseudospin index which can be represented by a spinor-like wave function in addition to the spin and orbital index. Akin to the situation in QED, the comparative contributions of the A and B sublattices in graphene are defined in terms of the spinor wave functions of the chiral pseudospin quantum number. However, in the case of graphene, the spin index specifies the sublattice rather than the real spin of the electrons.

Fig. 4 The linear branches of the conical *E–K* plot of graphene. Adapted from [16]

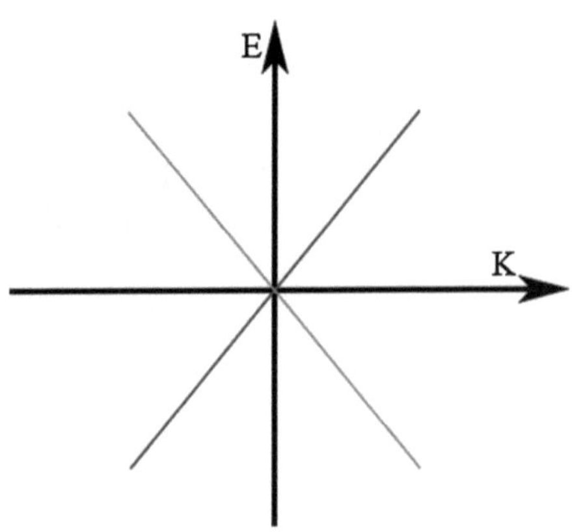

At low energies, the energy-momentum relation (dispersion relation) in graphene is linear near the six corners (Dirac points) of the hexagonal Brillouin zone, which leads to zero effective mass for electrons and holes, in contrast to the massive parabolic dispersion behavior of conventional semiconductors. Electrons and holes near the Dirac points behave like relativistic particles which are quantum mechanically described by the massless Dirac equation for spin ½ particles and hence are termed Dirac fermions. It is quite interesting and exciting to see that in graphene the low-energy carrier dynamics can be well formulated in terms of massless, chiral, Dirac fermions. However, the Dirac fermions move with the Fermi velocity v_F, which is 300 times smaller than the velocity of light. A host of the unusual surprises of QED can hence be expected to show up in graphene but at much smaller velocities. Theoretically speaking, the most distinguishing feature of graphene, in addition to its strict 2D crystalline nature, is this exquisite, long wavelength, Dirac dispersion with a Fermi velocity close to 10^6 ms^{-1} [18, 19].

The unique band structure and the presence of charge carriers mimicking massless Dirac fermions have endowed graphene with exceptional carrier transport properties. Graphene combines both semiconductor and metal properties and hence offers prospects as a replacement for the currently used semiconductors in computer chips. It has thermal conductivity around ten times higher than that of common metals like copper and aluminum. The single atom thick two dimensional planar structure of graphene, representing a system with fewer atoms and much better thermal conductivity compared to many common metals, naturally should be much faster in electrical conduction. The electrons in graphene are found to travel faster by virtue of the much smaller effects of thermal vibrations on the conduction electrons in graphene, compared to many common metals and semiconductors. This could be the reason for the much smaller value of electrical resistivity in graphene, around 10^{-6} Ω-cm, which is about 35 % smaller than that of the lowest resistivity metal, silver, at room temperature. In suspended graphene, the limiting mobility at room temperature comes around 200,000 cm^2/Vs, which is more than twice that of the high mobility conventional semiconductors [20, 21]. At lower temperatures of around 240 K, carrier mobilities of up to 120,000 cm^2/Vs have been practically achieved in suspended graphene samples.

The charge carriers in graphene are endowed with the pseudospin, as mentioned earlier, and this aspect has much to do with the exceptionally high room temperature charge carrier mobility observed in graphene. The backscattering of charge carriers in graphene is suppressed, since the back scattering involves the reversing of the pseudospin of the charge carriers in addition to the momentum, which is not allowed for low energy defect states.

For graphene monolayers supported on substrates like silicon dioxide the mobility drops down to around 10,000 cm^2/Vs as a result of the transfer of electron vibration directly from the substrate to the graphene electrons. However, there is ample scope for improvement and the room temperature mobility of graphene supported on substrates can be significantly enhanced by working out procedures for choosing the right substrates devoid of atomic scale dirts and thus reducing the effects of scattering due to charged impurities and remote interfacial phonons.

The current carrying capacity of graphene is about 5×10^8 A/cm^2 which corresponds to about 1×10^{-6} A per atomic row of carbon, which is astonishingly high [22]. Graphene has astounding mechanical strength which is about 200 times greater than that of steel, mainly due to its robust network of sp^2 bonds while being stretchable and flexible at the same time [23]. Its thermal properties are also quite surprising, with extremely high thermal conductivity around 5,000 W/mK at room temperature, which is twenty times higher than that of copper. Its thermal expansion coefficient is large and negative which is of much significance in thermal stress management in graphene based devices. All these miraculous properties of graphene are highly favored for a variety of microelectronic device applications [17, 24]. As the thinnest material ever synthesized, the one atom thick graphene membranes are impermeable to even the lightest gas molecules and are the strongest and the stiffest among man-made materials.

In graphene one has the practical realization of two-dimensionality on an atomic length scale and this two-dimensionality is stronger in the sense that the electrons in graphene remain two dimensional up to room temperature and even up to the melting point of graphene. Pure graphene is a truly two dimensional metal even at room temperature and here one has the unique experience of elevating the two dimensional electron physics from its low temperature realms to the warmth of room temperature ambience.

It is an enticing experience to work with the group of phenomena that one encounters rarely in condensed matter physics, which are exclusively specified in terms of the fundamental constants and do not depend of material parameters. In this context, graphene has another magical offer to replenish the present awareness in many related phenomena, both from the theoretical and experimental viewpoints. The optical transparency of suspended graphene can be defined in terms of the fine structure constant, given by

$$\alpha = e^2/\hbar c \tag{2}$$

where e is the electronic charge, c is the velocity of light and h is the Planck's constant. The optical or dynamical conductivity G is generally used to express the optical properties of low dimensional structures like thin films. In the case of graphene exhibiting the conical dispersion relation of the zero rest mass Dirac fermions, the energy can be expressed as

$$\varepsilon = \hbar v_F |k| \tag{3}$$

where $\mathbf{k}$ is the wave vector and $\mathbf{v_F}$ is the Fermi velocity.

G has been theoretically predicted to have a universal value

$$G_0 = e^2/4\hbar \tag{4}$$

if the photon energy is much larger than the Fermi energy $\boldsymbol{\varepsilon_F}$. This universal value of G_0 indicates that all the three optical parameters of graphene, the transmittance T,

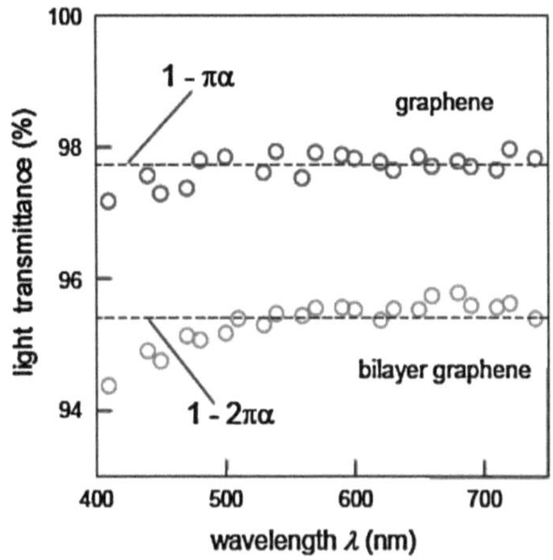

Fig. 5 Transmission spectrum of graphene, illustrating transmittance close to 98 % for monolayer graphene in the entire visible range. Adapted with permission from [24]. Copyright (2008) American Chemical Society

absorption A and the reflection R can be expressed in terms of fundamental constants. For normal incidence of light [25], the transmittance can be expressed as

$$T = (1 + 2\pi G_0/c)^{-2} \tag{5}$$

which on substitution of the value of G_0 can be approximated as $\approx 1 - \pi\alpha$.

Both R and T are explicitly related to G in the 2D case, which can be directly measured using graphene membranes. Here one has the exquisite experience of estimating the universal constant like G, using simple spectroscopic techniques, in the light of the unique electronic structure graphene can offer. Experimentally, G has been estimated to be close to $1.01G_0$ in the wavelength range from 450 to 800 nm and the corresponding transmittance T ≈ 97.7 %, which corresponds to an absorbance of about 2.3 % for graphene in this wavelength range [25, 26]. The transmittance spectrum of single layer and bilayer graphene is shown in Fig. 5. It can be seen that the transmittance decreases with the increase in the number of layers.

3 Graphene's Entry into Organic Optoelectronics

Graphene is a fascinating, truly two-dimensional material the present century has offered, whose electrons are confined to the two dimensional plane and exhibit characteristics akin to relativistic fermions. The extremely high charge carrier mobility, truly two dimensional crystalline order and the associated high electron mean free path, exceptionally high current carrying capacity, thermal conductivity, mechanical strength and flexibility and the high transparency over the entire visible range make it a highly preferred material for exotic applications in high speed electronics and flexible optoelectronics. The retention of the truly two dimensional nature of electrons even at room temperature and above is the underlying factor, enriching the vast realm of its application prospects.

Single layer and multilayer graphene can be synthesized by a variety of techniques ranging from the mechanical exfoliation of graphite flakes, as illustrated by Geim and co-workers in their pioneering work in 2004 [1] to the more sophisticated techniques including epitaxial growth of large area graphene films on metal and semiconductor substrates [27] and the chemical vapor deposition of graphene on polycrystalline nickel and copper metallic substrates [28, 29]. The realms of two dimensional condensed matter electron physics have been made accessible to ordinary room temperature laboratories with the possibilities of graphene synthesis using mechanical exfoliation. This technique, however, has the disadvantage that the graphene sheets so generated are randomly placed without proper order. In order to utilize the prospects of graphene as an exotic electronic material for device applications, one has to resort to large area graphene synthesis on suitable substrates using more sophisticated techniques.

The central theme of the present chapter is to reflect upon the versatile role of graphene in realizing flexible organic photovoltaic devices with higher power conversion efficiency and long term stability. In general pristine graphene, being a zero gap semiconductor, cannot be utilized as is for device applications. A variety of doping, intercalation and striping schemes [30, 31] have been developed to open a band gap in pristine graphene and formulate band gap engineering procedures to tailor the band gap suitably for device fabrication. Methods have also been developed for introducing various functionalities by attaching suitable functional groups and moieties onto the graphene structure using both covalent and non-covalent approaches [32–39]. A very common and easily synthesized form of functionalized graphene is the graphene oxide (GO) which is bonded with a large number of hydroxyl, carboxyl and epoxy groups. It can be obtained by simply sonicating the graphite oxide, prepared by Hummers, Brodie or Staudenmaier methods [40], in organic or aqueous solvents [41,42]. Pristine, well defined graphene sheets (both single and multilayer) can be obtained by micromechanical or chemical exfoliation of graphite [1, 43] chemical vapor deposition (CVD) technique [44, 45] or by the reduction of graphene oxide (GO) by electrochemical, chemical or thermal means [46–48].

Graphene and its derivatives have emerged as the most promising materials for the building blocks of present-day optoelectronic devices by virtue of their stable yet tunable electronic and optical properties. Graphene can be tailor-made and functionalized to fit in various roles as required for a variety of applications; it can exist as highly conducting, semiconducting and insulating based on the number of layers, the type of functionalization, the type of doping and the type of morphologies. The use of graphene and its derivatives as window layers, transparent electrodes, hole transport layers, electron transport layers and active material has paved the way for developing a new category of stable and efficient organic photovoltaic devices.

Over the past three decades, organic photovoltaic cells have emerged as the third generation solar cells with quite advantageous prospects to dominate the power generation industry in the near future. They are endowed with the inherent meritorious aspects of flexibility and cost-effective synthesis routes associated with organic molecules and polymers. The basic structure of an organic solar cell consists of the active material layer (donor + acceptor) capable of absorbing solar radiation sandwiched between the anode and the cathode, one of which has to be transparent enough to admit light transmission. In conventional type solar cells, the anode serves as the transparent electrode and in inverted type solar cells the cathode serves as the transparent electrode. Donor materials are usually low band gap polymers like poly (3-alkyl thiophenes) (P3AT) or organic molecules like the pthalocyanines, capable of efficient light harvesting [49–51].

Excitons are created, on light incidence on the donor material through the transparent electrode, which diffuse within the donor material and move towards the acceptor material. The strong electronegativity of the acceptor, combined with the built-in potential due to the work function difference between the two electrodes, results in effective carrier separation. Once the carriers are separated, they should be effectively transported to the electrodes to generate current in the external circuit. Suitable electron transport layers (ETL) and hole transport layers (HTL) are also necessary to get good power conversion efficiency (PCE) for the solar cell. In a typical solar cell, the anode is usually chosen as a transparent conducting material like indium tin oxide (ITO) which has a higher work function and the cathode is a metal like calcium or aluminum with a lower work function. In the earlier stages of development, organic solar cells generally had a bi-layer architecture represented by transparent anode/donor/acceptor/cathode as shown in Fig. 6a. The donor and acceptor materials in this structure remain as two separate layers sandwiched between the electrodes of dissimilar work function. Here, the thickness of the light absorbing donor layer is critical in such a way that the thickness should be of the order of the absorption length required for sufficient light absorption, which comes to about 100 nm. This thickness is quite high compared to the exciton diffusion length ($\sim$ 10 nm) in disordered organic semiconductors. Hence, it remained a big challenge to attain good power conversion efficiencies (PCE) until the introduction of bulk heterojunction (BHJ) concept in the 1990s. The BHJ concept [52, 53] could successfully address both the thickness requirements of the photoactive layer and the shorter exciton diffusion lengths. In these types of solar cells, the donor and acceptor materials interpenetrate each other forming a blend so that the interfaces between the donor and acceptor are spatially distributed as depicted in Fig. 6b. This can be achieved either by spin coating the donor-acceptor blend in the case of a polymer: polymer or polymer:organic small molecule active layer or by co-evaporation of conjugated molecules [52]. Further improvement in the solar cell characteristics has been achieved by introducing aligned (Fig. 6c) and non-aligned nanostructures, like zinc oxide nanorods, in the active layer, thereby improving the interfaces between the donor and acceptor layers [54–56]. The processability remains a difficult affair when inorganic nanomaterials are used as acceptor layers for fabricating solution processable solar cells. One of the best alternatives for nanostructured acceptor layer materials is graphene, which combines the flexibility and processability of polymers and the high mobility and thermal stability of the inorganic materials.

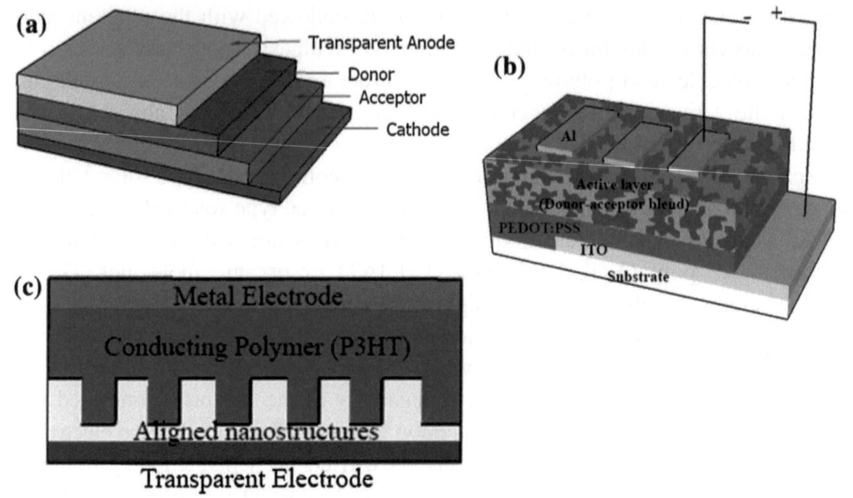

Fig. 6 Schematic of **a** planar heterojunction, **b** BHJ and **c** BHJ with aligned nanostructures

The enthralling features of graphene open the possibilities of its applications in versatile roles as the transparent conductive electrode, the hole transport layer (HTL), the electron transport layer (ETL) and the donor and acceptor layers to fabricate efficient, cost-effective, flexible and stable organic photovoltaic devices.

4 Graphene's Versatile Roles

The highly transparent graphene thin films are viable alternatives for the brittle and costly indium tin oxide (ITO) layers used in most of the optoelectronic devices. Pristine single layer graphene sheets obtained by mechanical exfoliation [57] exhibit a transmittance of about 97.7 % in the visible range and thus it absorbs only 2.3 % of the incident light. Pristine graphene has a typical work function of 4.4–4.5 eV which is very close to the highest occupied molecular orbitals (HOMO) of most of the p-type organic small molecules and polymers. This enables an effective and faster charge transfer between graphene and the organic molecules/polymers. The very high transparency, suitable work function, high electrical conductivity comparable to common metals and charge carrier mobility close to 10^5 cm^2/Vs have motivated many research groups to use graphene as the transparent conducting electrode instead of ITO in organic solar cells and obtained efficiencies comparable to that of ITO based ones [58]. It has been observed that the fill-factor of ITO-based solar cells fabricated on flexible substrates like polyethylene terephthalate (PET) drops to zero on bending the cell from 0°–60° whereas the CVD (chemical vapor deposition) graphene-based flexible OPVs are found to show negligible variation when bent up to 138°. The SEM analysis of the bent CVD graphene and ITO sheets on flexible substrates provides sufficient evidence of the crack-free nature of graphene sheets compared to the micro-crack embedded morphology of the bent ITO sheets [59]. This result is quite significant in realizing flexible photovoltaic devices without compromising the efficiency and stability. The organic solar cells fabricated on multilayer films of graphene as electrodes with 84.2 % transparency are found to give a fill-factor (FF) of 32.6 % and power conversion efficiency (PCE) of 1.17 % [60]. On comparison with the PCE of 3.43 % obtained with ITO-based cells, although the PCE is smaller for the graphene based cells, they have the advantages of low cost and flexibility. In practice, the lower conductivity of the graphene sheets compared with that of ITO films is one of the major issues affecting the overall efficiency of the fabricated devices. This challenge can be met by using doped CVD grown graphene as transparent electrodes in organic solar cells and the PCE values of these devices are found to be comparable to their counterparts with ITO electrodes [61]. Another method to improve the conductivity of graphene is to use layer-by-layer molecular doping of graphene with p-type materials like the tetracyanoquinodimethane (TCNQ) [62]. Sandwiched graphene/TCNQ films stacked structure has been used as an anode in polymer solar cells with a poly (3-hexylthiphene) (P3HT): Phenyl-C61-butyric acid methyl ester (PCBM) active layer. Among these, the cells constructed with multilayer anodes with 2 TCNQ layers sandwiched by 3 graphene layers are found to give the maximum PCE of $\sim$2.58 % which is quite higher than that of devices with acid-doped multilayer graphene as the transparent anode.

Another major issue with pristine graphene electrodes is the poor surface wetting with many of the hole transporting layers like PEDOT:PSS which makes pristine graphene-based photovoltaic devices inferior to the ITO-based devices in their overall performance. This problem has been resolved by doping the graphene layer using $AuCl_3$ which can alter the surface wetting properties and enable the formation of a uniform coating of hole transporting layer over it. Other metal chlorides like $IrCl_3$, $MoCl_3$, $OsCl_3$, $PdCl_2$ and $RhCl_3$ have been used to dope CVD-grown graphene sheets and an increase in work function has been observed with increase in dopant concentration due to spontaneous charge transfer from the specific energy level of graphene to the metal ions [63]. The sheet resistance and transmittance are found to decrease with increase in dopant concentration. From these studies, $RhCl_3$ has been identified to be the strongest p-dopant and doping with this metal chloride and modifies the work function of graphene from 4.2 to 5.14 eV. This doping method also improves the overall conductivity of the graphene electrode thereby shifting the work function of graphene to desired values which further improves the PCE values [61].

Transition metal oxides like MoO_3 have also been successfully used to address the surface immiscibility issues between graphene sheets and PEDOT:PSS layer [64]. The bonding between PEDOT:PSS and graphene improves considerably as a result of this modification. It has been inferred that surface wettability plays an important role in improving the overall device performance. A detailed study on the effect of graphene morphology, the various combinations of hole transport layers and counter electrodes on the PCE and stability has also been performed to give a better understanding in utilizing graphene as anode in organic solar cells. A significant increase in PCE has been obtained in the devices assembled using MoO_3 modified graphene as the transparent electrode [64].

Low-pressure chemical vapor deposited graphene layers with improved surface wettability have been introduced as a replacement for the ITO electrode in organic solar cells using a 15 nm PEDOT film as the hole transport layer on the graphene surface. In conventional methods, spin coated PEDOT:PSS films are used as the HTLs which have the disadvantages of poor surface wettability with graphene sheets, lower charge mobility compared to the PEDOT films and poor uniformity in thin film form. To avoid these issues between PEDOT:PSS and graphene layer, the very sophisticated vapor printing method, which includes shadow masking in combination with oxidative chemical vapor deposition (oCVD), has been used to get a very uniform, smooth, complete and pure PEDOT layer. This method yields organic photovoltaic devices of graphene/vapour printed PEDOT/DBP (tetraphenyldibenzoperiflanthene)/C_{60} (fullerene)/BCP (bathocuproine)/Al (aluminum) architecture having comparable efficiencies with the ITO based devices [65].

Another drawback of using pristine graphene as electrodes in photovoltaic applications is the significant reduction in the open circuit voltage when compared to the devices based on conventional semiconductors [66]. This problem is usually rectified by introducing a band gap in graphene and tailoring the Fermi level. Tailoring the optical and electrical properties of graphene can be achieved basically through four methods: (a) chemical modification (b) electrostatic field tuning, (c) hetero atom doping and (d) formation of graphene nanoribbons [66, 67]. Chemical modification of graphene can be done either by attaching various functional groups or by grafting organic small molecules/polymers to the graphene edges or basal planes [67]. These processes will introduce a band gap as well as modify the optical properties due to the introduction of new bands in the electronic structure of the obtained hybrid materials. The attached structures, or molecules, and the chemistry used determine the band gap and associated electrical and optical properties of the modified graphene.

Even though, pristine graphene sheets as such are not very promising as transparent electrodes in organic photovoltaic cells using polymers as the HTLs and polymer/organic small molecules as the active layers, in a recent study, 4–5 layers of graphene have been used as the transparent conductive electrodes for fabricating hybrid hetero junction solar cells based on silicon nanostructures and P3HT in which the PCE has attained values as high as 9.94 and 10.34 % [68]. Among the silicon nanostructures used, silicon nanohole arrays based cells show the highest PCE values (10.34 %) when compared to those cells based on silicon nanowire arrays (9.94 %) due to larger surface area offered and better support to graphene without dropping the capability of light absorption. The cell parameters, investigated by replacing graphene with metallic electrodes, are found to be comparatively inferior. The very high PCE values in these cells can be attributed to the efficient suppression of charge recombination and improved light harvesting capability of the graphene/silicon nanostructure combinations with P3HT. Interestingly, it has been found that solar cells with 5 layered-graphene electrodes out-perform those cells with metallic electrodes due to the higher optical transparency of graphene over the metallic films.

Interfacial/buffer layers are used to match the work function and energy levels between subsequent layers in organic photovoltaic devices. The use of interfacial layers enables efficient charge separation that results in excellent power conversion efficiencies. Graphene oxide (GO) can be used as the anode interfacial layer in organic bulk heterojunction solar cells with ITO as the transparent electrode. Solution processed GO, nickel oxide (NiO) and GO/NiO bi-layers have been used as anode interfacial layers in bulk heterojunction organic solar cells which exhibit excellent PCE and fill factor values [69]. Of these, ITO/GO/NiO/P3HT:PCBM/LiF/Al devices are found to be the best, having cell efficiency of 3.48 % with the J_{SC} of 8.71 mA/cm_2, V_{OC} of 0.602 V and FF of 66.44 % (Fig. 7, Table 1). These enhanced cell parameters have been attributed to the well-matched energy levels between the various layers of the device, achieved as a result of the incorporation of GO/NiO bilayer. A 49 % improvement in efficiency has been achieved in these devices when compared to the corresponding devices without the interfacial GO layer, which acts as the hole transporting layer.

In a recent innovative approach, GO has been used as the interfacial layer with ITO electrode and aluminum doped zinc oxide (AZO) cathode buffer layer to improve the PCE values of conventional and inverted type solar cells [70]. A remarkable improvement in the solar cell parameters has been observed in the

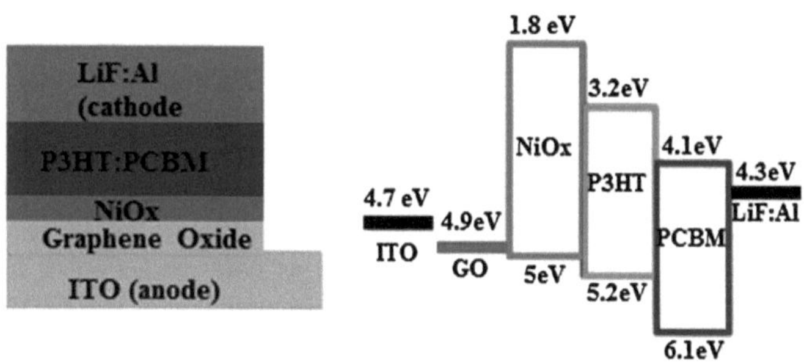

Fig. 7 Schematic of the GO/NiO_x bilayer solar cell and the corresponding band diagram. Reprinted from [69]. Copyright (2011), with permission from Elsevier

Table 1 Solar cell parameters of various devices with different combinations of HTLs

HTL configuration	J_{SC} (mA/cm^2)	V_{OC} (V)	FF (%)	PCE (%)
Without HTL	8.65	0.520	51.76	2.33
GO	8.85	0.565	54.94	2.75
Nickel oxide	9.08	0.604	56.58	3.10
GO/Nickel oxide	8.71	0.602	66.44	3.48
Nickel oxide/GO	9.11	0.602	54.91	3.01

Reprinted from [69]. Copyright (2011), with permission from Elsevier

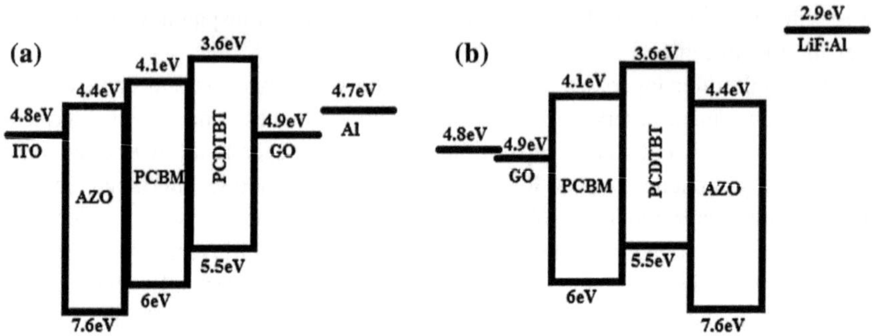

Fig. 8 Band diagrams of **a** inverted and **b** conventional type solar cells. Reprinted from [70]. Copyright (2012), with permission from Elsevier

inverted type cells. Inverted solar cells fabricated using solution processed AZO and tri-layer GO as cathode and anode buffer layers, the polymers PCDTBT, PBDTTPD as electron donors and PCBM as acceptor, are found to exhibit open-circuit voltages of 0.74 and 0.70 V, short-circuit current densities of −12.09 and −12.06 mA/cm2, fill factors (FFs) of 60.73 and 60.03 %, with overall power conversion efficiencies of about 5.46 and 5.07 %, respectively (Fig. 8). These values are quite higher compared to their conventional analogues as shown in Table 2.

The stability of thin film organic photovoltaic devices depends on the thermal and environmental stabilities of each layers used in the cell. The inhomogeneous electrical properties, high acidity suspension and hygroscopic properties of the commonly used PEDOT:PSS are some of the other major issues which affect the long term stability of the organic cells. A method to overcome these issues is to use graphene-based hole transport layers (HTL) instead of PEDOT:PSS. Improved stability has been observed in organic photovoltaic cells fabricated using UV/ozone treated graphene sheets instead of PEDOT:PSS as hole extraction or hole transporting layers. In such cells, comparable efficiencies with that of PEDOT:PSS based cells has been observed and the cells are capable of withstanding humid conditions up to 26 h of continuous working [71]. There has been a drastic decrease in the

Table 2 Solar cell parameters of conventional and inverted type solar cells using GO as anode interfacial layer and AZO as the cathode buffer layer

Device architecture	J_{sc} (mA/cm^2)	V_{oc} (V)	FF (%)	PCE average (%)
ITO/AZO/PCDTBT:PCBM/GO/Al	12.09	0.74	60.73	5.46
ITO/AZO/PBDTTPD:PCBM/GO/Al	12.06	0.7	60.03	5.07
ITO/GO/PCDTBT:PCBM/AZO/LiF:Al	7.18	0.65	60.01	2.81
ITO/GO/PBDTTPD:PCBM/AZO/LiF:Al	7.18	0.6	56.12	2.41

Reprinted from [70]. Copyright (2012), with permission from Elsevier

efficiency of PEDOT:PSS based devices which are found to work for only 14 h on exposure to the same humid atmosphere. There are also many other advantages of using UV/ozone treated graphene sheets instead of GO and reduced GO (r-GO) sheets as HTLs. The major disadvantage of using GO is the thickness of the layer, which, if higher, can make the GO layer more insulating and can affect adversely the overall efficiency of the cell. It is also very difficult to get proper band alignment with the various active layers when r-GO is used as the HEL.

The application of modified graphene is not limited to the transparent anode and the hole transporting layers in organic solar cells. It can also be used as efficient transparent cathode material as reported by a few research groups. The higher sheet resistance of the graphene layer, when compared with the conventional cathode materials, poses a major difficulty in overcoming the PCE values offered by the conventional cells. It has been reported that single layer graphene can be used as cathode material and the performance of the cell can be made much better if contact doping can be induced in the graphene layer, which reduces the sheet resistance considerably. The high transparency of the graphene also adds to the improvement in performance [72] (Fig. 9). Single layer graphene has been found to exhibit a very interesting property by which its work function can be made tunable when kept in contact with strong electron donating materials, which is a consequence of graphene's small density of states around the Dirac point. This process is termed as contact doping of graphene, since the contact with suitable electron donors

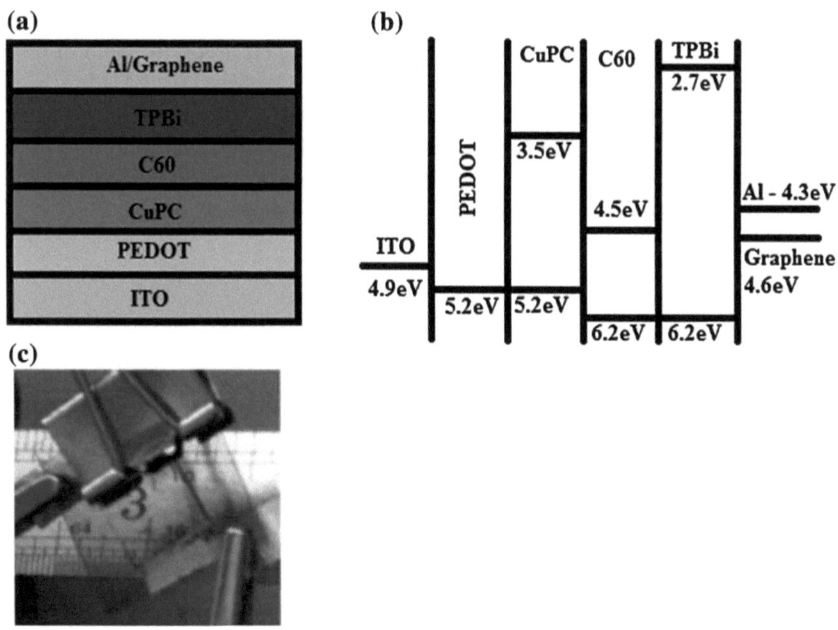

Fig. 9 **a** Device architecture of organic solar cells with graphene cathode, **b** band diagram of the device with graphene cathode and **c** photograph of the transparent devices with graphene cathode. Reprinted with permission from [72]. Copyright [2011], AIP Publishing LLC

facilitates work function tuning [73–75]. These studies highlight the possibility of using graphene as both cathode and anode materials in metal-free, flexible and light-weight, all-organic solar cells.

The major challenges for using graphene as cathode materials are concerned with achieving the required surface wettability, work function tuning and carrier transport. These issues can be solved to some extent by incorporating Al-TiO_2 nanocomposite to modify the single layer graphene [76]. The evaporated aluminum nanoclusters deposited on graphene improve the surface wettability for the subsequent deposition of TiO_2 in the composite, thereby reducing the work function of graphene to suit the energy levels. The solution processed TiO_2 being a good electron transporter helps graphene to extract charges more efficiently, thereby enhancing the overall efficiency of the solar cell. The self-assembly method adopted to deposit TiO_2 on the aluminum coated graphene offers better charge extraction when compared to the conventional spin coating process, which results in non-aligned structures. The main advantages of the self-assembly method are excellent uniformity and thickness control. The inverted organic solar cells with Al-TiO_2 modified graphene cathode exhibit twice the PCE values (2.58 %) compared to those with un-modified graphene cathodes (Fig. 10, Table 3). The PCE values of these organic solar cells have reached almost 75 % of those made using indium tin oxide.

In a recent study, graphene mesh transparent electrodes, prepared by photolithography and O_2 plasma etching, have been used as one of the transparent, conducting electrodes for organic bulk hetero junction solar cells with a power conversion efficiency of 2.04 % [77]. This is one of the highest PCE values for an organic solar cell based on solution-processed graphene electrode. A blend of poly (3-hexylthiophene) and phenyl-C_{61}-butyric acid methyl ester ($PC_{61}BM$) has been used as the active layer for this solar cell. The O_2 plasma treatment improves the hydrophilic nature of the graphene mesh and aids in the formation of uniform films on the PEDOT:PSS layer. This improved hydrophilicity is due to the introduction of some oxygen containing functional groups on the graphene surface.

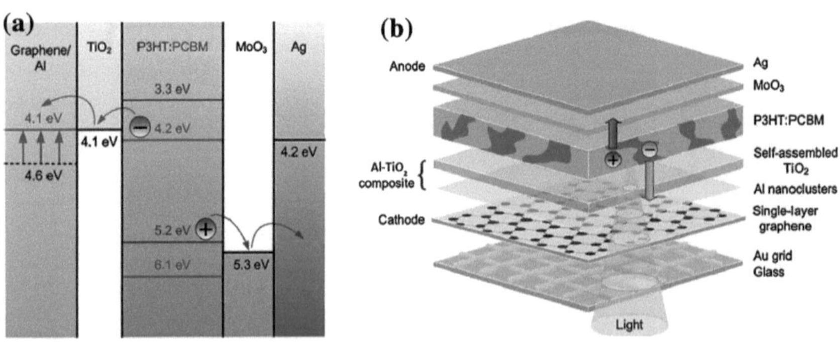

Fig. 10 **a** Band diagram of Al-TiO_2 modified, graphene based organic photovoltaic devices and **b** Architecture of device with graphene cathode. Adapted with permission from [76]. Copyright ©2013 American Chemical Society

Table 3 Device characteristics of inverted type polymer solar cells with Cathode/P3HT:$PC_{61}BM$/MoO_3/Ag architecture

Cathode	J_{SC} (mA/cm^2)	V_{OC} (V)	FF (%)	PCE (%)
SLG/Al-TiO_2	7.85 ± 0.24	0.58 ± 0.02	35.0 ± 3.2	1.59 ± 0.08
Grid/SLG/Al-TiO_2	8.55 ± 0.62	0.60 ± 0.01	50.1 ± 2.5	2.58 ± 0.09
ITO/Al-TiO_2	9.11 ± 0.25	0.63 ± 0.00	60.1 ± 0.3	3.45 ± 0.09
SLG/Al/spin-coated TiO_2	1.80 ± 0.65	0.30 ± 0.04	25.2 ± 0.5	0.14 ± 0.07
Grid only/Al-TiO_2	1.31 ± 0.56	0.61 ± 0.01	33.2 ± 0.5	0.27 ± 0.11

SLG single layer graphene
Adapted with permission from [76]. Copyright ©2013 American Chemical Society

5 Functionalization of Graphene

Functionalization of nanostructured materials refers to the attachment of functional groups at desired sites for suitable modification of the material characteristics. It is possible to design materials with tailor-made properties by adopting the most appropriate functionalization procedures.

The basal plane of graphene is comprised of strong covalent bonds of the sp^2 hybridized carbon atoms, whereas the edge sites contain dangling bonds and are comparatively more reactive. These dangling bonds at the edge sites are quite appropriate for covalently attaching different types of functional groups and molecules to achieve better processability, solubility, film forming properties and reactivity suitable for the desired type of chemical modifications. The graphene basal plane can also be covalently or non-covalently functionalized to impart modifications as required.

Functionalization of graphene in the basal planes and edges are excellent methods to design efficient hole and electron extraction materials with proper energy level matching with the various layers in organic energy harvesters. Edge functionalization of graphene can be used to synthesize cesium-neutralized graphene oxide (GO-Cs) by effecting charge neutralization of the –COOH groups attached to the periphery of GO with Cs_2CO_3 and the process is illustrated in Fig. 11a. The work function of GO gets modified upon this functionalization and GO-Cs has been identified to be a very good electron transporting material [78]. It is quite interesting to see that GO and GO-Cs can be used, respectively, as hole and electron extraction layers in conventional and inverted types of organic solar cells with a blend of P3HT and PCBM as the active layers [78–80]. These solar cells, with device architectures as depicted in Fig. 11b, c exhibit PCE values as high as 3.67 % and outclass the analogous conventional type BHJs with state-of-art hole and electron extraction layers. Recent research in edge functionalization of graphene has provided highly promising and effective ways to get functionalized graphene with controlled work function and optical properties with high prospects of improving the efficiency of organic solar cells [79–82].

Polymer solar cells with appreciable efficiency have also been realized based on basal-plane functionalized sulphated graphene oxide, GO-OSO_3H with a remarkable

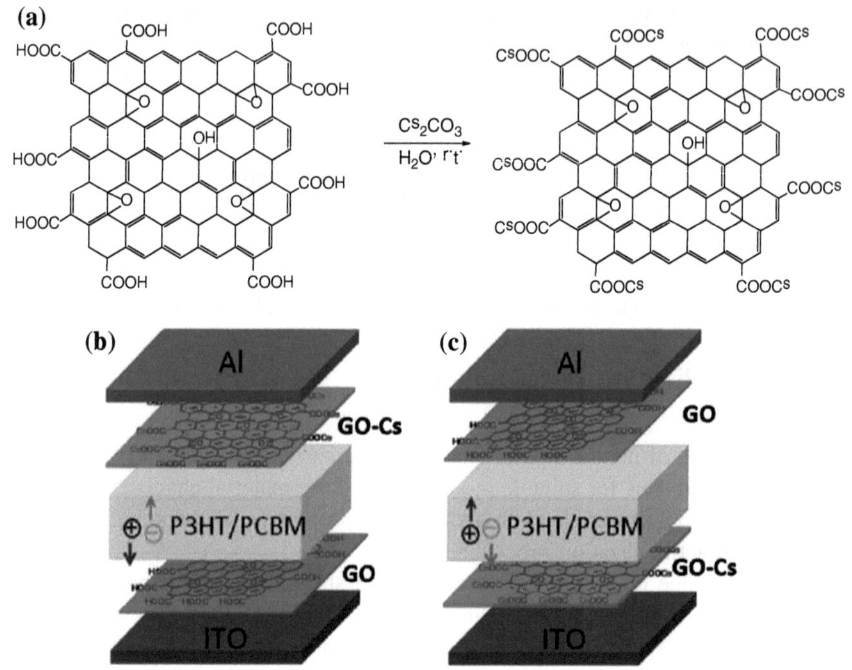

Fig. 11 **a** Synthesis route for Cs functionalization and **b**, **c** architecture of devices with edge functionalized GO as ETL. Adapted with permission from [79]. Copyright ©2013 American Chemical Society)

PCE of 4.37 % which is amongst the highest values obtained for solar cells with P3HT:PCBM active layers. Sulphated graphene oxide can be synthesized by substituting the hydroxyl or the epoxy groups of the carbon basal plane of graphene oxide (GO) with $-OSO_3H$ groups, by treating with fuming sulphuric acid, while keeping its –COOH edge groups intact, as illustrated in Fig. 12. The sulphated graphene oxide functions as the hole extraction material in this polymer solar cell which exhibits excellent performance characteristics [80].

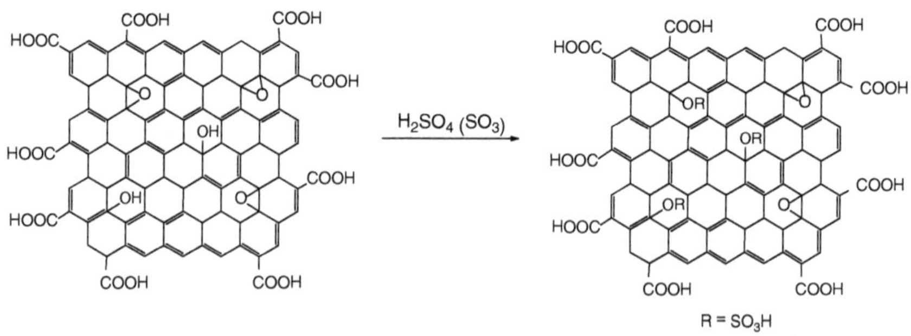

Fig. 12 The synthesis route for GO-OSO_3H

Graphene based acceptor layers can be used to improve the efficiency and thermal stability of organic/hybrid bulk hetero junction solar cells. Several modifications on the graphene surface have been adopted to improve the charge transport and separation in solar cells. The fullerene based acceptors with very low stability can be replaced by these modified stable counterparts. An electron transporting material developed from hydrazine-reduced graphene oxide by a simple lithiation reaction has been found to yield a covalent attachment of monosubstituted C_{60} on the graphene sheet [83]. A 2.5 fold improvement in the PCE values has been observed in P3HT based solar cells when C_{60} is replaced by this electron acceptor due to enhanced electron transport or efficient carrier separation. The C_{60} grafted graphene sheet will be a very attractive inclusion for hybrid all-carbon devices.

Graphene quantum dots (GQDs) represent a new type of nanomaterials which have very unique and distinctive electrical and optical properties and are being widely investigated for a variety of applications. The GQDs can be synthesized by the very simple, cost-effective and high-yield hydrothermal method. Although graphene quantum dots are used in hybrid/organic solar cells as electron acceptors, the power conversion efficiencies of these cells are not very promising. Graphene quantum dots can be effectively used as cathode buffer additives with cesium carbonate (Cs_2CO_3) in bulk hetero junction inverted polymer solar cells and 22 % enhancement in the PCE values (2.59–3.17 %) has been observed, when compared to that with Cs_2CO_3 alone [84]. This can be attributed to the improved exciton dissociation and suppressed electron–hole recombination at the cathode-polymer active layer interface. In this architecture, depicted in Fig. 13, GQDs act as excellent electron transfer and hole blocking material (Table 4).

In most of the bulk hetero junction organic solar cells, fullerene and fullerene derivatives have become indispensable as the electron acceptors owing to the higher charge mobilities compared to the polymer-based acceptor molecules. The overall stability of the cell depends upon the thermal stability of the electron acceptors also and hence, it is desirable to replace fullerene and its derivatives with suitable

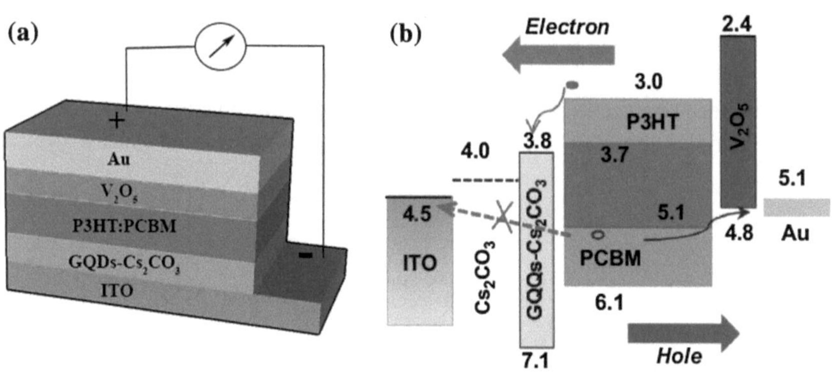

Fig. 13 **a** Device structure of the GQD based inverted organic solar cell and **b** Energy band diagram of inverted polymer solar cells with Cs_2CO_3 or GQDs-Cs_2CO_3 buffer layers. Reprinted from [84]. Copyright (2013), with permission from Elsevier

Table 4 Device characteristics of inverted organic solar cells with different cathode buffer layers

Cathode buffer	J_{SC} (mA/cm^2)	V_{OC} (V)	FF (%)	PCE (%)
GQDs	4.7	0.509	33.2	0.79
Cs_2CO_3	8.37	0.57	54.2	2.59
GQDs-Cs_2CO_3	9.04	0.585	60	3.17

materials with higher charge mobility and thermal stability. In a typical study, aqueous-dispersible graphene (a-dG) sheets have been presented as the substitutes for the acceptor materials that offer efficient charge transfer between the donor polymer, poly[3-(potassium-6-hexanoate) thiophene-2, 5-diyl] (P3KT) and the a-dG acceptor in a 'green' polymer solar cell with ITO/PEDOT:PSS/P3KT:a-dG/ZnO/Al architecture [85]. The a-dGs can be synthesized by reducing graphene sheets in the presence of the water soluble P3KT by non-covalent functionalization. An efficient charge transfer occurs between P3KT and a-dG, whose chemical structures are depicted in Fig. 14a, b. The charge transfer can be confirmed from the fluorescence quenching observed in P3KT due to the presence of a-dG. The performance of the a-dG modified polymer solar cell has also been found to be superior when compared to that of the one without a-dG in the P3KT film. This study has extended the role of graphene as excellent electron acceptors in easily processable, environmental friendly and low-cost polymer solar cells. In addition to the higher thermal stability and charge mobility, functionalized graphene has the advantages of low cost and ease of synthesis when compared to the other widely used organic acceptor materials.

Functionalized graphene blended with donor polymers such as P3HT has been used by many researchers as the active layer in their bulk hetero junction photovoltaic devices to enhance the stability and improve the charge separation. It has been observed that there is an optimal graphene content and a moderate annealing treatment is required to obtain the maximum efficiency out of the device.

The present PCE values of single bulk hetero junction (BHJ) photovoltaic devices have approached 9 % owing to the advances in the appropriate band gap tailoring and energy level modifications of the photo active layers and the controlled interface engineering strategies for efficient charge separation and collection by the electrodes [86–88]. The low charge carrier mobilities and narrow optical

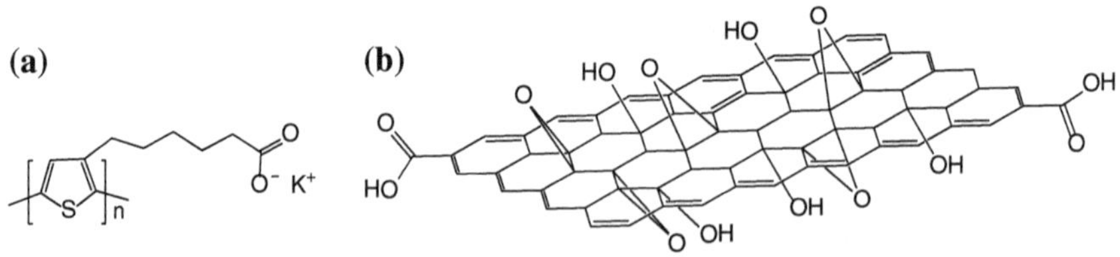

Fig. 14 Chemical structure of **a** P3KT and **b** a-dG

absorption band width associated with organic molecules and conjugated polymers are the prime challenges encountered for enhancing the PCE values well above 10 % for single junction devices [89–91]. The multi-junction or the tandem concept is a versatile approach to enhance the efficiency of photo voltaic devices, by which the harvest of a wide spectral solar radiation can be made viable. The solar radiation absorption window can be broadened by stacking together wide and low band gap photo active materials capable of light absorption over a much broader wavelength region. In the tandem solar cell configurations, where two or more sub-cells with matching light absorption are stacked and connected in series or parallel, PCE values above 10 % have been achieved [92, 93].

The adaptability of graphene has crossed another milestone with the realization that graphene can function effectively as the intermediate layer (IML) in tandem solar cells to boost the overall efficiency. In the tandem configuration with sub cells connected either in series or parallel, the IML should act as the protective layer between sub cells to prevent intermixing of any two sub cells. The material acting as the IML should also have quite high electrical conductivity and minimum light absorption. Compared to the series connected tandem cell, which does not demand stringent material perfection in terms of electrical conductivity and transmittance for the IML, the parallel connected tandem configuration is essentially in need of highly transparent and conductive IML without structural discontinuities. With most of the metallic IMLs the main hurdle is the light transmission loss close to 40 %, which can inhibit the light harvesting capability of the parallel tandem device [94]. Carbon nanotubes (CNTs), in spite of the high electrical conductivity are not suitable as efficient IML materials, owing to the large contact resistance between CNTs and the organic molecules [95]. In parallel connected tandem cells, any two sub cells can operate individually and higher efficiency can be achieved more easily, compared to the series connected tandem cells, and the demand for developing parallel connected tandem cells is on the soar all through the photovoltaic industry.

Graphene has recently been identified as a highly pursued material as the IML in both series and parallel connected tandem cells, owing to the excellent transparency and high electrical conductivity, combined with the possibility that the soft graphene membranes can be conveniently transferred to a variety of substrates, according to the requirement. Graphene films coated with MoO_3 offer high prospects as IMLs between each sub cell in tandem polymer (parallel and series configurations) solar cells and considerable improvement in the open circuit voltage and short circuit current values has been achieved [94]. The deposition of a MoO_3 layer of particular thickness over CVD grown graphene films modifies the work function of graphene layers to match with the energy levels of subsequent layers. It has also been found that the work function of graphene layers increases with the increase in thickness of MoO_3 layer; this offers the possibility to tune the work function of MoO_3 coated graphene layers, in terms of the MoO_3 layer thickness, to match with that of the layers in contact, to achieve efficient charge extraction. The work function of graphene interface has to be suitably controlled to minimize charge build- up between sub cells in tandem configuration. It has been established experimentally that graphene based IMLs are highly efficient in joining sub cells of tandem cells so that the open circuit voltage and short circuit current density get multiplied, in proportion to the number of sub cells involved [94]. This is a new leap towards developing organic tandem cells with much pronounced thermal and environmental stability, efficiency, and design flexibility.

5.1 Towards All-Carbon Devices

Electric power generation from inorganic semiconductor based photovoltaic technology is still alarmingly expensive compared to fossil fuel technology, even when intense research is on the soar globally for identifying strategies towards cost reduction. A host of new materials have been developed over the past two decades as alternatives to replace silicon and gallium arsenide like expensive photoactive inorganic materials, which include conjugated polymers, organic small molecules and carbon-based nanostructures. These organic analogues with excellent light absorption and tunable electrical conductivity offer the possibility of assembling thin film structures on light weight and flexible substrates. The cost reduction can be genuinely realized by adopting roll to roll, large area device processing technology, employing cost effective techniques like solution processing or the methods based on paintable inks, at low temperatures and ambient pressure conditions, which are feasible with organic materials [96, 97]. Solution deposition based roll to roll processes can be suitably adapted to install organic photovoltaic devices on any types of surfaces without worrying about the shape constraints, on textiles, automobiles and buildings, so as to make solar cells affordable and accessible on a general basis. Although there is a long way to go for realizing these visions, the recent advances in carbon-based nanostructures are very promising to expect breathtaking giant leaps in this direction within the next 5–10 years.

Carbon, known from time immemorial, constitutes one of the most abundant materials on the Earth's crest and in tune with the widespread technological demand, is produced in huge quantities to the extent of 9 Giga tons per year [98]. Along with the naturally occurring graphite, diamond and coal, carbon nanotubes, fullerenes and graphene, which are the nanostructured carbon allotropes, are also being extensively utilized for many technological applications. These carbon nanostructures and their substituted derivatives and functionalized forms offer excellent prospects in terms of charge carrier mobility and optical absorption for organic photovoltaic applications. Both conducting and semiconducting properties can be envisaged in these carbon nanostructures based on their chemical structure and are the prime factors for their suitability to be used in combination to realize devices consisting entirely of carbon-based materials. Though carbon-based devices are in the infant state of development, there is high probability that the carbon-based photovoltaic devices may dominate the power industry, as the next generation cost effective solar cells, within the next decade. Such devices have many attractive features to offer, the most important one being the possibility of production at much cheaper rates and in large quantities due to the abundance and availability of carbon. The carbon-based nanomaterials and their functionalized forms can be dispersed and also grown on suitable substrates using cost effective solution processes which makes their integration into roll to roll manufacturing processes highly feasible and scalable [99–101]. Many of the carbon-based active materials suitable for device fabrication are also endowed with unparalleled, long term temperature and chemical stability [102–104] which makes them highly sought after materials of the present century.

The recent surge of research enthusiasm towards carbon nanostructures has resulted in utilizing them as the transparent electrodes, interconnects, electron/hole transport layers, buffer layers [105] and as the acceptor layers in combination with conjugated polymers or small organic molecules as donor layers [96, 97] in organic photovoltaic devices. As the initial leap towards "all-carbon devices" attempts have been carried out to replace the conjugated polymers and small molecules with carbon nanostructures in combination, to function as the main active layer components, and bilayer photovoltaic devices with PCE values close to 0.85 % have been realized using fullerenes in combination with single walled carbon nanotubes (SWCNTs) or composites of SWCNTs with reduced graphene oxide (rGO) and fullerenes [106, 107]. Further modifications in device structure have resulted in efficiencies around 1.3 %, based on entirely solution deposited and carbon-based active layers consisting of semiconducting SWCNTs, the fullerene derivative PCBM and rGO in the bulk- heterojunction architecture [108]. The photo active layer is completely devoid of conjugated polymers or small molecules and the active layer consisting entirely of carbon nanomaterials provides atomic carbon concentration around 80–90 %, compared to the much lesser carbon concentration around 40–50 % in polymer based solar cells [108]. Carbon-based nanomaterials offer all the meritorious aspects of conjugated polymers and organic molecules including solution processability, chemical tunability and mechanical flexibility with the added benefits of much higher photo and thermal stability.

Although these advancements towards carbon-based photovoltaic devices are quite promising, these devices are not entirely carbon-based, and in these devices, indium tin oxide and silver/aluminum function as the anode and the cathode respectively. A breakthrough has been achieved [109] with the recent reports on the fabrication of the first, all-carbon photovoltaic device by Stanford University scientists, in which all the components are made entirely of carbon-based materials, including the anode, the active layer and the cathode. In this approach, the active layer consists of a bilayer film of solution processed and sorted semiconducting SWCNTs functioning as the light absorbing donor, and fullerene C_{60} as the acceptor, and reduced graphene oxide and doped (n-type) SWCNTs function respectively as the anode and the cathode [109]. This all-carbon solar cell has been found to have a PCE much less than 1 % for near infrared illumination. Although the PCE value is quite small for these all-carbon solar cells, there is ample scope for improvement. By suitably choosing the active layer components, especially the semiconducting SWCNTs, capable of light absorption in a broader range of the solar spectrum including the visible spectrum, the efficiency can be considerably improved. The use of unidirectionally aligned SWCNTs in the active layer can significantly reduce the exciton trapping and enhance the exciton diffusion length and contribute towards achieving higher efficiency [109]. The Stanford research team is also working out strategies for using smoother and more conductive and more transparent, graphene based anode layer to facilitate better charge collection and achieve more incident light intensity. It is also equally challenging to directly deposit a much smoother SWCNT cathode film with better contact to the semi-conducting SWCNT and C_{60} active layer without damaging the active layer which lies underneath to improve efficiency and device performance [109].

The dream of realizing all-carbon solar cells has already spread wings to conquer the heights of perfection. Though there is still a long way to go for this novel concept to flourish into marketable device technology, the ambience is all set for giant leaps towards miraculous achievements. Carbon-based materials are quite robust and are stable up to air temperatures around 1,100 °F. According to the Stanford research team working on all-carbon solar cells, the need for higher efficiency of these novel devices is balanced by the capability of these devices to outperform the conventional solar cells under extreme conditions of high temperature, pressure and physical stress. The manufacturing costs are also considerably reduced, since carbon materials can be coated on any substrate, employing solution deposition techniques, which do not require sophisticated and expensive machinery, contrary to the expensive processing technology of silicon-based devices. Considering the accessibility and abundance of carbon materials with potentials not yet utilized for energy harvesting technologies and the innovations emerging in material processing and device fabrication concepts, the future holds bright promises for the all-carbon photovoltaic devices. May be the days are not far away, to witness the flexible carbon solar cells adorning the surfaces and windows of buildings and automobiles, and generating electric power, in tune with the richness and simplicity of green chemistry.

References

1. Novoselov KS, Geim AK, Morozov SV, Jiang D, Zhang Y, Dubonos SV, Grigorieva IV, Firosov AA (2004) Science 306:666
2. Wallace PR (1947) Phys Rev 71:622
3. Ajiki H, Ando T (1993) J Phys Soc Jpn 62:1255
4. Mermin ND, Wagner H (1966) Phys Rev Lett 17:1133
5. Fasolino A, Los JH, Katsnelson MI (2007) Nat Mater 6:858
6. Geim AK (2009) Science 3:1530
7. Novoselov KS, Jiang D, Schedin F, Booth TJ, Khotkevich VV, Morozov SV, Geim AK (2005) Proc Natl Acad Sci USA 102:10451
8. Geim AK, Novoselov KS (2007) Nat Mater 6:183
9. Meyer JC, Geim AK, Katsnelson MI, Novoselov KS, Booth TJ, Roth S (2007) Nature 446:60
10. Blake P, Hill EW, Castro Neto AH, Novoselov KS, Jiang D, Yang R, Booth TJ, Geim AK (2007) Appl Phys Lett 91:063124
11. Meyer JC, Geim AK, Katsnelson MI, Novoselov KS, Booth TJ, Roth S (2007) Nature 446:60

12. Geim AK, MacDonald AH (2007) Phys Today 60:35
13. Avouris P (2010) Nano Lett 10:4285
14. Wilson M (2006) Phys Today 59:21
15. Castro Neto AHF, Guinea N, Peres MR, Novoselov KS, Geim AK (2009) Rev Mod Phys 81:109
16. Sharma SD, Shaffique A, Hwang EH, Rossi E (2011) Rev Mod Phys 83:407
17. Fuhrer MS, Lau CN, MacDonald AH (2010) MRS Bull 35:289
18. Novoselov KS, Geim AK, Morozov SV, Jiang D, Zhang Y, Katsnelson MI, Grigorieva IV, Dubonos SV, Firsov AA (2005) Nature 438:197
19. Zhang Y, Tan Y-W, Stormer HL, Kim P (2005) Nature 438:201
20. Chen JH, Jang C, Xiao SD, Ishigami M, Fuhrer MS (2008) Nat Nanotechnol 3:206
21. Morozov SV, Novoselov KS, Katsnelson MI, Schedin F, Elias DC, Jaszczak JA, Geim AK (2008) Phys Rev Lett 100:016602
22. Standley B, Bao W, Zhang H, Bruck J, Lau CN, Bockrath M (2008) Nano Lett 8:3345
23. Lee C, Wei XD, Kysar JW, Hone J (2008) Science 321:385
24. Balandin AA, Ghosh S, Bao W, Calizo I, Teweldebrhan D, Miao F, Lau CN (2008) Superior thermal conductivity of single-layer graphene. Nano Lett 8:902
25. Nair RR, Blake P, Grigorenko AN, Novoselov KS, Booth TJ, Stauber T, Peres NMR, Geim AK (2008) Science 320:1308
26. Sheehy DE, Schmalian J (2009) Phys Rev B 80:193411
27. Berger C, Song ZM, Li XB, Wu XS, Brown N, Naud C, Mayo D, Li TB, Hass J, Marchenkov AN, Conrad EH, First PN, de Heer WA (2006) Science 312:1191
28. Reina A, Jia XT, Ho J, Nezich D, Son HB, Bulovic V, Dresselhaus MS, Kong J (2009) Nano Lett 9:30
29. Li XS, Cai WW, An JH, Kim S, Nah J, Yang DX, Piner R, Velamakanni A, Jung I, Tutuc E, Banerjee SK, Colombo L, Ruoff RS (2009) Science 324:1312
30. Park J, Lee WH, Huh S, Sim SH, Kim SB, Cho K, Hong BH, Kim KS (2011) J Phys Chem Lett 2:841
31. Park J, Jo SB, Yu YJ, Kim Y, Yang JW, Lee WH, Kim HH, Hong BH, Kim P, Cho K, Kim KS (2012) Adv Mater 24:407
32. Sinitskii A, Dimiev A, Corley DA, Fursina AA, Kosynkin DV, Tour JM (2010) ACS Nano 4:1949
33. Hossain MZ, Walsh MA, Hersam MC (2010) J Am Chem Soc 132:15399
34. Nemes-Incze P, Osváth Z, Kamarás K, Biro LP (2008) Carbon 46:1435
35. He H, Gao C (2010) Chem Mater 22:5054
36. Liu Y, Zhou J, Zhang X, Liu Z, Wan X, Tian J, Wang T, Chen Y (2009) Carbon 47:3113
37. An X, Butler TW, Washington M, Nayak SK, Kar S (2011) ACS Nano 5:1003
38. Bai H, Xu Y, Zhao L, Li C, Shi G (2009) Chem Commun (13):1667 doi:10.1039/B821805F
39. Liu H, Gao J, Xue M, Zhu N, Zhang M, Cao T (2009) Langmuir 25:12006
40. Park S, Ruoff R S (2009) Nature Nanotech 4:217
41. Stankovich S, Piner RD, Chen X, Wu N, Nguyen ST, Ruoff RS (2006) J Mater Chem 16:155
42. Stankovich S, Dikin DA, Piner RD,Kohlhaas K A, Kleinhammes A, Jia Y, Wu Y, Nguyen S T, Ruoff R S(2007) Carbon 45:1558
43. Lotya M, Hernandez Y, King PJ, Smith RJ, Nicolosi V, Karlsson LS, Blighe FM, De S, Wang ZM, McGovern IT, Duesberg GS, Coleman JN (2009) J Am Chem Soc 131:3611
44. Kim KS, Zhao Y, Jang H, Lee SY, Kim JM, Kim KS, Ahn J-H, Kim P, Choi J-Y, Hong BH (2009) Nature 457:706
45. Li X, Cai W, An J, Kim S, Nah J, Yang D, Piner R, Velamakanni A, Jung I, Tutuc E, Banerjee SK, Colombo L, Ruoff RS (2009) Science 324:5932
46. Park S, Ruoff RS (2009) Nat Nanotechnol 4:217
47. Shao YY, Wang J, Engelhard M, Wang CM, Lin YH (2010) J Mater Chem 20:743
48. Zhu YW, Stoller MD, Cai WW, Velamakanni A, Piner RD, Chen D, Ruoff RS (2010) ACS Nano 4:1227

49. Kroon R, Lenes M, Hummelen JC, Blom PWM, De Boer B (2008) Polymer Rev 48:531
50. Vanlaekea P, Swinnenb A, Haeldermansb I, Vanhoylandb G, Aernoutsa T, Cheynsa D, Deibela C, D'Haena J, Heremansa P, Poortmansa J, Manca JV (2006) Sol Energ Mat Sol 90:2150
51. Yoon SM, Lou SJ, Loser S, Smith J, Chen LX, Facchetti A, Marks T (2012) Nano Lett 12:6315
52. Pagliaro M, Palmisano G, Ciriminna R (2008) Flexible solar cells. Wiley-VCH Verlag, GmbH & Co. KGaA, Weinheim, pp 85
53. Yu G, Gao J, Hummelen JC, Wudl F, Heeger AJ (1995) Science 270:1789
54. Liang Z, Gao R, Lan J-L, Wiranwetchayan O, Zhang Q, Li C, Cao G (2013) Sol Energ Mat Sol 117:34
55. Ahmadi M, Mirabbaszadeh K, Ketabch M (2013) Electron Mater Lett 9:729
56. Kang Y, Park N-G, Kim D (2005) Appl Phys Lett 86:113101
57. Wan X, Huang Y, Chen Y (2012) Acc Chem Res 45:598
58. Wang H-X et al (2013) Small 9:1266
59. Zhang Y, Zhang L, Zhou C (2013) Acc Chem Res 46:2329
60. Choi Y-Y et al (2012) Sol Energ Mat Sol 96:281
61. Park H et al (2010) Nanotechnology 21:505204 (6 p)
62. Hsu C-L (2012) ACS Nano 6:5031
63. Kwon KC et al (2012) Adv Funct Mater 22:4724
64. Park H et al (2012) Nano Lett 12:133
65. Park H et al (2012) ACS Nano 6:6370
66. Wang H-X (2013) Small 9:1266
67. Georgakilas V et al (2012) Chem Rev 112:6156
68. Yiming W et al (2013) J Phys Chem C 117:11968
69. Ryu MS and Jang J (2011) Sol Energy Mat Sol 95(10):2893
70. Yusoff ARBM et al (2012) Org Electron 13(11):2379
71. Kwon KC et al (2013) Sol Energy Mat Sol 109:148
72. Cox M et al (2011) Appl Phys Lett 98:123303
73. Ryu S, Brus LE, Kim KS, Kim P, Yu Y-J, Zhao Y (2009) ACS Nano 9:3430
74. Lee DS, Krauss B, Patthey L, von Klitzing K, Smet JH, Starke U, Coletti C, Riedl C (2010) Phys Rev B 81:235401
75. Rusu C, Brocks G, van den Brink J, Kelly PJ, Khomyakov PA, Giovannetti G (2009) Phys Rev B 79:195425
76. Zhang D et al (2013) ACS Nano 7:1740
77. Zhang Q, Wan X, Xing F, Huang L, Long G, Yi N, Ni1 W, Liu Z, Tian J, Chen Y (2013) Nano Res 6:478
78. Liu J, Xue Y, Gao Y, Yu D, Durstock M, Dai L (2012) Adv Mater 24:2227–2231
79. Dai L (2013) Acc Chem Res 46:31–42
80. Liu J, Xue Y, Dai L (2012) J Phys Chem Lett 3:1928
81. Jeon I-Y, Shin Y-R, Sohn G-J, Choi H-J, Bae S-Y, Mahmood J, Jung S-M, Seo J-M, Kim M-J, Chang DW, Dai L, Baek J-B (2012) Proc Natl Acad Sci USA 109:5588
82. Choi EK, Jeon IY, Bae SY, Lee HJ, Shin HS, Dai LM, Baek JB (2010) Chem Commun 46:6320
83. Yu DS, Park K, Durstock M, Dai LM (2011) J Phys Chem Lett 2:1113
84. Yang HB et al (2013) Sol Energy Mat Sol 117:214
85. Liu Z et al (2012) Sol Energy Mat Sol 97:28
86. Chen HY, Hou JH, Zhang SQ, Liang YY, Yang GW, Yang Y, Yu LP, Wu Y, Li G (2009) Nat Photonics 3:649
87. You JB, Li XH, Xie FX, Sha WEI, Kwong JHW, Li G, Choy WCH, Yang Y (2012) Adv Energy Mater 2:1203
88. Price SC, Stuart AC, Yang LQ, Zhou HX, You W (2011) J Am Chem Soc 133:4625
89. Yu G, Gao J, Hummelen JC, Wudl F, Heeger AJ (1995) Science 270:1789
90. Li G, Shrotriya V, Huang J, Yao Y, Moriarty T, Emery K, Yang Y (2005) Nat Mater 4:864

91. Guenes S, Neugebauer H, Sariciftci NS (2007) Chem Rev 107:1324
92. You J, Chen C-C, Hong Z, Yoshimura K, Ohya K, Run X, Shenglin Y, Gao J, Gang L, Yang Y (2013) Adv Mater 25:3973
93. You JB, Dou LT, Yoshimura K, Kato T, Ohya K, Moriarty T, Emery K, Chen CC, Gao J, Li G, Yang Y (2013) Nat Commun 4:1446
94. Tong SW, Wang Yu, Zheng Y, Man-Fai N, Loh KP (2011) Adv Funct Mater 21:4430
95. Tanaka S, Mielczarek K, Ovalle-Robles R, Wang B, Hsu D, Zakhidov AA (2009) Appl Phys Lett 94:113506
96. Mayer AC, Scully SR, Hardin BE, Rowell MW, McGehee MD (2007) Mater Today 10:28
97. Sun Y, Welch GC, Leong WL,Takacs CJ, Bazan GC, Heeger AJ (2011) Nat Mater 11:44
98. Cox PA (1989) The elements: their origin, abundance, and distribution. Oxford University Press, New York
99. Premkumar T, Mezzenga R, Geckeler KE (2012) Small 8:1299
100. Bae S, Kim H, Lee Y, Xu X, Park JS, Zheng Y, Balakrishnan J, Lei T, Ri Kim H, Song YI (210) Nat Nanotechnol 5:574
101. Lee CW, Han X, Chen F, Wei J, Chen Y, Chan-Park MB, Li LJ (2010) Adv Mater 22:1278
102. Lee C, Wei X, Kysar J, Hone W (2008) J Science 321:385
103. Amanda SB, Phys J (2009) Condens Matter 21:144205
104. Gao D, Helander MG, Wang ZB, Puzzo DP, Greiner MT, Lu ZH (2010) Adv Mater 22:5404
105. Zhu H, Wei J, Wang K, Wu D (2009) Sol Energy Mater Sol Cells 93:1461
106. Tung VC, Huang JH, Tevis I, Kim F, Kim J, Chu CW, Stupp SI, Huang J (2011) J Am Chem Soc 133:4940
107. Tung VC, Huang JH, Kim J, Smith AJ, Chu CW, Huang J (2012) Energy Environ Sci 5:7810
108. Bernardi M, Lohrman J, Kumar PV, Kirkeminde A, Ferralis N, Grossman FC, Ren S (2012) ACS Nano 6(10):8896
109. Ramuz MP, Vosgueritchian M, Wei P, Wang C, Gao Y, Wu Y, Chen Y, Bao Z (2012) ACS Nano 6(11):10384

Nanomaterials in Nanomedicine

Francis Mensah, Hailemichael Seyoum and Prabhakar Misra

Abstract Nanomedicine refers to the applications of nanotechnology to the field of medicine. Nanomaterials have led to the development of novel devices for the early detection of malignant tumors, as well as significant enhancements in efficient drug, gene and protein delivery mechanisms to targeted sites in the human body. As nanoparticles become increasingly smaller in size, they also present the potential for harming certain organs of the body. Safety issues involving nanoparticles need to be solved using in vivo techniques. Research in nanomedicine has improved biological therapies, such as vaccination, cell therapy and gene therapy. A particular kind of colloidal nanoparticle, called the liposome, which has properties similar to a red blood cell, has viscoelastic properties that make it extremely useful for a variety of applications in the pharmaceutical and consumer product sectors of the global market. Liposomes have been clinically established as efficient nanosystems for targeted drug delivery. Their efficacy has been demonstrated in reducing systemic effects and toxicity, as well as in attenuating drug clearance. The Maxwell Spring-Dashpot model has been reviewed for liposomes and the viscoelastic exponential equation shown to fit the liposome data. The relevance of this study is to the increasing use of viscoelastic characteristics of liposomes for efficient drug delivery and targeted destruction of malignant tumors. Nanobiotechnology has the potential to facilitate the integration of diagnostics with therapeutics, and in turn lead to personalized medicine tailored for a specific individual.

F. Mensah
Department of Natural and Physical Sciences, Virginia Union University, Richmond, VA, USA

H. Seyoum
Department of Chemistry and Physics, University of the District of Columbia, Washington DC, USA

P. Misra (✉)
Department of Physics and Astronomy, Howard University, Washington DC, USA
e-mail: pmisra@howard.edu

P. Misra (ed.), *Applied Spectroscopy and the Science of Nanomaterials*,
Progress in Optical Science and Photonics 2, DOI 10.1007/978-981-287-242-5_11

Keywords Nanoparticles · Tumor · Cancer · Gold nanoparticles · Drug delivery · Protein · Gene delivery · Biosensor · Bioimaging · Quantum dot · Liposome · Hemoglobin · Red blood cell · Vesicles · Phospholipids

1 Introduction

Nanotechnology is the marriage of technology with matter primarily in the size range 1–100 nm (Fig. 1) that leads to a variety of applications useful to the world at large. Nanomedicine refers specifically to the application of nanotechnology to medicine. It utilizes very tiny materials to develop new and novel therapeutic and diagnostic techniques [1–3]. Nanomaterials have useful physicochemical properties; they are ultra-small in size and have large surface area-to-mass ratio with high reactivity. The development of nanotechnology has led to the construction of powerful new devices for early detection of cancerous tumors. It has also led to the improvement and efficiency of drug, gene and protein delivery mechanisms, primarily because nanoparticles can be constructed by various techniques so that specific material(s) of choice can be delivered to targeted site(s). It is worthwhile to mention that as nanoparticles become smaller in scale, they also have the potential of harming some organs, which is another reason why their study in medicine is important. Safety issues involving nanoparticles can be solved effectively using in vivo techniques.

Nanoparticles have been used to target the Central Nervous System (CNS), which is challenging owing to the necessity of crossing the Blood Brain Barrier (BBB). They are also used in the treatment of Alzheimer's Disease (AD), which is a neurodegenerative disorder prevalent in the aging senile population. It is characterized by severe neuronal loss and proliferation of plaques composed of β-amyloid peptide (A_β) and τ-protein deposits. An imbalance between production and clearance leads to the aggregation of A_β peptides, especially in neurotoxic forms, and

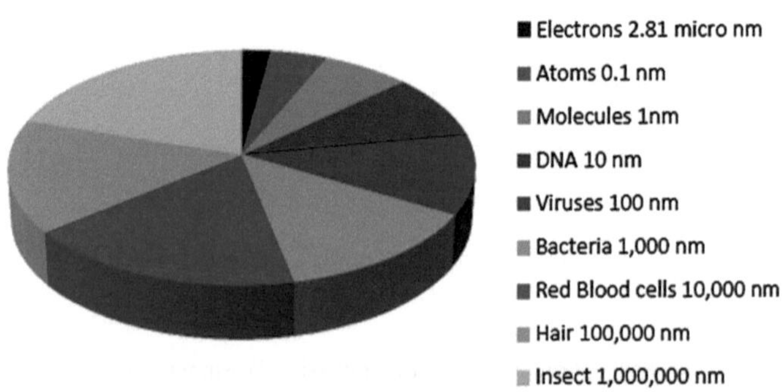

Fig. 1 Size comparison of a variety of small scale objects

may be the initiating factor in AD. Even today, the dearth of deeper knowledge regarding the biophysiology of such processes is one of the main reasons for the lack of early tumor detection.

Research in nanomedicine has improved biological therapies, such as vaccination, cell therapy and gene therapy. Nanobiotechnology forms the basis of many new devices being developed for medicine and surgery (e.g. nanorobots, nanostars, etc.). It has applications in several branches of medicine at the nanoscale, such as: nanooncology—relating to cancer research; nanoneurology—the study of neurological disorders; nanocardiology—the study of cardiovascular disorders; nanoorthopedics—the study of diseases of bones and joints; nanoophthalmology—the study of eye diseases; and also other infectious diseases. Nanobiotechnology can help facilitate the integration of diagnostics with therapeutics and also the development of personalized medicine, i.e. prescription of specific therapeutics best suited for a particular individual.

A particular type of nanoparticles, called liposomes, which are colloidal in nature, find extensive applications in the pharmaceutical and consumer product sectors. Liposomes are viscoelastic and deformable in nature, similar to red blood cells. Recent scientific literature has shown that liposomes can be used to incorporate red blood cells for delivery into the human body. Understanding rheological properties of liposomes is vital to their utilization. In this chapter, we use the liposome as a protypical nanoparticle to present an overview regarding their applications in nanomedicine, and discuss the viscoelastic and deformation properties of liposomes in conjunction with the red blood cells' evolution equation in mathematical biology.

2 Nanoparticles in Therapeutics

Nanotechnology has many applications in therapeutics. Nanoparticles can be used as platforms or carriers for insoluble or poorly soluble drugs. Nanoparticle-carried drugs often have altered pharmacokinetics (PK) and disposition profiles compared to their native forms [4–6]. Nanoparticles can permeate the leaky vasculature surrounding tumors and areas of inflammation, which is referred to as the enhanced permeability and retention effect (EPR). Additionally, many nanoparticle-based anticancer drug strategies now exploit this "passive targeting" mechanism to concentrate drugs in tumors of soft tissue and epithelial cells [7, 8]. The preferential delivery of drugs to tumors makes lower dosages effective and in turn reduces the undesirable side effects of chemotherapeutics. Furthermore, the nanoparticle delivery approach has reduced side effects and improved efficacy because the nanoparticles specifically 'target' the drug to parts of the body where it is needed [9].

Nanoparticles can serve as a platform for the attachment of chemical moieties that execute a variety of medical functions. Ligands for particular cellular receptors can be attached to a nanoparticle and facilitate 'active targeting' to tissues representing those receptors [10, 11]. Hydrophilic molecules, such as polyethylene glycol (PEG), can be bound to a nanoparticle surface to increase solubility and biocompatibility [12]. Image contrast agents, such as chelated gadolinium, can also be conjugated to nanoparticles for diagnostics. The resulting nanoparticle therapeutic is a multipart, multifunctional entity with much greater complexity than the conventional small-molecule drug. Nanoparticles are unlike conventional pharmaceuticals in that they serve as multicomponent, multifunctional systems.

The surface of a nanoparticle can be functionalized with hydrophilic polymers to improve solubility or help the particle elude uptake by the immune system, targeting molecules (e.g., antibodies), drugs and imaging contrast agents for diagnostic purposes. The interior core of a nanoparticle can be solid (quantum dots), liquid (liposomes) or contain an encapsulated drug. Nanoparticles are also of importance in pharmaceutical pipelines—in clinical trials or in preclinical development. For several of the nanoparticle-based therapeutics, preclinical characterization has been a rate-limiting and time-consuming phase of the commercialization process. Nanoparticles represent a broad range of materials (e.g. liposomes, dendrimers, nanocrystals, metal colloids and fullerenes) that have been envisioned for a broad spectrum of clinical applications.

Some methods are uniquely expedient for nanoparticles or are particularly revealing of particle toxicity or efficacy. For physicochemical characterization, such techniques include size characterization by transmission-electron microscopy (TEM) and dynamic light scattering (DLS), and localization in tissue by chemical composition detection using energy dispersive x-ray (EDX) spectroscopy.

3 Nanoparticles in Diagnostics

Cancer represents a high rate of mortality in the U.S. and around the globe and early diagnostic of any dormant metastasis sites is exceedingly important for the patient's survival. Cancer accounts for ¼ of all deaths in the USA. Breast cancer is prevalent in women who carry the high-risk gene and if diagnosed at an early stage (0 or I) can have a survival rate as high as 98 % over a 5 year period. However, that survival rate drops to 85 % when diagnosed later at stage II and is only 20 % for a stage IV tumor. In the case of pancreatic cancer, the death rate is about 75 % within 1 year of detection. The pancreatic cancer associated high death rate is due to the lack of effective and efficient early detection screening. Thus, it is important that research be improved toward the invention of new tools for early detection of the tumor even at the pre-syndrome stage.

3.1 Nanoparticles in Tumor Detection

Tumor biomarkers are substances that can be divided into several categories and can be found in tissue, blood, urine, or stool of cancer patients. Measurements from these tumor detectors are used in the detection of cancer. In imaging, cell membrane protein biomarkers can easily be targeted, while intracellular proteins are difficult to target because of the relatively large size of the nanoparticles, which prevent them from crossing the cell's membrane barrier. Several biomarkers are known today, among them are the epidermal growth factor receptors (EGFR), which can form a dimer with another EGFR of a human epidermal receptor (HER). The EGFR is also intimately linked in the development of cancerous tumors. There are other members of the HER family, such as HER2, HER3 and HER4. HER2 is known as a breast cancer biomarker and can dimerize with itself or with an EGFR or with HER3 [13]. Some other tumor biomarkers are the vascular endothelial growth factor (VEGF), which plays an important role in angiogenesis, the integrin molecules, the folate receptors; the Matrix metalloproteinases (MMPs) representing a family of zinc-dependent endopeptidases; the prostate-specific membrane antigen (PSMA); and phosphatidylserine.

3.2 Quantum Dots

Quantum dots are nanoparticles made of semiconductor materials, such as chalcogenides (selenides or sulfides) of metals like cadmium or zinc, for example CdSe or ZnS [14], as illustrated in Fig. 2. Due to their small size, they exhibit unique optical and electrical properties that are different from the bulk material. One example is the emission of photons following excitation that are visible to the human eye. The ability to precisely control the size of a quantum dot enables the manufacturer to determine the wavelength of the emission, which in turn

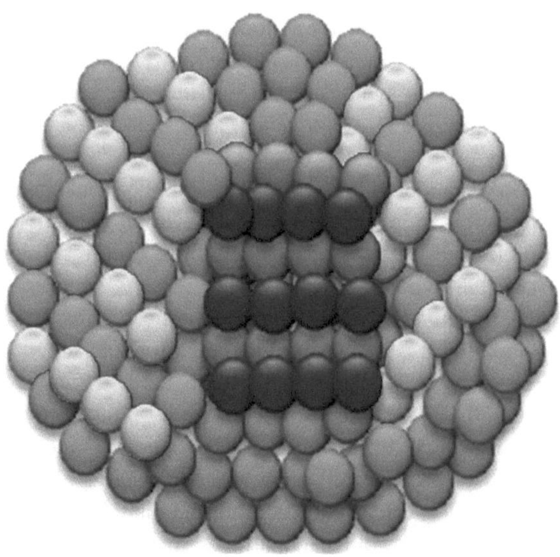

Fig. 2 Schematic of quantum dots showing an inner spherical core of CdSe surrounded with a shell of ZnS

determines the color of light the human eye perceives. Quantum dots can therefore be "tuned" during production to emit any color of light preferred by the user. The ability to control or tune the emission from the quantum dot by changing its core size is termed the "size quantization effect".

The smaller the quantum dot, the closer it is to the blue end of the spectrum, and the larger the dot, the closer it is to the red end. Dots can even be tuned beyond the visible light spectrum, into the infra-red or into the ultra-violet. At the end of the production process, quantum dots physically appear either as a powder or in solution [14]. Quantum dots can serve as biomarkers, emitting light of the desired color to represent a particular state.

3.3 Gold Nanoparticles

Metal nanoparticles are used extensively in various biomedical applications due to their small size to volume ratio and because of their extensive thermal stability. Gold nanoparticles (GNPs) are a good example because of their amenability to synthesis and functionalization, which includes toxicity and ease of detection. GNPs are colloidal suspensions of gold particles of nanometer size and have been synthesized by an array of methods, mainly based on the approach that uses the reduction of chloroauric acid in the presence of a stabilizing agent. The most commonly used method, the citrate synthesis method, includes reduction of chloroauric acid using trisodium citrate, resulting in the formation of GNPs [15, 16]. Besides their usual spherical form, GNPs have been synthesized in various other shapes, affecting their physical and biochemical properties. For example, hexagon and boot shaped GNPs show different surface enhanced Raman scattering (SER), which in turn can be used to detect molecules conjugated to GNPs, such as avidin, and thereby making these functionalized GNPs (fGNPs) useful for biolabelling, bioassay, clinical diagnosis and therapy [17]. Gold nanocages of six and eight facets have also been synthesized [18]. Similarly, gold nanorods have been synthesized, which find usage in biomedical applications for cancer imaging and photothermal therapy [19].

Gold nanoparticles have many applications in nanomedicine. One important application is in the treatment of ovarian cancer. Positively charged gold nanoparticles are usually toxic to cells, but cancer cells manage to avoid nanoparticle toxicity. Mayo Clinic researchers [5] have determined why this occurs and developed a method to make the nanoparticles effective against ovarian cancer cells. Nanoparticles supposedly kill cells by causing cellular calcium ion levels to increase but researchers discovered that a regulatory protein in the mitochondria essentially buffers the rising calcium by transporting it into the mitochondria, thus subverting cell death. Cancer cells have an abundance of this transporter and may thus be protected from nanoparticle toxicity. The research team discovered that, if they inhibit calcium uptake into the mitochondria, sufficient cellular stress builds up, making the gold nanoparticles more effective in destroying cancer cells [5].

Fig. 3 Illustration showing tetraethyleneglycol functionalized gold

As a result of these studies, it can be expected that the understanding of how mitochondrial transport mechanisms work will help in the design of targeted therapies against cancer.

In yet another recent study, the GNPs were grown in a lysozyme crystal, which can potentially serve as a bifunctional molecule (Fig. 3) for specific catalytic activity [20]. Methods are being devised to synthesize GNPs with functional moieties to increase their affinity to biological molecules and use as drug-carriers in cells with increased specificity.

3.3.1 Oligonucleotide Functionalized Nanoparticles

Several research groups have devised methods to functionalize gold and other nanoparticles using oligonucleotides either alone or with some modifications. It is also known that DNA conjugated nanostructures can be synthesized in a controlled manner, either by attaching a specific number of single-stranded DNA molecules through thiol caps or by saturating the surface of the GNPs by single-stranded DNA molecules [21]. Kinetic and thermodynamic studies on DNA hybridized to GNPs have shown that single-stranded DNA (ssDNA) first reaches the GNPs and then slowly diffuses on its surface [22]. Secondary structure of a DNA hairpin inhibits interaction between GNPs and DNA, thereby increasing the stability of adhered DNA. Aptamer-GNP conjugation has been exploited to target prostate cancer cells [23], which was accomplished by attaching GNPs with an oligonucleotide that is complementary to the sequence of the anti-PSMA (prostate specific membrane antigen). It also facilitates the attachment of PSMA-GNPs to anti-PSMA antibodies. The results show a promising role of such fGNPs in the detection and imaging of cancerous cells. Moreover, DNA functionalized GNPs (Fig. 3) were employed to design a chip-based DNA bio bar code sensor to detect target DNA sequences [24]. In this case, the bio bar code amplification of the target DNA is assessed using a complementary DNA attached to GNPs and the subsequent detection of the amplified DNA, instead of the original target DNA.

3.3.2 PEGylation

PEGylation is one of the most commonly used functionalization methods for GNPs. GNPs are coated with a layer of only PEG or in conjunction with other molecules such as biotin, peptides or oligonucleotides, thereby helping the internalization of these GNPs to the target cells. Due to their ability to bind the cell membrane, these functionalized GNPs can serve as good drug-carriers. PEGylated GNPs functionalized with biomolecules, such as lectin, lactose and biotin, have been synthesized [25–32]. PEGylated GNPs are one of the most commonly used nanoparticles for gene delivery. Overall, PEGylated GNPs are useful in cellular and intracellular targeting of biological materials.

3.3.3 Peptide/Amino Acid Conjugation

Functionalization of nanoparticles with amino acids and peptides helps enhance specificity and efficacy of nanoparticle-based delivery systems. GNPs functionalized with amino acids, such as lysine, polylysine and glycine, bind DNA with higher efficiency for gene delivery without toxicity. Primary ammonium groups of these amino acids contribute to a higher binding capacity to the cationic groups on the DNA. Also lysine dendrons were found to be superior to polylysine for expression of the reporter β-galatosidase gene [33]. In addition, peptide functionalized GNPs are known to activate macrophages, holding promise to be used as adjuvants for vaccine delivery.

The GNPs functionalized with an amyloid growth inhibitory peptide (AGIP) associated with Alzheimer's disease have been found useful for intracellular drug delivery. Peptide-conjugated GNPs are also being used to devise a protein kinase assay using secondary-ion mass spectrometric imaging technique and change in the mass of the peptide substrate following kinase action [34]. Such an approach is much simpler compared to traditional methods using radioactive or fluorescent labeling. Thus, peptide conjugated nanoparticles hold promise for use in bioimaging, diagnosis, and therapeutic applications. GNPs are also being functionalized using both peptides and oligonucleotides for perinuclear localization for a variety of functions, such as cell imaging, target- specific internalization, etc. [35].

Functionalized GNPs have been used for targeting drugs and biomolecules to specific cell types and to organelles, such as the nucleus or mitochondria. GNPs, functionalized with PEG and 3-mercaptopropionic acid, were shown to penetrate the nucleus of HeLa cells without causing severe cytotoxicity and hence can be used as a nuclear drug delivery carrier [36]. Similarly, GNPs encapsulated by liposomes have been studied for their cellular targeting and uptake capacity while carrying drugs or other cargos [37]. Intracellular uptake of GNPs, as small as 1.4 nm, has been shown to enhance internalization 1000-fold. Such nanoparticles harbor significant potential to be used as gene delivery vehicles, drug-carriers and carriers for other biomolecules.

3.3.4 Drug Delivery

GNPs are suitable for the delivery of drugs to cellular destinations due to their ease of synthesis, functionalization and biocompatibility. GNPs functionalized with targeted specific biomolecules can effectively destroy cancer cells or bacteria [38]. Large surface to volume ratio of GNPs offer a large number of drug molecules being carried by the GNPs [39]. GNPs have been used for the co-administration of protein drugs due to their ability to cross cellular membranes [40], possibly due to the interaction of GNPs with cell surface lipids.

3.3.5 Detection

GNPs are also used in the detection of various biological molecules, which include proteins, enzymes, DNA, antigens and antibodies, etc. Detection of microorganisms can be achieved by several biochemical, microbiological, and molecular methods. Recent advances in the field of nanotechnology have made it possible to detect microorganisms using nanoparticles functionalized with oligonucleotides complementary to the gene tags of the microorganisms. In one such study, oligonucleotides complementary to the unique sequences of the heat shock protein 70 (HSP 70) of Cryptosporidium parvum were used to functionalize GNPs, which could be used to detect the oocytes of Cryptosporidium in a colorimetric assay, offering a simple and robust method of molecular detection [41].

3.3.6 Metal Sensors

Development of an easy colorimetric assay to detect uranium has been achieved by using DNAzyme-GNPs system [42]. Traditionally, uranium in the environment is detected using complex biophysical techniques, such as via fluorimetry, ICP-MS, and atomic absorption spectroscopy. However these methods are difficult to use on-site. The DNAzyme-GNP system provides an alternative to these traditional methods. DNAzymes are catalytic DNA molecules developed in vitro with specific affinities to metal cofactors, such as Uranyl (UO_2^{2+}), which is the most common bioavailable form of uranium. These biosensors are able to detect uranium in two ways, either by disassembly of DNAzyme functionalized GNPs in the presence of uranyl ions, causing a visible color change from purple to red ("turn-on" method), or by the "turn-off" method, which is based on different adsorption properties of single- and double-stranded DNA on GNPs in the presence of uranyl ions. The above-cited technique is significant, as it can detect uranyl below the maximum contamination limits specified by the U.S. environmental protection agency (EPA).

GNPs functionalized with aza-crown ether acridinedione were developed as a fluorescent chemosensor for metal ions based on the shift in the surface plasmon resonance of GNPs with aggregation of nanoparticles via sandwich complexation [43]. GNPs functionalized with L-cysteine were used for the detection of mercury ions (Hg^{2+}). In the presence of UV light and mercury ions Hg^{2+}, these GNPs tend to aggregate in their detection, making them useful biosensors for on-site applications [44]. A similar biosensor for Hg^{2+} detection was developed using oligonucleotide fGNPs [45].

3.3.7 Toxicity of GNPs in Biological Systems

In spite of their extraordinary capacity to bioconjugate to various molecules, there have been studies showing GNPs to be cytotoxic due to their inherent physio-chemical properties. GNPs have been found to induce death response. In a recent report, an embryonic stem cell test (EST) was developed to check the embryo-toxicity of the GNPs [46]. The EST is an in vitro standard assay that is used to classify substances as strongly, weakly or non-embryotoxic. The embryonic stem cells (ESCs) were exposed to GNPs for 5 days to assess the cytotoxicity, which exhibits the following order: gold salt ($HAuCl_4{\cdot}3H_2O$) > cobalt ferrite salt ($CoFe_2O_4$) > cobalt ferrite nanoparticles coated with silanes (Si–CoFe) > GNPs coated with hyaluronic acid (HA–Au). The $\sim$5 nm gold nanoparticles have been shown to induce oxidative stress and toxicity in blue mussel Mytilus edulis, at 750 ppb concentration after 24 h [47]. Similarly, the presence of sodium citrate residues (stabilizing agent for synthesis of GNPs) on the surface of GNPs elicited toxicity in alveolar cell lines in vitro [48]. Sodium citrate not only compromised cell viability but also affected cell proliferation. However, these nanoparticles remained localized in the membrane-bound vesicles and were not freely dispersed in the cytoplasm. The effect of polycaprolactone (PCL) coating on internalization and cytotoxicity of GNPs has been evaluated on ECV-304 cells [49, 50]. In comparison to PCL coated GNPs, bare GNPs were shown to exhibit significant changes in cell morphology and the cytoskeleton. Additionally, PCL-coated GNPs were shown to be less cytotoxic as compared to bare GNPs.

4 Applications of Liposomes in Nanomedicine

4.1 Background

Many different types of nanoparticles are currently being studied for applications in nanomedicine. These nanoparticles can be carbon-based skeletal-type structures, such as the fullerenes, or micelle-like lipid-based liposomes, which are already in use for numerous applications in drug delivery and in the cosmetic industry. Colloids, typically liposome nanoparticles, selected for their solubility and suspension properties are used in cosmetics, creams, protective coatings and stain-resistant clothing. The structure of a liposome is made of phospholipids and cholesterol. Because of their nanoscale size, liposome can be classified as nanoparticles. Indeed, the diameter of the SUVs (Small Unilamellar Vesicles) ranges from 20 to about

100 nm and those of the LUVs (Large Unilamellar Vesicles), MLVs (MultiLamellar Vesicles) and MVVs (MultiVesicular Vesicles) range from 100 nm to several microns. In addition, the thickness of the membrane is approximately in the range 5–6 nm. Modified liposomes at the nanoscale have been shown to have excellent pharmacokinetic profiles for the delivery of DNA, antisense oligonucleotide, siRNA, proteins and chemotherapeutic agents. Examples of marketed liposomal drugs with higher efficacy and lower toxicity than their nonliposomal analogues are now known.

Doxorubicin is an anticancer drug that is widely used for the treatment of various types of tumors. It is a highly toxic compound that affects not only the tumor tissue, but also the heart and kidney, a fact that limits its therapeutic applications. However, the development of doxorubicin enclosed in liposomes culminated in an approved nanomedical drug delivery system. Such a novel liposomal formulation has resulted in reduced delivery of doxorubicin to the heart and renal system, while elevating the accumulation in tumor tissue by the EPR effect. Furthermore, a number of liposomal drugs are currently being investigated, including anticancer agents, such as camptothecin and paclitaxel (PTX), as well as antibiotics, such as vancomycin and amikacin. Liposomes are now being targeted for use as a delivery system for both medicinal and vaccine purposes [51].

Liposomes have both a head and a tail group (Fig. 4). The head is hydrophilic and attracted by the aqueous medium, while the tail is hydrophobic and repelled by water [52]. Hydrophilic molecules can easily be incorporated into the aqueous center of the liposomes, while lipophilic molecules can be incorporated into the lipid bilayer. The phospholipids, when in the presence of an aqueous solution, organize in such a way as to minimize the interaction between their hydrocarbon chains and water molecules. Liposomes have a prolonged half-life in blood plasma, which allows for their accumulation in pathological areas with compromised vasculature, and their intracellular penetration is suitable for targeted delivery of drugs or magnetic resonance contrast agents [53].

Phospholipid vesicles have been widely used as model cell membranes and drug carriers. Liposomes, which are used as delivery systems, may encapsulate hydrophilic substances in their aqueous core [Fig. 4]. Amphiphilic and lipophilic substances, such as oil soluble UV filters, can be incorporated into the lipid bilayer. Loaded liposomes as well as non-loaded empty liposomes are used in cosmetics and in skin treatment.

4.2 Aggregation and Properties of Liposomes

Solutions of liposome samples show propensity for aggregation [54]. Particles tending to agglomerate exhibit properties that obey the laws of population dynamics. This can be verified by the observations in solution of liposomes [5] shown in Figs. 5, 6 and 7.

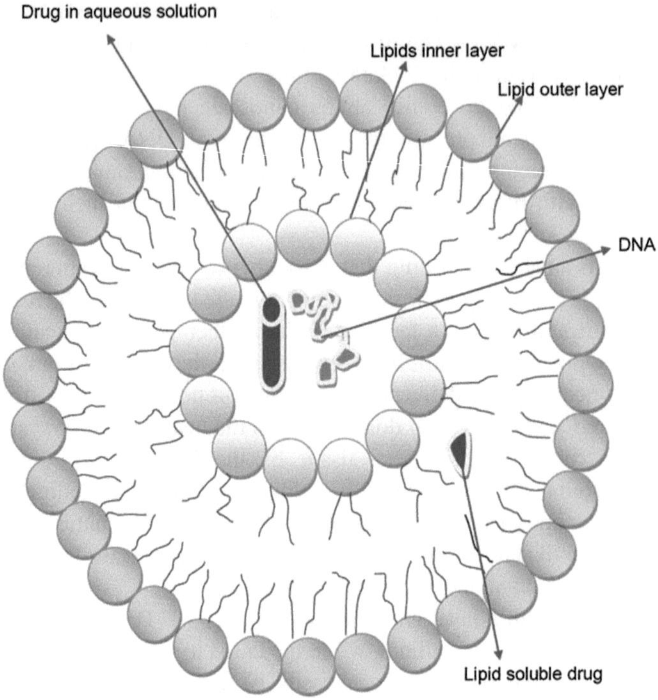

Fig. 4 Liposome image and structure showing wrapping of drugs and DNA in aqueous solution

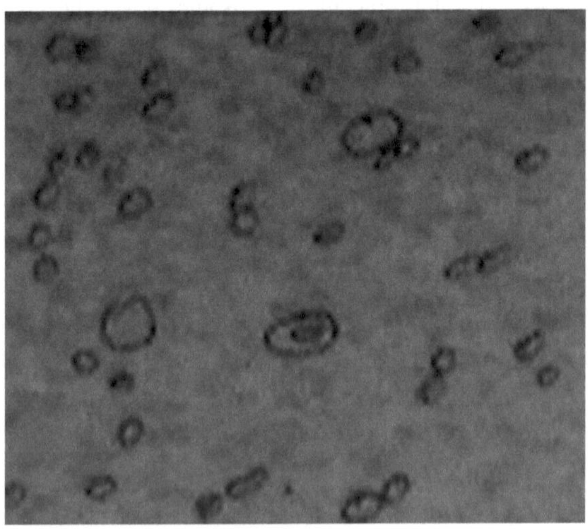

Fig. 5 Liposome solution exhibiting negative staining with uranylacetate

As can be seen from Figs. 5, 6 and 7, negative staining with uranylacetate shows tendency towards agglomeration and cooperative phenomena (Fig. 6); and the same situation occurs with freeze fracture (Fig. 7). However, administration of liposome

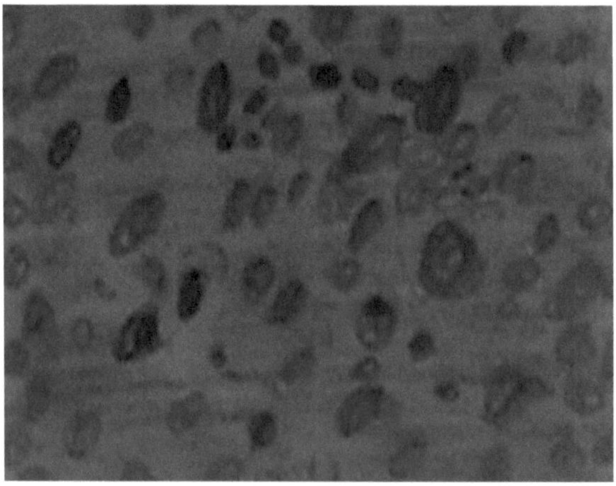

Fig. 6 Liposome solution showing negative staining with uranylacetate in agglomeration

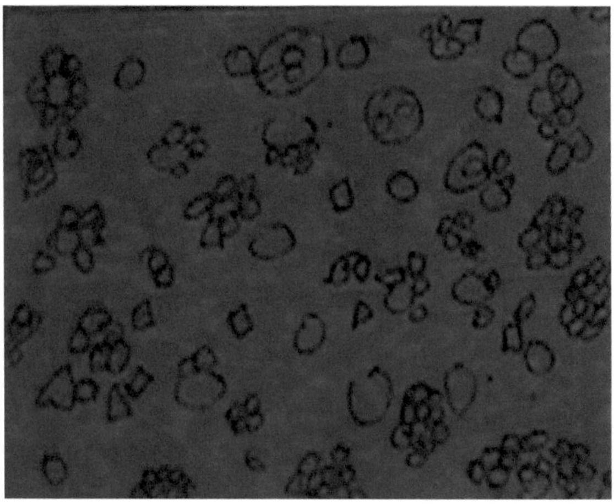

Fig. 7 Liposome solution exhibiting freeze fracture

particles to a specific site in the body still remains problematic [55]. One possible solution is to immobilize liposomes in a hydrogen network and inject the liposome gel to the affected part. Thus, critical understanding of rheological properties of liposomes can help in their use and application.

Rao et al. [56] have added Chol-PEG-Chol to the liposome solution, which transforms the liposome into an electric gel-like solid that can hold its own weight under gravity. Chol-PEG-Chol was prepared using diamino-polyethylene glycol and cholesteryl chloroformate [57]. Rao et al. [56] investigated the thermo-reversibility of the liposome gel upon repetitive heating and cooling processes. They confirmed the intactness of the liposome particles in the gel by using dynamic light scattering measurements and also demonstrated that the liposome gel exhibits viscoelastic properties similar to that of the Maxwellian model. The liposome gel was significantly weakened when exposed to relatively low stress and recovered rapidly after removal of the stress. The dynamic shear modulus of the gel was dependent upon the liposome concentration and the polymer concentration. Moreover, in another report, Harland et al. [58] confirmed that the data associated with the phospholipid bilayer membranes, a key component of liposomes, fitted well into the Maxwellian model of viscosity. These studies showed that each phospholipid component shows both a viscous-dominant and an elastic-dominant regime response.

A liposome is a deformable material and can be described by the theory of viscoelasticity. Viscoelasticity of liposomes is significant because it aids this biomaterial to perform its duties as an effective transport agent and in turn the elasticity of liposomes can help in transdermal drug delivery. Nipper et al. [59] have performed research relating to the detection of liposome membrane viscosity using radiometric molecular rotors, and have investigated the viscosity changes in liposome and cells in general.

4.3 Aggregation and Stability of Liposomes

Liposomes can be prepared with their surface carrying a positive or a negative charge or with no charge, based on the lipid composition and chemical components, such as stearylamine or dicetylphosphate. According to electrostatics, these charges on the surface of the liposome can create an electric potential between the surface and the bulk ionic medium. For a single charge, the electric potential created at a given distance r is given by the relation,

$$V(r) = k\frac{q}{r}$$

where $k = 9 \times 10^9\ \mathrm{N \cdot m^2/C^2}$ is the Coulomb constant, q the value of the charge in the System International (SI) unit Coulomb (C) and $V(r)$ has the SI unit of Volt (V) when the distance r is in meters.

The electric potential plays a key role in the physical stability of the liposome and participates in the aggregation or disaggregation (flocculation or deflocculation) of the liposome particles. It is important to think of the surface charge distribution, since these particles occur on the surface. The magnitude of the surface charge distribution density is given by the equation

$$\sigma = \frac{dq}{dA}$$

for any elemental charge dq on an elemental surface dA. The surface charge distribution increases the physical stability of the liposome, where the liposome particles are shielded by an electrical barrier to prevent disturbance by other particles, so that the liposomes can stay in a dispersed state.

The aggregation of liposomes when used as a drug carrier system can lead to some undesirable effects as well. If the aggregation results, for example, in the coalescence or fusion of the liposome, it can cause leakage of the entrapped drugs. Embolization can also happen in vivo on intravenous administration. Aggregation can cause increase in particle size and lead to changes in the in vivo clearance of the liposomes. Physically—stable liposomes can be obtained by increasing their surface potential, which in turn causes an increase in the electrostatic repulsive forces and a reduction in aggregation. Also, an increase in the ionic strength of the medium—from hypotonic to isotonic—may cause a reduction in the electric double layer thickness and a decrease in the surface potential. Hence, aggregation of liposomes may be induced to result in an increase in particle size, an effect which is significant for liposomes with entrapped markers.

4.4 Maxwell Spring-Dashpot Model for Liposomes

The Maxwell model is a combination of two mechanical analogs: (1) Hookean spring model and (2) Newtonian dashpot model. The Hookean model is given by

$$\sigma = k\,\varepsilon \tag{1}$$

where k is the constant of proportionality between the stress (analogous to the force acting on the spring) and the strain (analogous to the displacement of the spring). The Newtonian dashpot model is represented by the equation

$$\sigma = \eta\,\dot{\varepsilon} \tag{2}$$

where η is the constant of proportionality between the stress (analogous to the force acting on the dashpot) and the strain rate (analogous to the displacement of the dashpot per unit of time). The constant η is analogous to the viscosity of a fluid. The dashpot models the instantaneous deformation within the fluid-like material and its magnitude is related to the fraction of energy stored in the dashpot as energy due to viscosity.

Figure 8 displays the spring-dashpot Maxwell model for a viscoelastic material in which a Hookean spring and a Newtonian dashpot are connected in a series configuration.

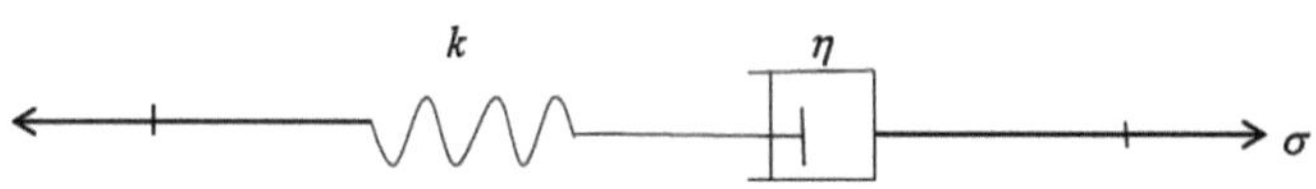

Fig. 8 The spring-dashpot Maxwell model for a viscoelastic material

In this model, the stress on each element will be the same, while the strain will add up:

$$\sigma_S = \sigma_d = \sigma \tag{3}$$

$$\varepsilon_s + \varepsilon_d = \varepsilon \tag{4}$$

where the subscript s stands for spring and the subscript d for dashpot.

In order to find the evolution equation of the Maxwell model we take the first time derivative of the strain ε in the previous equation. We obtain:

$$\dot{\varepsilon}_s + \dot{\varepsilon}_d = \dot{\varepsilon} \tag{5}$$

Using Eqs. (1) and (2), Eq. (5) can be written as:

$$\frac{\dot{\sigma}}{k} + \frac{\sigma}{\eta} = \dot{\varepsilon} \tag{6}$$

By multiplying Eq. (6) by k on both sides, it follows:

$$\dot{\sigma} + \frac{\sigma}{\eta/k} = k\dot{\varepsilon} \tag{7}$$

The constant that appears in the denominator of the second term of the left side of Eq. (7) can be denoted by $\tau = \frac{\eta}{k}$, which has the dimension of time and represents the material's viscoelastic response time.

Eq. (7) is now written:

$$\dot{\sigma} + \frac{\sigma}{\tau} = k\dot{\varepsilon} \tag{8}$$

and represents the constitutive equation of a Maxwell material—it is an equation that relates the stress and strain for a material.

Eq. (8) can also be written:

$$\tau\dot{\sigma} + \sigma = \eta\dot{\varepsilon} \tag{9}$$

If a particular strain rate $\bar{\gamma}$ is suddenly applied at t = 0 and held at that value for subsequent times, it can be proved that for t > 0,

$$\sigma(t) = \eta\bar{\gamma}(1 - e^{-t/\tau}) \tag{10}$$

On the other hand, if the strain rate, which has the constant value $\bar{\gamma}$ for t < 0, is suddenly removed at t = 0, it can be shown that for $t \geq 0$,

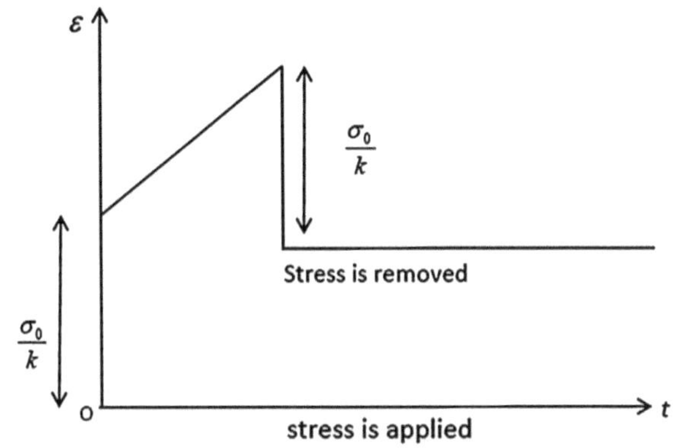

Fig. 9 Creep recovery response in the Maxwell model

$$\sigma(t) = \eta\bar{\gamma}\, e^{-t/\tau} \tag{11}$$

Hence, the stress relaxes exponentially from its equilibrium value to zero. The constant τ represents the relaxation time and can be denoted as $\tau = \tau_M$.

4.5 Creep Recovery Response

Let us now consider a creep test. It is known that for the Maxwell model, when the stress $\sigma_0(t=0) = \eta\bar{\gamma}$ is applied, the spring will stretch immediately, but the dash-pot will take time to react. However, the initial strain is $\varepsilon(0) = \frac{\sigma_0}{k}$ and the solution to the Maxwell model represented above by Eq. (6) is $\dot{\varepsilon} = \frac{\sigma_0}{\eta}$, whose solution yields $\varepsilon = \sigma_0\left(\frac{1}{\eta}t + \frac{1}{k}\right)$.

It is interesting to mention that the creep response can be expressed in terms of a creep compliance function:

$$\varepsilon(t) = \sigma_0 J(t) \text{ with } J(t) = \frac{t}{\eta} + \frac{1}{k}$$

After the removal of the load, the spring reacts immediately, but the dash-pot has no tendency to recover. Therefore, there is a residual creep strain due to the dash-pot as shown in Fig. 9.

4.6 Liposomes and Red Blood Cells

Liposomes composed of phosphatidylcholine (PC), sphingomyelin and ganglioside, and resembling the surface characteristics of red blood cells, are found to show substantially reduced uptake by the reticuloendothelial system and exhibit an

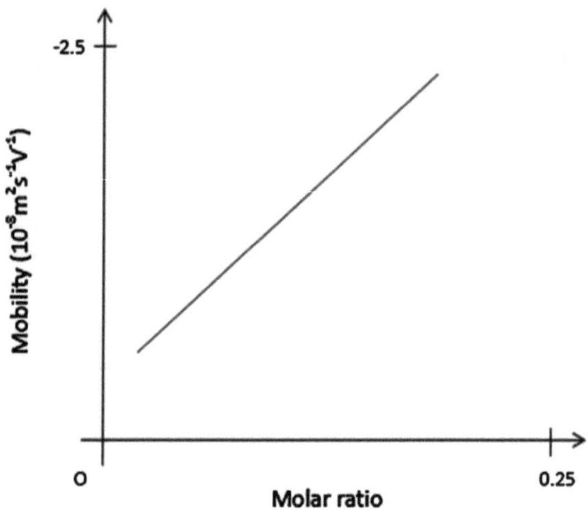

Fig. 10 Liposome mobility after incorporation of the charge-induced agent versus molar ratio of dicetylphosphate (PS)

increase in the half-life of liposomes in blood circulation. It is important to note that liposomes made of PC, cholesterol and dicetylphosphate, at a molar ratio of 1.6:1:0.15 are similar to the lipid composition of the membrane of red blood cells whose cholesterol content is about 23 % by weight of the total membrane lipid. Such liposomes show negative mobility of $-1.3 \times 10^{-8} m^2 s^{-1} V^{-1}$ in the presence of 1/15 M phosphate buffer at pH = 7.4 and temperature 25 °C, which is almost the exact red blood cell electrophoretic mobility under the same conditions of the phosphate buffer. This type of situation provides great stability to the liposomes in the retention of entrapped drugs. Figure 10 shows a theoretical graph of the mobility of liposomes as a function of the molar concentration of dicetylphosphate (PS) or monosialylganglioside (GMI) incorporated into the liposomes. As can be seen from Fig. 10, the graph exhibits a negative mobility on the y-axis.

4.7 Red Blood Cells Incorporated Within Liposomes

In an important study, Sakai et al. [60], studied red blood cells (RBCs) incorporated within liposomes. They studied the flow patterns of RBCs suspended in viscous Newtonian and non-Newtonian fluids. The investigators observed the complete axial alignment and near-wall excess of RBCs in the noncapillary size tube under extreme conditions.

Liposomes, because of their carrier efficacy, are being used for cancer treatment. Additionally, liposomes can be used to encapsulate drugs. Improvements have been made in recent studies to prevent drug leakage from liposomes. Liposomes affect the rate of clearance of blood cytotoxic contents, protecting the latter from inactivation and determining their fate in blood [61, 62]. As expected, liver and spleen take up portions of the injected dose of liposomes. Localization and the knowledge of the total possible uptake of liposomal agents by tissues, such as mucosa, skeletal and cardiac muscle, lungs and kidney, is often inaccurate [54]. The primary reason is that many drugs are lost during their circulation in blood. Also, they spill out of the cells of tissues in the form of metabolites [63]. Encapsulation of hemoglobin inside a lipid bilayer has also been the subject of some research, but needs to mature.

There is a significant connection between liposomes and cells. Experiments on the control of cell metabolism by the use of liposomes have been carried out [64] and the data suggested that cultured cells were able to take up liposomes, the contents of which were delivered into the liposomes and after the disruption of the carrier, the agents were set to act freely. Liposomes can fuse with cells [65–67] in such a way as to inhibit cell growth when composed of unsaturated phospholipids and not when solid dipalmitoyl phosphatidycholine liposome is used. Significant amount of data has also been accumulated regarding the fate of liposomes after injection into animals. It has been observed that, after the intravenous injection of liposomes composed of one or more layers, and containing radiolabels in both their lipid and aqueous phase, the ratio of the two labels in the blood plasma often remains similar to that in the injected preparation. Therefore, it has been inferred that the carrier retains its structural integrity in the circulation, although the possibility exists that a labeled agent can form a complex with the labeled lipid marker and circulate in the blood even after disruption of the liposomes [54].

4.8 Red Blood Cells (RBC) and the Polymerization Equation

The hemoglobin molecule is composed of four globular protein subunits, each of which is composed of a protein chain tightly associated with a non-protein heme group. The proteins are folded chains of a large number of different amino acids called polypeptides. The role of hemoglobin in the RBC is to carry oxygen and release it to the tissues and organs. However, deoxy-hemoglobin molecules lose this property because of the lack of oxygen and tend to polymerize. Liposomes, which are artificially prepared vesicles and are composed of lipid bilayers, can be used as a drug, RBC or DNA carrier in the blood. Let us examine here the theoretical situation of the deformation of liposomes by using the evolution equation for the RBCs. We propose to use the same evolution equation for the complex {Liposome, RBC}, where the RBC is incorporated within the liposome, which in turn leads us to consider the important rheological properties of liposomes in such a model.

The mathematical model for the polymerization of sickle hemoglobin S is given by the integro-differential equation [68, 69]:

$$\ddot{Q}(t) + \eta\dot{Q}(t) - \frac{1}{2}\frac{\dot{Q}^2(t)}{Q(t)} + \omega^2 Q(t) = 0 \tag{12}$$

where, Q(t) is the time-dependent deformation and represents the extra stretch with respect to the equilibrium position, ω is the frequency of vibration of the polymer and η, the coefficient of friction per unit of mass. $\eta\dot{Q}$ is the viscous constraint per unit of mass due to the frictional force of the medium, $\frac{1}{2}\frac{\dot{Q}^2}{Q}$ represents the constraint of inertia per unit of mass due to the force inertia. The last term on the left hand side represents the spring force.

Defining a new variable, $f = \frac{\dot{Q}}{Q}$, the above equation is expressed in terms of the Riccati equation, $\dot{f} = -\frac{1}{2}f^2 - \eta f - \omega^2$, which gives the solution of the evolution equation in a steady state regime as follows:

$$Q(t) = Q_\infty[1 + l\exp(-\alpha t)]^{-1/v} \tag{13}$$

where l, α, v are coupling parameters which are related to η and ω as follows:

$$\alpha = \eta\,\delta; \quad \delta = \left[1 - \frac{2\omega^2}{\eta^2}\right]^{1/2} \tag{14}$$

$$l = \frac{f_0}{f_1 - f_0}; \quad f_1 = -\eta - \eta\delta; \quad v = \frac{\alpha}{f_1} \tag{15}$$

where f_1 represents one of the roots of the Ricatti equation, and $f_0 = \lim_{t \to 0} f(t)$.

Consider now that the strain is the deformation per unit of the original length; the strain, ε can therefore be written as follows:

$$\varepsilon(t) = \varepsilon_\infty[1 + l\exp(-\alpha t)]^{-1/v} \tag{16}$$

The viscosity follows from the same expression as the deformation and can be written as

$$\eta(t) = \eta_\infty[1 + l\exp(-\alpha t)]^{-1/v} \tag{17}$$

where η_∞, l, α, v are defined by Eqs. (14) and (15).

4.9 Model for the Viscosity of Liposomes

The model represented by Eq. (13) is used to fit the viscosity versus the shear rate of the graph in Fig. 11 [70]. All of the parameters describing the model were found using the software Easy Plot: $l = -1$, $\alpha = 0.0746$, $v = 1.101$, which shows the success of such an exponential model in representing liposome viscosity data.

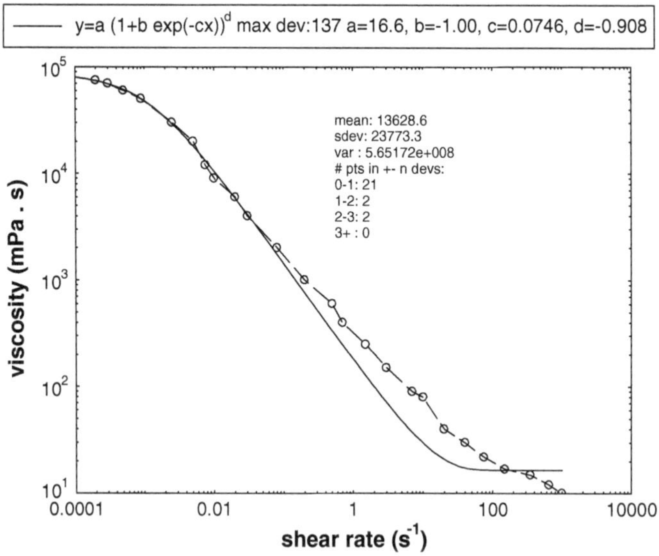

Fig. 11 Fitting of the graph of viscosity versus shear rate of the RBC at 25 °C for the liposome-dextran model

5 Conclusion

In conclusion, GNPs can be considered as extraordinary molecular carriers for targeting, intracellular trafficking, and the delivery of a huge array of biomolecules that include DNA, RNA, proteins, peptides, drugs, genes and other molecules of therapeutic significance. They do not cause significant cytotoxicity owing to their physiochemical properties. Efforts need to be undertaken for designing GNPs to enhance the bioavailability of fGNPs with less immunogenicity and cytotoxicity that can be effectively used in vivo. A judicious choice between the size and functionalization method of the GNPs is indeed a prerequisite for the effective use of GNPs in a variety of important biomedical applications.

Liposomes are currently one of the most clinically established nanosystems for drug delivery. Their efficacy has been demonstrated in reducing systemic effects and toxicity, as well as in attenuating drug clearance. Liposomes are also subject to certain limitations, including low encapsulation efficiency, fast burst release of drugs, poor storage stability and lack of tunable triggers for drug release. Furthermore, since liposomes cannot usually permeate cells, drugs are released into the extracellular fluid. As such, many efforts have focused on improving their stability and increasing their circulation half-life for effective targeting or sustained drug action. Surface modification is one method of conferring stability and structural integrity against a harsh bioenvironment after oral or parenteral administration. Surface modifications can be achieved by attaching polyethylene glycol units, which form a protective layer over the liposome surface (known as stealth liposomes) to slow down liposome recognition, or by attaching other polymers, such as poly(methacrylic acid- co-cholesteryl methacrylate) and poly(acrylic acid), to improve the circulation time of liposomes in blood. To overcome the fast burst release of the chemotherapeutic drugs from liposomes, drugs such as doxorubicin may be encapsulated in the liposomal aqueous phase via an ammonium sulfate gradient. This kind of strategy enables stable drug entrapment with negligible drug leakage during circulation, even after prolonged residence in the blood stream. Further efforts to improve control over the rate of release and drug bioavailability have been made by designing liposomes whose release is environmentally triggered. Accordingly, the drug release from liposome-responsive polymers, or hydrogel, is triggered by a change in pH, temperature, radiofrequency or magnetic field. Liposomes have also been conjugated with active-targeting ligands, such as antibodies or folate, for target-specific drug delivery.

The Maxwellian model described above for liposomes has convincingly shown that a viscoelastic exponential model is able to fit the liposome data and the relevance of this formalism is seen in the ever increasing utilization of liposome viscoelastic properties in drug carrying for the detection and destruction of tumors. However, more detailed studies are needed to better understand the rheology of liposomes and their interactions with drugs and other encapsulated molecules.

Acknowledgments This paper was partially supported by the STEM Research Mini-Grant UDC 05-16-2011/07-16-2011. We would like to thank Ms. Kafayat Olayinka for putting together the liposome structural illustration represented by Fig. 4 and Ms. Kristyn Miller, M.Ed., of Virginia Union University, for reviewing and proofreading a portion of this chapter.

References

1. Zhang L, Gu FX, Chan JM, Wang AZ, Langer RS, Farokhzad OC (2008) Nanoparticles in Medicine: therapeutic applications and developments. Clin Pharmacol Ther 83:761–769
2. Navalakhe RM, Nandedkar TD (2007) Application of nanotechnology in biomedicine. Indian J Exp Biol 45(2):160–165
3. Jain KK (2008) Nanomedicine: application of nanobiotechnology in medical practice. Med Prince Pract 17(2):89–101. doi: 10.1159/000112961. (Epub 2008 Feb 19. PharmaBiotech, Basel, Switzerland)
4. Andrieux K, Couvreur P (2013) Nanomedicine as a promising approach for the treatment and diagnosis of brain diseases: the example of Alzheimer's disease. Ann Pharm Fr 71(4):225–233
5. Arvizo RR, Moyano DF, Saha S, Thompson MA, Bhattacharya R, Rotello VM, Prakash YS, Mukherjee P (2013) Probing novel roles of the mitochondrial uniporter in ovarian cancer cells using nanoparticles. J Biol Chem. doi:10.1074/jbc.M112.435206
6. Hall JB, Dobrovolskaia MA, Patri AK, McNeil SE (2007) Characterization of nanoparticles for therapeutics. Nanomedicine 2(6):789–803
7. Fang J, Sawa T, Maeda H (2003) Factors and mechanism of 'EPR' effect and the enhanced antitumor effects of macromolecular drugs including SMANCS. Adv Exp Med Biol 519:29–49
8. Maeda H, Wu J, Sawa T, Matsumura Y, Hori K (2000) Tumor vascular permeability and the EPR effect in macromolecular therapeutics: a review. J Control Release 65:271–284

9. Allen TM, Cullis PR (2004) Drug delivery systems: entering the mainstream. Science 303:1818–1822
10. Sahoo SK, Labhasetwar V (2003) Nanotech approaches to drug delivery and imaging. Drug Discov Today 8:1112–1120
11. Patri AK, Kukowska-Latallo JF, Baker JR Jr (2005) Targeted drug delivery with dendrimers: comparison of the release kinetics of covalently conjugated drug and non-covalent drug inclusion complex. Adv Drug Deliv Rev 57:2203–2214
12. Harris JM, Chess RB (2003) Effect of pegylation on pharmaceuticals. Nat Rev Drug Discov 2:214–221
13. Cai W, Niu G, Chen X (2008) Multimodality imaging of the HER-kinase axis in cancer. Eur J Nucl Med Mol Imaging 35:186–208
14. http://www.nanocotechnologies.com/content/AboutUs/AboutQuantumDots.aspx
15. Connor EE, Mwamuka J, Gole A, Murphy CJ, Wyatt MD (2005) Gold nanoparticles are taken up by human cells but do not cause acute cytotoxicity. Small 1:325–327
16. Turkevitch J, Stevenson PC, Hillier J (1951) A study of the nucleation and growth process in the synthesis of colloidal gold. Faraday Soc 11:55–75
17. Hu M, Qian L, Brinas RP, Lymar ES, Kuznetsova L, Hainfeld JF (2008) Gold nanoparticle-protein arrays improve resolution for cryo-electron microscopy. J Struct Biol 161:83–91
18. Sun Y, Xia Y (2002) Shape-controlled synthesis of gold and silver nanoparticles. Science 298:2176–2179
19. Huang Y, Yu F, Park YS, Wang J, Shin MC, Chung HS, Victor C, Yang VC (2010) Co-administration of protein drugs with gold nanoparticles to enable percutaneous delivery. Biomaterials 31:9086–9091
20. Vekilov PG (2011) Gold nanoparticles: grown in a crystal. Nat Nanotech 6:82–83
21. Pellegrino T, Sperling RA, Allvisatos AP, Parak WJ (2007) Gel electrophoresis of gold nanoconjugates. J Biomed Biotech 2007:26796:1–26796:9
22. Chen C, Wang W, Ge J, Zhao XS (2009) Kinetics and thermodynamics of DNA hybridization on gold nanoparticles. Nucl Acid Res 37:3756–3765
23. Javier DJ, Nitin N, Levy M, Ellington A, Richards-Kortum R (2008) Aptamer-targeted gold nanoparticles as molecular specific contrast agents for refelectance imaging. Bioconjugate Chem 19:1309–1312
24. Chang TL, Tsai CY, Sun CC, Uppala R, Chen CC, Lin CH, Chen PH (2006) Electrical detection of DNA using gold and magnetic nanoparticles and bio bar-code DNA between nanogap electrodes. Microelectron Eng 83:1630–1633
25. Takae S, Akiyama Y, Otsuka H, Nakamura T, Nagasaki Y, Kataoka K (2005) Ligand density effect on biorecognition by PEGylated gold nanoparticles: regulated interaction of RCA (120) lectin with lactose installed to the distal end of tethered PEG strands on gold surface. Biomacromolecules 6:818–824
26. Ishii T, Otsuka H, Kataoka K, Nagasaki Y (2004) Preparation of functionally PEGylated gold nanoparticles with narrow distribution through autoreduction of auric cation by alpha-biotinyl-PEG-block-[poly(2-N, N-dimethylamino)ethyl methacrylate)]. Langmuir 20:561–564
27. Khalil H, Mahajan D, Rafailovich M, Gelfer M, Pandya K (2004) Synthesis of zerovalent nanophase metal particles stabilized with poly(ethylene glycol). Langmuir 20:6896–6903
28. Otsuka H, Akiyama Y, Nagasaki Y, Kataoka K (2001) Quantitative and reversible lectin-induced association of gold nanoparticles modified with alpha-lactosylpomega-mercapto-poly (ethylene glycol). J Am Chem Soc 123:8226–8230
29. Olivier JC, Huertas R, Lee HJ, Calon F, Pardridge WM (2002) Synthesis of PEGylated Immunonanoparticles. Pharm Res 19:1137–1143
30. Otsuka H, Nagasaki Y, Kataoka K (2003) PEGylated nanoparticles for biological and pharmaceutical applications. Adv Drug Deliv Rev 55:403–419
31. Shimmin RG, Schoch AB, Braun PV (2004) Polymer size and concentration of effects on the size of gold nanoparticles cappped by polymeric thiols. Langmuir 20:5613–5620
32. Tshikhudo TR, Wang Z, Brust M (2004) Biocompatible gold nanoparticles. Mater Sci Tech 20:980–984

33. Ghosh PS, Kim CK, Han G, Forbes NS, Rotello VM (2008) Efficient gene delivery vectors by tuning the surface charge density of amino acid-functionalized gold nanoparticles. ACS Nano 2:2213–2218
34. Kim YP, Oh E, Oh YH, Moon DW, Lee TG, Kim HS (2007) Protein kinase assay on peptide–conjugated gold nanoparticles by using secondary-ion mass spectrometric imaging. Angew Chem Int Ed Engl 46:6816–6819
35. Patel PC, Giljohann DA, Seferos DS, Mirkin CA (2008) Peptide antisense nanoparticles. Proc Natl Acad Sci USA 105:17222–17226
36. Gu YJ, Cheng J, Lin CC, Lam YW, Cheng SH, Wong WT (2009) Nuclear penetration of surface functionalized gold nanoparticles. Toxicol Appl Pharmacol 237:196–204
37. Chithrani DB, Dunne M, Stewart J, Allen C, Jaffray DA (2010) Cellular uptake and transport of gold nanoparticles incorporated in a liposomal carrier. Nanomedicine 6:161–169
38. Duncan B, Kim C, Rotello VM (2010) Gold nanoparticle platforms as drug and biomacromolecule delivery systems. J Contr Release 148:122–127
39. Grace NA, Pandian K (2007) Antibacterial efficacy of aminoglycosidic antibiotics protected gold nanoparticles—a brief study. Colloids Surf A 297:63–70
40. Huang Y, Yu F, Park YS, Wang J, Shin MC, Chung HS, Victor C, Yang VC (2010) Co-administration of protein drugs with gold nanoparticles to enable percutaneous delivery. Biomaterials 31:9086–9091
41. Javier DJ, Castellanos-Gonzalez A, Weigum SE, White AC, Richards-Kortum R (2009) Oligonucleotide-gold nanoparticle networks for detection of Cryptosporidium parvum heat shock protein 70 mRNA. J Clin Microbiol 47:4060–4066
42. Lee JH, Wang Z, Liu J, Lu Y (2008) Highly sensitive and selective colorimetric sensors for uranyl (UO22 +): Development and comparison of labeled and label-free DNAzyme-gold nanoparticle systems. J Am Chem Soc 130:14217–14226
43. Velu R, Ramakrishnan VT, Ramamurthy P (2010) Colorimetric and fluorometric chemosensors for selective signaling toward Ca^{2+} and Mg^{2+} by aza-crown ether acridinedione-functionalized gold nanoparticles. Tetrahedron Lett 51:4331–4335
44. Li D, Wieckowska A, Willner I (2008) Optical analysis of Hg^{2+} ions by oligonucleotide–goldnanoparticle hybrids and DNA-based machines. Angew Chem Int Ed Engl 120: 3991–3995
45. Chai F, Wang C, Wang T, Ma Z, Su Z (2010) L-cysteine functionalized gold nanoparticles for the colorimetric detection of Hg^{2+} induced by ultraviolet light. Nanotechnology 21:025501: 1–025501:6
46. Guglielmoa CD, López DR, de Lapuente J, Mallafre JM, Suàrez MB (2010) Embryotoxicity of cobalt ferrite and gold nanoparticles: a first in vitro approach. Reprod Toxicol 30:271–276
47. Tedesco S, Doyle H, Blasco J, Redmond G, Sheehan D (2010) Oxidative stress and toxicity of gold nanoparticles in Mytilus edulis. Aquat Toxicol 100:178–186
48. Uboldi C, Bonacchi D, Lorenzi G, Hermanns MI, Pohl C, Baldi G, Unger RE, Kirkpatrick CJ (2009) Gold nanoparticles induce cytotoxicity in the alveolar type-II cell lines A549 and NCIH441. Part Fibre Toxicol 6:18
49. Mao Z, Wang B, Ma L, Gao C, Shen J (2007) The influence of polycaprolactone coating on the internalization and cytotoxicity of gold nanoparticles. Nanomed Nanotechnol Biol Med 3:215–223
50. Tiwari PM, Vig K, Dennis VA, Singh SR (2011) Functionalized gold nanoparticles and their biomedical applications. Nanomaterials 1:31–63. (ISSN 2079-4991)
51. Kersten GFA, Crommelin DJA (2003) Liposome and ISCOMs. Vaccine 21:915–920
52. Mensah F, Sridhar R, Misra P (2010) Simulation and modeling of laser-tissue interactions based on a liposome-dye system molecular and cellular biomechanics. CMC. Comput Mater Continua 7(4):203–212
53. Torchilin VP, Trubetskoy VS, Whiteman KR, Caliceti P, Ferruti P, Veronese FM (2006) New synthetic amphiphilic polymers for steric protection of liposomes in vivo. J Pharm Sci
54. Gregoriadis G, Alison AC (1980) liposome in biological systems. Wiley, New York

55. Brandl M, Bachmann D, Drechsler M, Bauer KH (1993) Liposome reparation using high—pressure homogenizers. Liposome Technology, 2nd ed. CRC Press, Boca Raton, vol I. Liposome in Preparation and Related Techniques, Chap. 3, pp 50–64
56. Rao Z, Inoue M, Matsuda M, Taguchi T (2011) Quick self-healing and thermo-reversible liposome ge. Colloids Surf B Biointerfaces 82(1):196–202
57. Carrion C, Domingo JC, de Madariaga MA (2001) Preparation of long-circulating immunoliposomes using PEG—cholesterol conjugates: effect of the spacer arm between PEG and cholesterol on liposomal characteristics. Chem Phys Lipids 113:97
58. Harland CW, Bradley MJ, Parthasarathy R (2010) Phospholipid bilayers are viscoelastic. In: Weitz DA (ed) Procedings of the National Academy of Sciences of the United States of America, vol. 107, no 45, pp 19146–19150. Harvard University, Cambridge, MA
59. Nipper ME, Dakanali M, Theodorakis E, Haidekker MA (2011) Detection of liposome membrane viscosity perturbations with ratiometric molecular rotors. Biochimie 93 (6):988–994. (Impact Factor: 3.14)
60. Sakai H, Sato A, Okuda N, Takeoka S, Maeda N, Tsuchida E (2009) Peculiar flow patterns of RBCs suspended in viscous fluids and perfused through a narrow tube (25 μm). Am J Physiol Heart Circ Physiol 297:H583–H589
61. Kimelberg HK (1976) Differential distribution of liposome-entrapped [3H] methotrexate and labelled lipids after intravenous injection in a primate. Biochim Biophys Acta 448(4):531–550
62. Freise J, Schäfer G, Schmidt FW, Magerstedt P (1977) The carrier potential of liposomes for methotrexate. Changing of the tissue levels of methotrexate in the organs of mice. Z Krebsforsch Klin Onkol Cancer Res Clin Oncol 90(2):187–195
63. Kimelberg HK, Mayhew E, Papahadjopoulos D (1975) Distribution of liposome-entrapped cations in tumor-bearing mice. Life Sci 17(5):715–723
64. Gregoriadis G, Buckland RA (1973) Enzyme-containing liposomes alleviate a model for storage diseases. Nature 244:170–172
65. Grant C, McConnell H (1973) Fusion of phospholipid vesicles with viable Acholeplasma laidlawii. Proc Natl Acad Sci U S A 70:1238–1240
66. Papahadjopoulos D, Mavhew E, Poste G, Smitrt S, VAIL WJ (1974) Incorporation of lipid vesicles by mammalian cells provides a potential method for modifying cell behaviour. Nature (Lond) 252(163):166
67. Papahadjopoulos D, Poste G, Mavhew E (1974) Cellular uptake of cyclic AMP captured within phospholipid vesicles and effect on cell-growth be- haviour. Biochim Biophys Acta 363:404–418
68. Mensah F, Grant JR, Thorpe AN (2010) The evolution of the molecules of deoxy-hemoglobin S in sickle cell anaemia: a mathematical prospective. Int J Phys Sci 4:576–583
69. Mensah F, Sridhar R, Misra P (2010) Molecular structure and rheological properties of hemoglobin molecules in sickle cell disease. Res Signpost. (978-81-308-0458-3)
70. Sakai H, Sato A, Okuda N, Takeoka S, Maeda N, Tsuchida E (2009) Peculiar flow patterns of RBCs suspended in viscous fluids and perfused through a narrow tube (25 μm). Am J Physiol—Heart Circulatory Physiol 297:H 583–H589. doi: 10.1152/ajpheart.00352

58. Brandl M, Bachmann D, Drechsler M, Bauer KH (1993) Liposome preparation using high-pressure homogenizers. Liposome Technology. 2nd ed. CRC Press, Boca Raton, vol 1. Liposome in Preparation and Related Techniques [illegible]

[illegible]

MIX
Papier aus verantwortungsvollen Quellen
Paper from responsible sources
FSC® C105338

If you have any concerns about our products,
you can contact us on
ProductSafety@springernature.com

In case Publisher is established outside the EU,
the EU authorized representative is:
Springer Nature Customer Service Center GmbH
Europaplatz 3, 69115 Heidelberg, Germany

Printed by Libri Plureos GmbH
in Hamburg, Germany